BOB BECKENHAUER

# EMBEDDED MICROCONTROLLERS AND PROCESSOR DESIGN

# EMBEDDED MICROCONTROLLERS AND PROCESSOR DESIGN

GREG OSBORN

**Prentice Hall**
Upper Saddle River, New Jersey
Columbus, Ohio

**Library of Congress Cataloging-in-Publication Data**

Osborn, Greg.
    Embedded microcontrollers and processor design / Greg Osborn.
        p. cm.
    ISBN-13: 978-0-13-113041-8
    ISBN-10: 0-13-113041-2
    1. Embedded computer systems—Design and construction.    2. Microcontrollers—Design and
construction.    3. Microprocessors—Design and construction.    I. Title.
    TK7895.E42O83 2010
    004.16—dc22

                                                                        2008043565

**Editor-in-Chief:** Vernon Anthony
**Acquisitions Editor:** Wyatt Morris
**Editorial Assistant:** Christopher Reed
**Project Manager:** Rex Davidson
**Senior Operations Supervisor:** Pat Tonneman
**Operations Specialist:** Laura Weaver
**Art Director:** Candace Rowley
**Cover Designer:** Rachel Hirshi
**Cover photo:** iStock
**Director of Marketing:** David Gesell
**Marketing Assistant:** Les Roberts

This book was set in Times Roman by Aptara®, Inc. It was printed and bound by Hamilton Printing Co.
The cover was printed by Phoenix Color Corp.

Pearson Education Ltd.
Pearson Education Singapore Pte. Ltd.
Pearson Education Canada, Ltd.
Pearson Education—Japan

Pearson Education Australia Pty. Limited
Pearson Education North Asia Ltd.
Pearson Educación de Mexico, S.A. de C.V.
Pearson Education Malaysia Pte. Ltd.

**Prentice Hall**
is an imprint of

www.pearsonhighered.com

10  9  8  7  6  5  4  3  2  1
ISBN-13: 978-0-13-113041-8
ISBN-10:    0-13-113041-2

# PREFACE

Microcontrollers have become ubiquitous elements of everyday life. Most of the electronic products we use in daily life have a microcontroller tucked inside. They are used in household appliances, automobiles, copiers, cell phones, and even used to control powerful locomotives. Where electricity is used, you will find a microcontroller!

There are many technical books on the subject of microcontrollers. Why develop a new book? All the popular microcontroller chips and architectures have "how-to" books in print. The focus of this text is a broader introduction to the student of microcontroller processor technology, both in single-chip and intellectual property form.

Many electronic engineering students are required to take a course in embedded system design with microcontrollers. Devices such as the Intel 8051, ZiLOG Z8, or Microchip PIC are most often the microcontroller of choice because of their widespread popularity. Also, they have extensive, inexpensive tools available to provide design support.

A wider variety of choices for an embedded microcontroller-based design face the engineer today, not only the popular single-chip devices, but also intellectual property cores for ASIC system-on-a-chip (SoC) design. Although the computer world has solidified around Intel-based architectures, the microcontroller world continues to evolve with innovative new designs.

This book is organized into three major sections: an introduction to microcontroller architectures, single-chip microcontrollers, and embedded IP cores. Each chapter within the section begins with learning objectives. Questions are provided at the end of the section to monitor progress.

Not only are specific chips and cores covered, but also this book introduces the student to the concepts of microcontroller architectures: for instance, how the concept of the computing devices evolved, and why different types of devices are used in design.

Single-chip microcontrollers as referenced in this book are typical commercial high-volume classical designs. Certainly, myriad parts are also available, particularly from "fabless" design houses. Microcontroller cores are in reference to established system-on-a-chip intellectual property cores and marketed as such. In architectural discussions in this book, the term "processor" incorporates both the "processor" element of "single-chip microcontrollers" and "IP cores."

This book is intended to provide the reader with an introduction to single-chip and embedded microcontroller processor design. The difference between architectures of the CISC- and RISC-based processors is discussed. Single-chip microcontroller design flows and embedded processor design flows are discussed.

The 16-bit Freescale MC9S12X family of single-chip microcontrollers is covered in detail. The RISC-based PIC18F4520 and the ZiLOG Z8 Encore! 8-bit microcontrollers are also discussed. The peripherals that are available with various members of the families are explained.

The concept of instruction set architecture (ISA) is introduced to develop an understanding of the commonality of the CISC and RISC processor families, respectively. This is expanded to the design of SoC embedded controllers-based core IP using the ISAs of ARM and MIPS. The ARM10TDMI and MIPS32 4KE™ IP cores are presented in some detail.

Configurable processor technology is increasingly important, particularly in the design of higher performing consumer products. It allows customization of the core processor, which can have both performance and power impact on the SoC embedded design. The Tensillica Xtensa LX2 Series configurable processor is covered.

A discussion of derivative RISC application-specific processors is presented. An overview of a digital signal processors (DSP), including the Texas Instruments' TMS320C55 and Analog Devices' ADSP-BR533 Blackfin is given. The methodology of the engineering design flow is also covered. Different tools available to the engineer for the design process are discussed. An example of using an integrated design environment (IDE) for single-chip microcontroller is presented.

Software programming for microcontroller design can be as simple as a program for controlling lawn sprinklers to a complex RTOS for robots. Programming techniques from simple polling loops to multilevel interrupt driven systems are discussed. Many single-chip microcontrollers include functional blocks for serial I/O. They are primarily used to communicate data. The UART, I$^2$C, I$^2$S, CAN/LIN SPI, and USB peripheral functions are discussed.

System-on-a-chip design requires a close relationship with the semiconductor foundry. As a fabless design technology, SoCs need specialized engineering techniques to integrate the functions needed for the chip. Combining IP functional blocks available from the foundry with those from independent companies to achieve a working chip is a complex process.

This book is intended as an introductory understanding of microcontrollers in single-chip and embedded forms. The concept of ISA is developed with the methodology for product design. System-on-a-chip design is introduced through the use of intellectual property.

Microcontroller design, at any level of abstraction, is based on a balancing of available technologies. The primary three technologies this book will focus on are processor, memory, and software: processor technology as it is defined in terms of semiconductor fabrication capability; memory technology as it is implemented in a hierarchical storage structure; and software technology as it is implemented in the form of assembler and optimizing compilers.

Within the scope of this book, generalizations are taken as they relate to characteristics of microcontroller-based design. In general, CISC-based processors have more complex instructions than RISC-based processors. In general, RISC registers sets are orthogonal when compared to CISC. In general, optimizing C compilers are more efficient for RISC than CISC.

RISC and CISC are considered in their global context of instruction set architecture (ISA). New innovations in architecture, such as VLIW and EPIC, are mentioned for comparison. The focus of this book is on microcontroller technology, which is predominately RISC- or CISC-based. This will provide the basic knowledge needed for the student to understand other derivative ISAs.

Microcontrollers have, at their heart, a microprocessor. In this book, the term *processor* is used in a broad sense. Whether implemented as core IP in an SoC, or in traditional single-chip form, the basic concept of the processor is the same. MIPS32 4KE™ IP can be incorporated as single chips from NEC or an SoC in a CISCO router. They are implemented differently but are the same architecturally.

This book is intended as an introduction to the topic of microcontroller technology for college engineering students. It is not a hardware reference manual. It is not intended as a series of application notes. The concepts presented are in general form. This will allow a broad group of engineering students to understand the basic concepts and apply them to real-world situations.

An online instructor's manual is available for instructors using this text for a course. To access supplementary materials online, instructors need to request an instructor access code. Go to **www.pearsonhighered.com/irc,** where you can register for an instructor access code. Within 48 hours after registering, you will receive a confirming e-mail, including an instructor access code. Once you have received your code, go to the site and log on for full instructions on downloading the materials you wish to use.

The author thanks the following reviewers of the manuscript: C. Richard G. Helps, Brigham Young University; James Streib, Illinois College; Chao-Ying Wang, DeVry University—Columbus; and Richard Warren, Vermont Technical College.

**Greg Osborn**

# CONTENTS

**CHAPTER 3**   **EMBEDDED MICROCONTROLLER TECHNOLOGY**   **30**

## CHAPTER 15    DIGITAL SIGNAL PROCESSORS    385

# CHAPTER 1

## Embedded Processors

### OBJECTIVE: INTRODUCTION TO MICROCONTROLLER CONCEPTS

The reader will learn:

1. Basic markets and applications for microcontrollers.
2. Concept of commercial and embedded processors.
3. System-on-a-chip design using intellectual property.
4. Concept of instruction set architecture.
5. Semiconductor technology trends.

## 1.0 MICROCONTROLLERS

Microcontrollers are primarily used in three ways: as stand-alone devices, part of an embedded processor system, or integrated into a system on a chip (SoC). At each level they may incorporate incremental numbers of features through logic blocks to enhance their utility. At the point of SoC design, all elements of the system logic are combined into a single integrated circuit (Figure 1.1). General-purpose, or stand-alone, chips typically incorporate the broadest set of application-specific features to increase potential market appeal.

**FIGURE 1.1** Zilog eZ8 Processor (Reprinted by Permission of ZiLOG Corporation.)

## 1.1    MICROCONTROLLER MARKETS

Most commercial microcontrollers are designed to support a wide variety of applications. This broadens the market for which they can be used and is a key way to keep their cost low. Twelve

**FIGURE 1.2** Major Market
Segments

| | |
|---|---|
| Aerospace and Defense Electronics | Digital Imaging |
| Automotive | Industrial Measurement & Control |
| Battery Powered | Medical Electronics |
| Broadcast and Entertainment | Server I/O |
| Consumer/Internet Applications | Telecommunications |
| Data Communications | Wireless |

major categories of microcontroller markets are identified in Figure 1.2. Each category contains a multitude of applications.

## 1.2          DATA PATH

Microcontrollers are generally referred to by the bit width of their data path. This is the number of parallel bits fetched from memory and processed by the arithmetic logic unit (ALU). The data path width of microcontroller and embedded processor cores typically range from 4, 8, 16, 32, or even 64 bits. Larger word sizes of 128 and 256 bits exist in specialized application specific processors (ASPs), such as those used in graphic controllers for video gaming.

## 1.3          COMMERCIAL MICROCONTROLLERS

The largest group of microcontrollers by volume are based on 8-bit data paths. Some commercial designs incorporate an 8-bit microprocessor core with the basic functional blocks. Several of the early chip designs have become de facto industry standards. The Intel i8051 and Freescale 68HC11 are examples.

Figure 1.3 shows the breadth of design possibilities using the PIC microcontrollers from Microchip Technologies. Just about any type of peripheral interface is available to design simple to complex embedded system controllers. SoC design takes it to another level by incorporating the CPU with the peripherals on a single die.

## 1.4          SoC CORE PROCESSORS

Relative shipment volumes for embedded processor cores used in SoC designs are shown in Figure 1.4. Shipment volumes for these cores can be extremely high, reflecting the large vertical markets for which they are targeted. Inexpensive electronic wristwatches using 4-bit microcontrollers ship in volumes of tens of millions, whereas more expensive 32-bit-based video game consoles ship in volumes of millions of units. The total worldwide shipments for 4- and 8-bit processors exceeds a billion units per year.

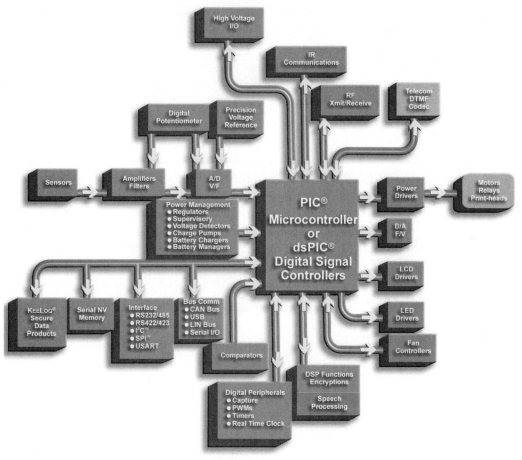

**FIGURE 1.3** Microchip PIC Embedded Control Solutions (Copyright 2004 Microchip Technology Incorporated. Reprinted with permission.)

**FIGURE 1.4**

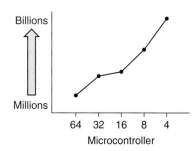

## 1.5    RELATIVE SoC UNIT VOLUMES

The overall market for commercial and embedded processors is shown in Figure 1.5. It is broken down into five major market segments, which are identified and shown as percentages in the pie chart. The relative share of market (SOM) for each application area can be seen. Microcontrollers are fairly evenly distributed across major market segments. This reflects their design utility over the wide range of applications.

**FIGURE 1.5**   Dollar Volume by Market Segment

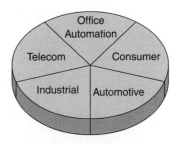

## 1.6          VERY-LARGE-SCALE INTEGRATION (VLSI) CHIP DESIGN TOOLS

Due to advances in semiconductor fabrication technology, microchips can be designed with millions of transistors. It has become possible for design engineers, using computer-aided design (CAD) tools (Figure 1.6), to create microchips tailored to specific applications. These types of microchips are referred to as application specific integrated circuits (ASICs).

**FIGURE 1.6**   VLSI Design Flow

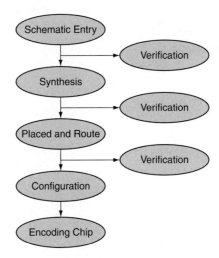

Advanced VLSI CAD tools make it possible to incorporate the processing core of popular microprocessors into the silicon chip design itself. The core is designed with appropriate functional blocks targeted to a specific application. These blocks, when combined with the processor core, emulate the basic functions of a computer system (Figure 1.7). Fundamentally, these devices are a computer SoC.

Designing an ASIC or SoC is done with the aid of sophisticated design automation (DA) software programs. The circuit is described using a high-level programming language (VHDL), as shown in Figure 1.8. A computer then executes the program to produce the actual chip design. It is a complex process, but essentially a microchip is derived from a software program.

## 1.7          INTELLECTUAL PROPERTY

Software engineers have developed standard protocols to define transistor circuits. This has allowed basic functional blocks of logic to be defined as independent blocks of code. These include microcontrollers and core processors. These blocks of code or logic programs are

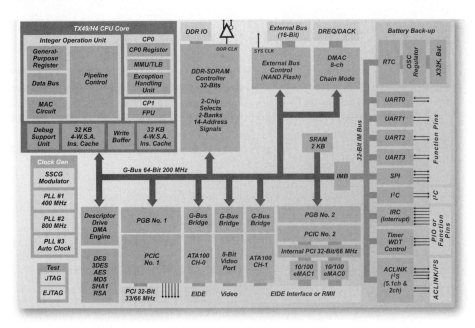

**FIGURE 1.7** Toshiba TX4939XBG SoC (Reprinted by permission of Toshiba America Electronics Corporation, TAEC.)

**FIGURE 1.8** Generic VHDL Code

```
library IEEE;
use IEEE.std_logic_1234.all;
entity and1 is
generic (gate_delay : Time: = 500 ps);
port(In1,In2 : in std_logic;
     Z : out std_logic;
end and1;

architecture behavioral of and1 is
begin
Z <= (In1 and In2) after gate_delay;
end behavioral;
```

referred to as intellectual property, or IP. By combining different IPs, it is possible to create an SoC with customized functions targeted to specific applications at an economical price. The following table lists broad categories of popular IP products showing the breadth of what is available.

| | |
|---|---|
| Bus interfaces | Digital signal processing |
| Analog/mixed signal | Processor & microcontrollers |
| Arithmetic & mathematic functions | Memory element |
| Test cores | Security & error detection |
| Video/image/audio | Wireline communications |
| Wireless communications | Platform level IP |
| Hardware OS/RTOS | Intercore communications |
| Embedded configurable logic | Machine tool controllers |

## 1.8        INSTRUCTION SET ARCHITECTURES

Complex instruction set computers (CISC) and reduced instruction set computers (RISC) are two basic classes of instruction set architectures. Both ISAs have emerged for the design of microcontrollers and microprocessors. Both types of processor architectures can be found in the microcontroller world, each with its own set of perceived advantages and disadvantages. Refer to Figures 1.9 and 1.10 for a comparison of the architectures.

**FIGURE 1.9**   CISC Architecture

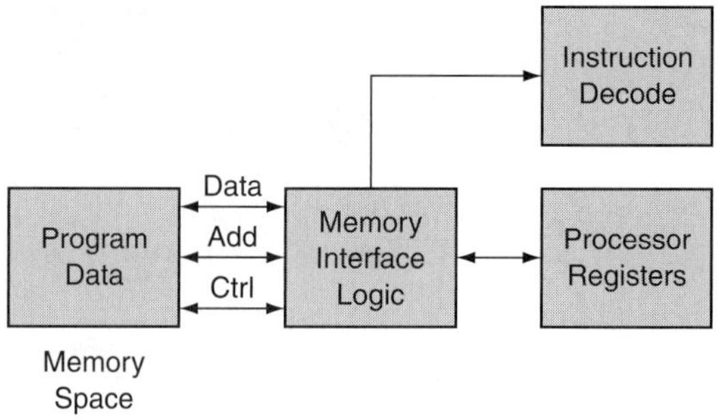

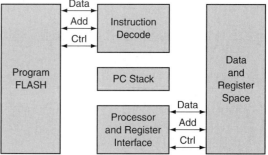

**FIGURE 1.10**   RISC Architecture

## 1.9        RETURN ON INVESTMENT

The design decision to use a commercial microcontroller for a product is the result of a combination of interrelated technical and economic factors. More often than not, economics plays the overriding role. Commercial microcontrollers reduce the front-end design costs resulting in a faster time to market. Selling the product is what generates company profits!

By contrast, custom SoC-based designs require a large up-front investment for the design of the chip. The time to market can easily be 18 months or more. The production cost for the chip is much lower than for a commercial microcontroller design. However, this comes at a significant increase in risk that the product will earn enough money for the company to be profitable.

The graph in Figure 1.11 shows the relative economics for designing a microcontroller-based product. This is a simple diagram of return on investment (ROI), but effectively demonstrates the concept of "risk." The area below the *x*-axis represents "cost," and the area above it "revenue."

Designing with a commercially available commercial microcontroller is a way to quick market penetration. The trade-off is that the end product cost will be higher. This may limit the potential revenue and hence the profitability of the product over time.

Using an SoC-based solution is more expensive in up-front design costs. Not only is the basic design cost greater, but also the intangible of time to market is at risk. It can easily take a

**FIGURE 1.11** ROI Commercial Microcontroller

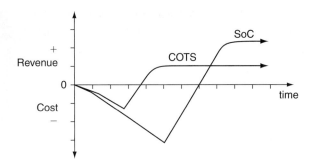

year or more to get a product out the door. In the meantime, other competitive products are reaching the market. Weighing against the risk is the potential for greater return (profit) on the investment, as shown in Figure 1.11.

Many times, there are no clear-cut technical reasons to use a commercial part or design an SoC. That is when the marketing department steps in to make a decision. Most often, the design chosen is the one with the lowest costs that gets the product to the market in the shortest amount of time.

## 1.10 SEMICONDUCTOR TECHNOLOGY DEVELOPMENTS

The pace of semiconductor, or more specifically complementary metal-oxide semiconductor (CMOS), technology is continually progressing at a rapid rate. Advances in semiconductor fabrication techniques have allowed continuous shrinking of the CMOS transistor size. Relative dimension of millimeter size hair to the size of an electron is shown in Figure 1.12.

Gordon Moore developed a law that says CMOS transistor density will double every 18 months. This has been more broadly applied to many semiconductor and related technologies. For microprocessors (and embedded controllers), this is interpreted as the number of CMOS transistors that can be fabricated on a single die. Moore's law is shown in Figure 1.13.

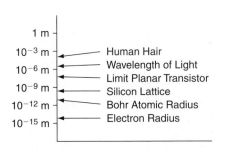

**FIGURE 1.12** Relative Sizes of Things

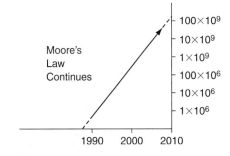

**FIGURE 1.13** Moore's Law

CMOS transistor scaling is shown in Figure 1.14. By the year 2010, the gate size of a transistor is projected to be in the 15 nanometer range. Remembering that the chip density is doubling every 18 months, this means billions of transistors on a chip. Of course, many technologies need to be refined to achieve the 15 nanometer gate. It is not simply a matter of drawing finer lines on a mask.

A basic CMOS transistor is shown in Figure 1.15. The gate, source, and drain are shown. The type of the CMOS transistor development needed to reach the technology goals of 2020 are shown in Figure 1.16. Note the evolution to a circular design structure. This demonstrates that the fundamental geometrical design of the CMOS transistor must also change with the shrinking size to the nanometer range.

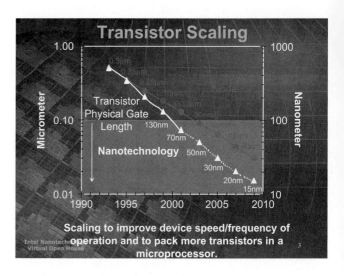

**FIGURE 1.14** Transistor Scaling (Reprinted by permission of Intel Corporation.)

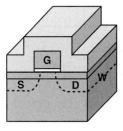

**FIGURE 1.15** CMOS Transistor

A chart from Intel Corporation, Figure 1.17, demonstrates that CMOS device scaling to meet Moore's law will require innovative new design techniques as we approach the year 2013. This includes the equipment and technologies needed to design and manufacture devices in the nanometer-sized range. Technological advances at this level require the convergence of multiple interrelated technologies.

Figure 1.18 shows the future direction of CMOS design technology. Device scaling will not in itself be sufficient to yield higher-performing CMOS transistors; new, integrated solutions

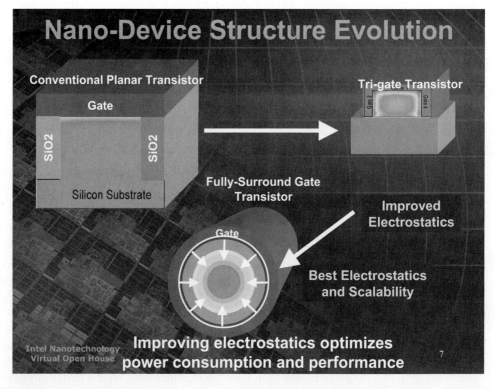

**FIGURE 1.16** CMOS Nano-Device Structure Evolution (Reprinted by permission of Intel Corporation.)

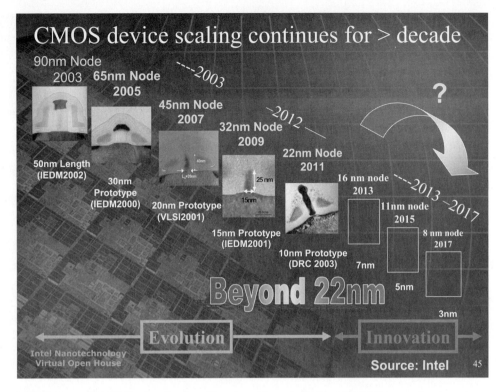

**FIGURE 1.17**    CMOS Device Scaling (Reprinted by permission of Intel Corporation.)

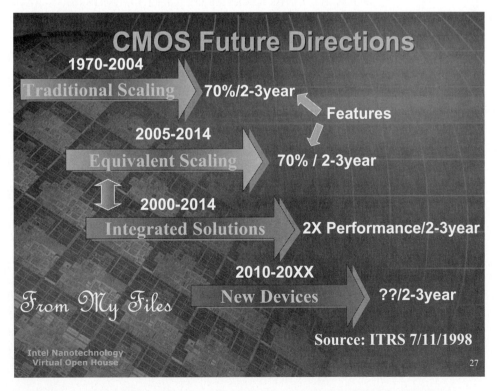

**FIGURE 1.18**    CMOS Future Directions (Reprinted by permission of Intel Corporation.)

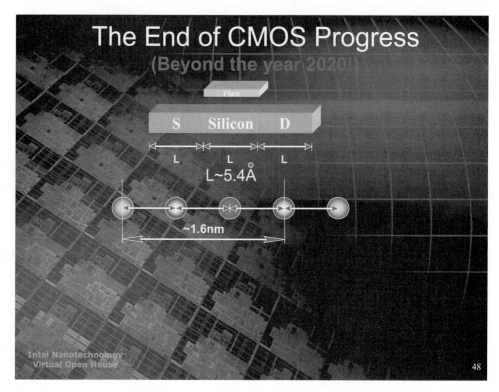

**FIGURE 1.19**   The End of CMOS (Reprinted by permission of Intel Corporation.)

will be required. As we can see in Figure 1.19, CMOS transistors will eventually reach a point where they can no longer be made smaller. Intel projects that this will happen after the year 2020. At that time, perhaps devices based on quantum effects will be introduced. There is still a long way to go for CMOS technology pioneered by Intel with the introduction of the i4004 microprocessor in 1970.

## SOURCES

Gargini, P. *Intel nanotechnology overview presentation*. International Nanotechnology Virtual Open House, Intel Corporation.
David, K. October 2004. *Silicon nanotechnology presentation*. Intel Corporation.

# CHAPTER 2

# Microcontroller Architecture

**OBJECTIVE: AN UNDERSTANDING OF THE EVOLUTION OF COMPUTER PROCESSOR DESIGN TO CURRENT RISC AND CISC PROCESSORS FOR EMBEDDED DESIGN**

The reader will learn:

1. Basic concepts of semiconductor technology.
2. Introduction to microprocessor concepts.
3. Microcontroller functional design.
4. RISC microcontroller introduction.
5. RISC and CISC processors as Intellectual Property.

## 2.0 COMPUTER ON A CHIP

Today's microcontrollers and embedded processors are direct architectural descendants from the earliest computers. They have evolved over time with improvements in silicon design and semiconductor fabrication technology to their current implementations. Containing the inherited functionality of their predecessors, they can truly be called computers on a chip.

The first commercial microcontrollers emerged in the mid 1980s from such firms as TI, Intel, Fairchild, and National Semiconductor. They were direct integrations of existing controllers designed from standard TTL (transistor–transistor logic) gate-level functions. A key driving force for their development was to replace hard-wired fixed designs with a much more flexible programmable logic-based design. This led to more cost effective controllers with reduced time to market (i.e., greater profitability).

Microcontroller design reflects the constantly changing functional requirements of electronically controlled products. With new products continually being introduced to the marketplace, the need for innovation in control logic is constantly being challenged. There is a continuous and increasing demand for greater product functionality at ever-decreasing prices.

The applicability of a microcontroller for any given design is dependent on many interrelated factors that are not all of a technical nature. Economics most often plays an important, if not overriding, role in the selection decision of a microcontroller in a product design. However, for the engineer, there remains the task of creating the best design for the given specifications.

Microcontrollers and embedded processors can and are used in virtually all electromechanical-based products; from everyday household appliances to cell phones to complex industrial equipment. Electronic-based controller design can eliminate or reduce complex mechanisms easily prone to failure. Solid-state designs enhance the functionality of a controller and increase reliability, while reducing maintenance and product cost.

## 2.1    JOHN von NEUMANN

John von Neumann is considered one of the founding fathers of modern computer design. As with many of the pioneers, he had a solid education in mathematics, receiving his doctorate degree in that subject in 1928. Due to the emerging war in Europe, like many scientists of his day, he immigrated to the United States, where he took a position at Princeton in the Institute for Applied Mathematics.

**FIGURE 2.1**  John von Neumann (www.lanl.gov)

World War II required everyone, including those involved with developing technologies, to aid in the war effort. An urgent requirement was placed on the academic community to convert their theoretical work to practical applications in support of national security. Von Neumann had a particular interest in applied mathematics. He focused enthusiastically on bringing mathematical rigor to the new computing machines.

### 2.1.1  von Neumann Architecture

Many groups of scientists worked on different and new technologies. Von Neumann was able to understand how the newly developed electronic computing machines could be applied to existing problems. His insight into how the new electronic technology could be implemented into a practical computing machine has been named the "von Neumann architecture."

It is fundamentally a concept whereby a program's instructions are organized in a serial fashion and stored separately. The computer has separate functional units that operate in a synchronized way. He devised a system of codes that would control the fixed wired logic of the machine. These instructions would determine the operations to be performed by the computer.

Up to that time, machines were hand wired, or hard coded, for a specific set of data operations. To change a sequence of instructions, or program, meant rewiring the connections between the functional units. It was a tedious, error prone, and time-consuming task. It also severely limited the capability of the computer to do useful work.

Von Neumann's approach resulted in high productivity by increasing the amount of time available for the computer to do useful work (uptime). With the ability to easily change the instruction codes, more complex programs could be developed and tested. It was a pioneering and powerful step in the development of the new field of electronic computing machines.

**FIGURE 2.2**   von Neumann Architecture

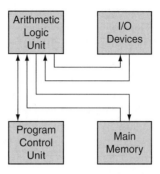

## 2.2      COMPUTER ARCHITECTURES

Today's microprocessors and microcontrollers share a common heritage. They have evolved over time with new technologies, new innovations in design, and new markets. However, underlying the change that has produced thousands of different chip designs is a basic commonality.

### 2.2.1  CISC and RISC

The two major groupings are considered to be CISC (complex instruction set computer) and RISC (reduced instruction set computer). These instruction set architectures (ISAs) define the computer system design. Newer types of processor designs have been enabled by the rapid progress in semiconductor technology. Putting one million transistors on a chip is a feat that was not possible before.

Within the scope of this book, the focus will be on embedded processors and microcontrollers with typical CISC or RISC ISAs. There are many ways to instantiate or embed the ISAs into silicon. However, philosophically, they have the same architectural constructs; they all execute instructions.

In the real world, maximizing product profitability determines which embedded processor is used for a particular product. Given that assumption, engineering design trade-off decisions still need to be made. The design goal is still to have the greatest functionality with the highest performance at the lowest production cost.

**FIGURE 2.3**   Market Forces

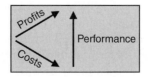

Both types of ISAs can be found as stand-alone commercial microcontrollers or as intellectual property in a wide variety of products. There is not a fixed set of design rules that dictates which ISA to use for a specific product design. Within many markets there is a broad overlap, and often the design selection is based on purely economic factors.

Single-chip microcontrollers are designed to be more general purpose in nature so they can appeal to a broader market. Intellectual property–based designs tend to be narrowly focused for a specific product. However, they still share the same basic functional design elements necessary to be computers on a chip.

## 2.3 SEMICONDUCTOR TECHNOLOGY

During the decade of the 1960s, semiconductor technology advanced to the point where multiple bipolar transistors could be put on a chip forming logic functions or integrated circuits (ICs). AND, NAND, NOR, FLIP-FLOPS, and many other basic logic functions became available as individual chips. By combining chips with the various logic functions onto printed circuit boards, larger more complex circuits could be designed.

An early application of this logic was to design controllers for devices connected to computers. Computers operated at very high speeds relative to the input and output machines connected to them. They were designed with specialized circuits that were company proprietary and not commercially available.

### 2.3.1 Small-Scale Integration

The introduction of SSI logic to the marketplace enabled the design of printed circuit board–level controllers. Engineers could create solid-state designs with commercially available affordable parts. The introduction of SSI logic enabled a new generation of innovation.

**FIGURE 2.4** Small-Scale Integration

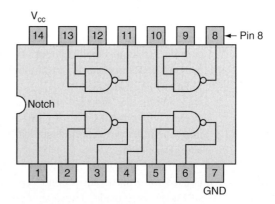

At the beginning of the decade, computers, or mainframes as they were popularly called, were very large and bulky. They required rooms fitted with special support equipment, often with locked doors and humorously called "temples." The commercial availability of logic circuits allowed engineers to create a new generation of computers that were more compact in size. They still maintained the basic functionality of the larger mainframes.

### 2.3.2 Hardware Bus

Computer design was also changing. Traditional techniques had the central processing unit itself execute the control functions needed to interface with peripheral devices. Central processing units (CPUs) could do only one task at a time. Controlling peripherals meant less time to execute user programs. A new design concept was introduced to partition the system by function. Rather than have all the controlling functions reside with the CPU, the peripherals themselves were enhanced with their own intelligence.

The enabling technology for this is a bus. A bus is an electronic "highway" that allows information (data, addresses, and control signals) to move between functional units (see Figure 2.5). The concept of a bus is a major innovation in computer architecture. All microcontroller designs incorporate the bus concept into their design.

**FIGURE 2.5** Basic Bus
Architecture

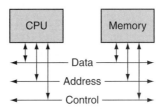

### 2.3.3 Intelligent Peripherals

The emergence of intelligent peripherals increased the overall performance of the computer. It allowed the CPU to spend more time processing programs rather than controlling the hardware. This was measured as the CPU time spent executing in user mode versus the time needed in supervisor mode, which included peripheral device control time.

### 2.3.4 Standardized I/O Interfaces

A market-driven demand was being created for having more flexible controller design. If manufacturers could develop more flexible or generic controllers, they could reduce costs. They would have to produce fewer printed circuit boards (PCBs) at the minimum. A way to accomplish this was to create a basic core design at the PCB level and then personalize it for a specific device with simple switches or jumpers. The PCB was still a single unit that reduced manufacturing costs, but it could now be customized for a specific product. This is similar to what is being done on PC motherboards to this day.

(a)

Up to this time, computers were designed as complete systems with all hardware and software integrated. I/O, memory, disks, and other functional units were incorporated intimately with the CPU into the overall hardware design. Although the computers were being designed logically into basic functional blocks, they were still electronically single tightly integrated units.

(b)

**FIGURE 2.6**
(a) Interface Specific;
(b) General Interface

The emergence of third-party electronics provided, for the first time, the ability for independently designed products. The first efforts were done in the I/O functions, such as terminals and printers. Although they provided the function, they still needed to be interfaced to the computer. To do this required a third-party manufacturer to match the electrical and timing characteristic of the specific computer manufacturer's interface.

Needless to say, each manufacturer had a different interface that they considered company proprietary. For an independent manufacturer, this meant designing a specific hardware controller for each interface. Although there was a move by manufacturers to reach a common standard, there were still company-specific design issues with which to deal.

What engineers were developing were intelligent peripherals. These were controllers that could be designed to comprehend different interfaces and be adjusted for the differences. As computer system design became more sophisticated, more and traditional CPU-controlled I/O was downloaded to the peripheral controllers. The computers were being constructed with partitioned design techniques.

**FIGURE 2.7** Partitioned Design Concept (Reprinted by permission of NXP Semiconductors.)

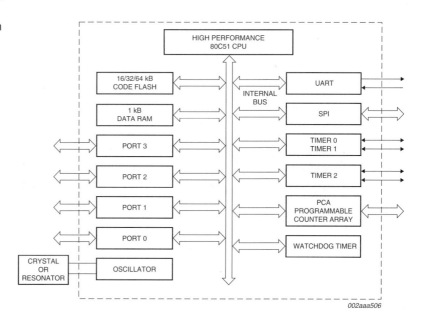

## 2.4        MSI AND LSI

In the late 1960s, medium-scale integration (MSI) was being introduced to the market. It was still based on bipolar transistor logic and extended the range of functions available with SSI as single chips. Octal latches, 4-bit ALUs, and 16-bit NAND gates reduced the number of SSI packages required for a design. This enabled more complex and intelligent designs than the previous generation of SSI-based controllers.

Specialized single-chip components were now possible. Standardized interface chips like RS-232 (Figure 2.8) for serial communication became available. Continuing to shrink the size of transistors, semiconductor fabrication process technology translated directly to more sophisticated integrated circuit components.

**FIGURE 2.8** RS-232 Serial I/O Interface

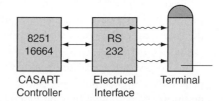

Specialized device controllers were being designed for a wide variety of peripheral functions that were more flexible, more reliable, and most important, less costly. And they were being designed not only for traditional keyboards, cathode ray tube (CRT) displays, and printers, but also many other computer-related functions. New components were being developed to enhance control functions for direct memory accessing, internal clocking, and timing circuits, as well as newly developing serial transmission protocols.

There was also another direction that could be taken. Higher levels of transistor integration meant fewer components were required to implement a specific design. With fewer components, lower costs could be attained for the controller part of the product design. The logical progression is to a single chip.

## 2.5             ELECTRONIC CALCULATOR

TI introduced a single four-function electronic calculator chip in 1968. It supported add, subtract, multiply, and divide. It caught the imagination of the public. A computer on a chip was born. Besides lowering the cost point of basic calculators, it also enabled the era of portable calculators. The reduced power consumption required for the single chip, along with lower power displays, meant the calculator could be battery powered.

Small electronic devices could now mimic the functionality of traditional computers. The desktop calculator had two or three forms of input and output, a keyboard, display, and an optional printer. It interfaced directly with a person and human speeds. The computer functionality was there; it was the data rate that was different. Fast computer execution speeds were not necessary. The fact that it took a long time to add two numbers electronically compared to the computer did not matter as long as it happened quickly for the person operating the machine. The perceived performance was what was important.

**FIGURE 2.9** Four Function Calculator (Courtesy of Texas Instruments.)

### 2.5.1 Programmable Calculator

Offering the user the ability to automate a series of calculations was the next logical step. HP introduced a portable programmable scientific calculator, the HP-35, to the market in early 1971. In many ways it was a miniature handheld computer. With the addition of software programming, all the basic functions of a computer were in the user's hand.

The HP-35, however, for all its calculation capabilities, remained just a sophisticated calculator. General-purpose computers have much larger instruction sets that are generic in nature to support a wide range of data processing requirements, not just mathematics. The HP-35 contains a very limited subset of instructions needed for its calculations. It is an application-specific product designed for a single purpose.

Semiconductor technology progressed rapidly in the late 1970s and early 1980s to large-scale integration (LSI). Newly developed N-channel and P-channel metal-oxide semiconductor (MOS) allowed thousands of transistors to be fabricated on a single die. A driving force for this technology was the replacement of slow, bulky, and error-prone core computer memory. The INTEL 1101, a device with 1 k bits of storage, was introduced in 1971.

This new technology enabled the design of significantly more complex and inherently more powerful components. Again, the driving force of the marketplace put demands on engineers for more advanced, sophisticated, and economical controllers. As the capabilities of the components increased, the applications they could be used for increased. The advances in technology not only met the needs for current requirements but created opportunities for newer designs.

## 2.6        MICROPROCESSORS

The demand for CRT terminals increased dramatically in the early 1980s as computers became more user friendly. Each manufacturer had specific requirements for what they wanted. Terminals had fixed specifications, and the computer I/O system directly controlled them. In addition, as terminal use grew, the ability to control them directly from a central I/O controller became increasingly difficult.

There was a need to move much of the standardized control functions from the I/O controllers to the terminal. Functions such as detecting depressions of keys on the keyboard, sending them to the computer, and displaying characters on the CRT were tasks that should be done by the terminal controller.

The ASCII character set was originally developed by industry to encode the typical characters found on a teletype terminal keyboard: upper/lowercase letters, numbers, punctuation, and so on. Seven data bits provide 128 "codes" ($2^7$ in binary), sufficient to incorporate this basic character set. This conveniently can be contained in a byte (8 bits).

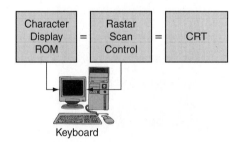

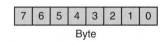

'A' $\Rightarrow$ '01000001'
'5' $\Rightarrow$ '00110101'

ASCII Codes

Byte

Keyboard

**FIGURE 2.10**   Terminal Design                   **FIGURE 2.11**   ASCII Byte Code

CRT makers began asking the semiconductor industry to provide a solution for them. With the increasing capability of LSI logic to integrate logic functions on a single chip, more complex controller designs became possible. What the CRT manufacturers needed was a programmable controller that was application specific to their needs, similar to a programmable calculator for mathematics.

### 2.6.1 Application-Oriented Processing

The semiconductor manufacturers were receiving requests from engineers designing a broad range of products. Central to their requests was the need for programming to their specific application. Everyone wanted a calculator for his or her product.

Increasing numbers of application-oriented LSI controllers were being introduced to the market to meet growing demands from a wide range of products. There was a commonality to the requirements for the chips. They all needed input and output and a control function, just like the calculator chips. The HP-35 was able to solve a broader range of mathematical problems because of its instructions and programming capability.

Semiconductor manufacturers knew that with the new LSI capability, they could create programmable devices for a wide range of applications. If they incorporated more instructions into the design, the chip could have a broader range of application. It would be more general purpose by design.

This had implications for the manufacturing cost and hence the retail price of the chip; however, the flexibility of its use would far outweigh any increase in cost. Instead of a fixed limited application-oriented instruction set, the new device could have many more general-purpose type instructions. Interfacing the chip to LSI memory and I/O components would allow it to be customized for a specific application.

### 2.6.2 Intel i4004

CRT manufacturers were asking for a controller that handled 4 bits of data for the ASCII character set. They wanted to manipulate, or process, the bits that were scanned from the keyboard, displayed as characters, and communicated with the central computer. Intel responded with a three-chip set of components. One of them, 4004, could process 4 bits of data. It was micro in size, but a processor nevertheless. The Intel 4004 was called a microprocessor.

**FIGURE 2.12**  Intel i4004 (Reprinted by permission of Intel Corporation.)

### 2.6.3 Intel i8080

The 8-bit i8080 microprocessor was limited to executing instructions to process data as its name implies. The "data" being the signals from other parts of the circuit translated to logical 1s and 0s and converted to byte format. It simply performed a transformation on the data passing through it, as shown in Figure 2.13.

**FIGURE 2.13**  Microprocessor as Transform Function

The i8080 needed a complement of support chips to function, such as program memory, data storage, and I/O ports, as well as the system control logic, as shown in Figure 2.14. There was a design overhead when using microprocessors. It increased the size and complexity of the printed circuit board for the controller, for example. It made the design process more complex.

Both were factors in the controller's cost. But the flexibility it provided to the engineer more than made up for it. The microprocessor replaced fixed logic with programmable logic.

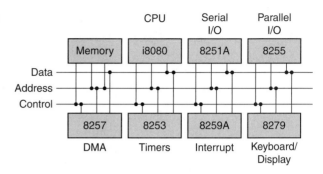

**FIGURE 2.14**   Intel i8080 System Chip Set

When used in the CRT terminal, the microprocessor acted as programmable logic. It dramatically increased the flexibility of the terminal to be configured for different interfaces. It allowed for a more robust feature set that could be tailored for individual customers. The perceived value to the customer more than made up for any additional cost in the controller design.

This early application of microprocessors to embedded controller design proved fantastically successful. Soon the use of microprocessors was expanded to all types of design where programmability brought enhanced functionality. Product designs became more market driven as customers explored their new power to add increasing numbers of features to their products.

## 2.7     MICROPROCESSOR PERIPHERALS

The success of the microprocessor concept brought with it the need for associated parts. The same problem that existed for complex fixed logic was being created for the circuitry to support the microprocessor. More LSI components were needed to simplify the additional logic requirements in the same way the microprocessor was needed to replace fixed logic designs.

There was a commonality for microprocessor designs based on the nature of the microprocessor itself. The i8080 was based on an 8-bit data word. It was natural to develop a chip with corresponding 8-bit data ports, the i8255. It allowed each port to be programmed for input or output.

CRT terminals had a necessary requirement for serial I/O with the computer. The i8251 USART (universal synchronous/asynchronous receiver/transmitter) was developed in response to this requirement. To support the need for counting and timing, an i8253 was designed. High-level data link control/synchronus data link control (HDLC/SDLC) protocol chips, data encryption standard (DES) encryption chips, and many other highly integrated programmable devices were designed to meet the engineer's appetite for additional functions. A partial list of popular Intel-compatible peripheral devices is given in the table shown in Figure 2.15.

### 2.7.1 Microcomputer

As the ability of engineers increased to make more complex controller designs, from the system level, the controllers began to take on the characteristics of computers. They incorporated all the basic functions: program and data storage, I/O, timers and counters, and system control logic as shown in Figure 2.16. In fact, because they were programmable, they could be used in other ways, like processing data. In a way, the first CRT terminal controllers could act like computers. They were physically small; it was natural to call them microcomputers.

**FIGURE 2.15**   Intel Peripheral
Chips

| | |
|---|---|
| 8251A | USART |
| 8257 | Direct memory access |
| 8253 | Timer/counter |
| 8255 | Parallel I/O |
| 8259 | Interrupt controller |
| 8271 | Multiprotocol serial controller |
| 8273 | Keyboard controller |
| 8275 | CRT controller |

**FIGURE 2.16**   Intelligent
Controller

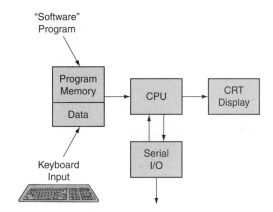

As microprocessor designs proved themselves in the marketplace, there was an increasing demand for engineers to use them for different design applications. Some products, like portable devices, had limits to how physically large the controller could be. Also, many applications only required simple I/O and a few functions.

Semiconductor manufacturers responded to the challenge. Taken to its logical conclusion, microprocessor-based systems could be reduced to the size of a single LSI chip. The capabilities of semiconductor fabrication technology of the day limited what could be accomplished. However, from a functional point of view, a basic single-chip microprocessor system should be possible.

## 2.8   INTEL i8051 MICROCONTROLLER

The peripherals used with the i8080 had become de facto industry standards. They had emerged as practical system building blocks. By reducing the memory and simplifying the peripheral functions, designing an 8-bit single-chip microcontroller would be feasible. The result, the Intel i8051, was introduced to the market in 1977. Figure 2.17 is a functional block diagram of the original i8051 (MCS-51).

Market demand continued to build for a lower-cost microcontroller. In response, Intel introduced a derivative of the i8051, the i8021. It had a subset of the i8051 functionality while maintaining the family architecture. To further reduce costs, it was offered in a 24-pin plastic dual in-line package (PDIP) versus the 40-pin version of the i8051.

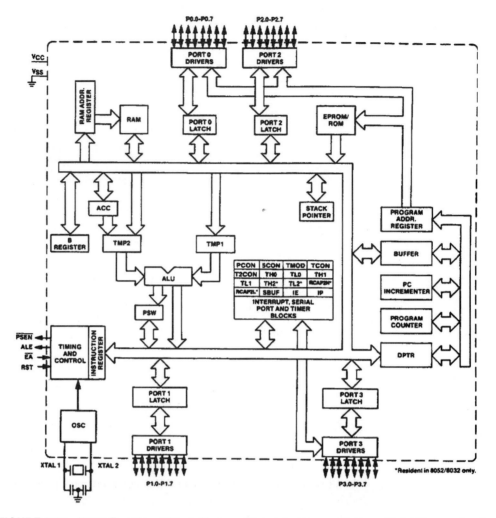

**FIGURE 2.17**    i8051 Functional Block Diagram (Reprinted by permission of Intel Corporation.)

## 2.9    RISC INTRODUCTION

During the early 1980s, the advancement in semiconductor fabrication technology enabled the design of RISC-based microcontrollers. The Harvard architecture, developed in the early 1950s had been impractical to implement because of the lack of cost-effective electronics. The integration level of transistors on a die, achieved by the early 1980s enabled RISC microcontrollers to be designed on a single chip.

An underlying concept of RISC is the overlap of instruction execution of the program. Multiple instructions are executed in parallel to maintain an effective throughput rate of one instruction execution per machine. This requires a tight coupling of the software with the processor. RISC processing demands optimized software compilers to achieve the maximum instruction execution rate.

### 2.9.1  RISC Processors

The initial RISC processors were composed of multiple chips in the same manner as the i8080. A separate RISC processor, data memory, and program memory were connected via busses. The

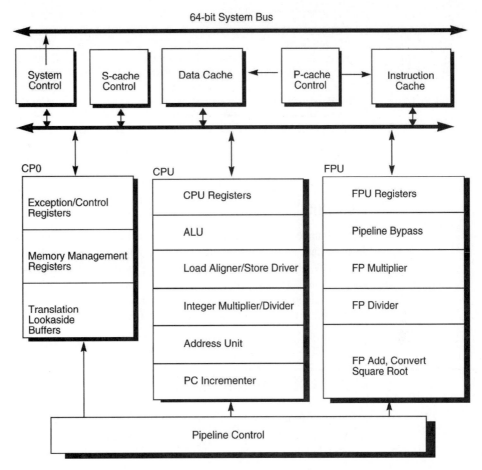

**FIGURE 2.18** RISC R4000-Based System Block Diagram (Reprinted by permission of MIPS Technologies Incorporated.)

RISC chip was then interfaced to standard peripheral I/O devices via a system bus. Figure 2.18 shows a typical system block diagram using the MIPS R3000 chipset. The command and memory bus are common to all devices similar to the concept used in the CISC system design.

The initial CISC design 8-bit microcontrollers were first introduced in the mid 1970s. The introduction of RISC-based chip sets with similar functional capability lagged CISC by almost 10 years. At the time the MIPS 2000 RISC chipset was introduced, its performance rating exceeded that of the i80286.

## 2.9.2 RISC Synergy

By the late 1980s, RISC processors had established themselves in the marketplace. There were competing designs from HP, SUN, and MIPS. There was a significant difference from the existing CISC products. These RISC chips (MIPS 2000, HP-PA, Sun SPARC) were not designed to be sold as end products to the semiconductor marketplace. Sun, HP, and MIPS all introduced RISC-based workstations.

RISC processors are more efficient for workstation design. They also used UNIX as an operating system and C-language for program generation. These are significant departures from CISC personal computers with the Windows operating system. These and other features of RISC would build a wall between the RISC and CISC communities.

| Actions Semi | ALi Corporation | AMD | ASUSTek |
|---|---|---|---|
| Atheros | Broadcom | Centillim | Chartered |
| Conexant Systems | ESS Technology | Infineon | IDT |
| LSI Logic | Marvell | Microchip | NEC |
| NXP | PMC-Sierra | Sony (SCEI) | Toshiba |

**FIGURE 2.19**   Example List of MIPS RISC Licensees

### 2.9.3 RISC Marketing

MIPS, founded in 1982, created a business model to meet the need for second sourcing. This was an important issue regarding the use of any type of semiconductor chip in a customer design. Customers wanted to have multiple sources for their chips. This competitive situation would encourage competitive pricing.

Semiconductor companies licensed their proprietary chips to other manufacturers. These were contractual agreements where the licensed company would pay royalties per chip shipped.

## 2.10   FABLESS SEMICONDUCTOR COMPANY

MIPS was formed as a fabless semiconductor company. They did not have a wafer fabrication facility to manufacture their chips. The parts were then sold by MIPS directly, as if they were a semiconductor manufacturer. This "fabless" business model is shown in Figure 2.20.

**FIGURE 2.20**   Fabless Business Model

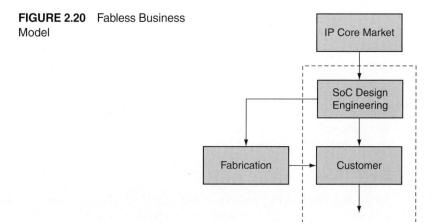

The first chip from MIPS was a 16-bit RISC processor chipset, the RS2000. It used standard nonvolatile program memory and RAM for data memory. This design was intended for MIPS's workstation. Additionally, as per the typical RISC model, they offered a C-compiler for software program generation.

To increase their business, MIPS chose to market their chip design technology to a limited number of silicon manufacturers. They developed an architectural licensing scheme. This allowed the manufacturers to incorporate the MIPS core into their own designs. Subsequently, a range of single-chip microcontrollers based on RISC architecture was introduced to the marketplace.

## 2.10.1  RISC as Intellectual Property

This innovative business model was the leading edge for intellectual property design utilizing microprocessor cores. The MIPS licenses soon introduced a variety of chips based on the same core technology. By using the same RISC ISA processor, common software could be used.

Compatible software was another important innovation. It allowed a broad range of chips to utilize compatible C-compilers. Additionally and of key significance was that independent companies could develop debug tools for the software and hardware. These integrated development environments encouraged a variety of products.

With this new business model, RISC-based processors were now able to compete with traditional CISC microprocessors. Many companies designing products with a common ISA allows for sharing of a common design environment. Listed following are common market segments used in the IP community. Note the variety and breadth of applications.

Physical library
Analog and mixed signal
Arithmetic and mathematic functions
Test cores
Peripheral cores
Bus interfaces
Digital signal processing
Processors and microcontrollers
Memory element
Security/error corr. det./modulation
Video/image/audio
Wireline communications
Wireless communications
Platform level IP
Hardware OS/RTOS
Intercore communications
Embedded configurable logic

The CISC monopoly on the microprocessor market was broken. A limited number of companies could no longer dominate the market. The limitation of chip design was overcome. Innovation was returned to the marketplace. From this time forward, the RISC ISA dominated the variety of designs in the marketplace.

It was the semiconductor technology limitations to integrate sufficient transistors on a single die that was the limiting factor in RISC development. The design concepts were firmly established in theory and experimental circuitry. It took the LSI technology to enable practical manufacture at economic pricing.

## 2.10.2  RISC Technology Curve

RISC processor architectural development continued rapidly, in step with the semiconductor fabrication technology. As the processors increased in speed, functions previously on separate chips were integrated with the processor core. RISC and CISC were now on similar semiconductor technology paths.

## 2.11        EMBEDDED CONTROLLER IP

A convergence of very-large-scale intregration computer-aided design (VLSI CAD) and semi-conductor fabrication technologies during the 1980s created a new microcontroller design paradigm. It enabled engineers to incorporate microcontroller functionality with broader system control functions onto a single chip. The microcontroller portion of the chip became known as an embedded CPU or core (refer to Section 2.24).

**FIGURE 2.21**   Infineon C166SV1 Functional Block Diagram (Reprinted by permission of Infineon Technologies, AG.)

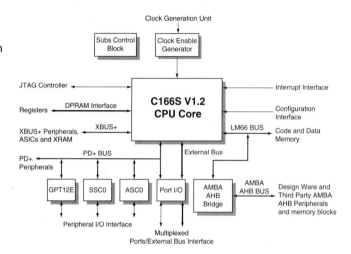

Logic functions are converted by the VLSI design tools to the die masks used in the manufacturing process. A high level description language (HDL) of the functions is converted to mathematical equations. The software then processes them, and a single physical layout of the die is created (refer to Figure 2.22).

**FIGURE 2.22**   VLSI Design Flow

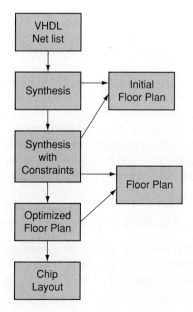

### 2.11.1  CISC IP

Intel licensed the i8051 to design companies for their use in system on chip designs. Once the logic has been converted to HDL, it can readily be incorporated into many designs. With an industry standardized design language, the i8051 as intellectual property (IP) can be easily transferred to third-party companies.

It is possible for skilled engineers from companies in many different industries to design ASICs and SoCs with the i8051. For writing software programs, development tools for the i8051 microcontroller are widely available at affordable prices. In addition, the i8051 is a design proven over time. This gives the engineer a high degree of confidence when using it in a chip project.

There were other widely used CISC microcontrollers also on the market such as the Freescale 68HC11 and the Zilog Z8. These companies, however, chose not to license their ISA technology to the open market. They chose a strategy to keep the design technology in house. As a result, the most widely used CISC ISA for embedded processor design remains the Intel i8051.

### 2.11.2  RISC IP

The number of companies offering RISC IP continues to increase. It is a tribute to the success of the "fabless" business model. Within the SoC concept, independent companies have the ability to design their own full-custom microcontrollers. For example, companies like ATMEL can produce a line of chips using the ARM 7 core.

### 2.11.3  Third-Party IP

By the early 1990s, a design model of incorporating IP into embedded processor designs became firmly established. New companies entered the emerging marketplace with RISC-based ISAs. Advanced RISC machines (ARM) introduced an ISA directed at the very lucrative personal electronics market. Their business model for generating revenue was licensing their processor core technology to companies and collecting royalties.

High-performance products such as internet routers and servers demanded the maximum computing processing power available. With the rapid advances in semiconductor technology (Moore's law), it now became feasible to incorporate these complex and powerful RISC processor cores in SoCs. This was not possible with the proprietary CISC microprocessors available on the market (Intel Pentium and AMD Athalon).

Taken as a whole, an SoC represents a tremendous amount of engineering design investment. However, by leveraging this design effort, SoC engineers could leverage this design manpower. Embedded RISC core SoCs do represent an enormous amount of design effort; however, it is spread over many companies. By purchasing IP functions, engineering design teams develop sophisticated SoCs for a reasonable monetary investment.

---

## 2.12          APPLICATION SPECIFIC PROCESSORS

An alternate path was emerging for SoC design utilizing the most powerful RISC IP available. By licensing established ISA IP from companies like MIPS or ARM, engineers could focus on application-oriented design functionality to differentiate their products. Application-specific processors (ASPs) were developed to meet specific product requirements. Although manufactured in relatively low volume, they were economical because of the use of existing IP.

Advances in VLSI CAD tools suites enables new processor design technology. The ISA itself can be configured to a specific application. This tight coupling between the instruction set

**FIGURE 2.23**   IP Core–Based
Design

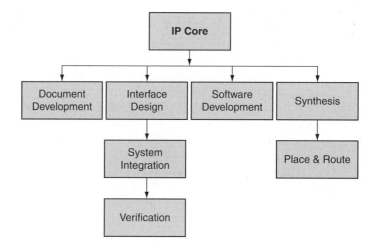

and the processor core further optimizes the processor performance (Figure 2.23). This is often an advantage to design smaller (in terms of die size) cores that can still maintain a high degree of performance.

Maintaining processor die size relative to the SoC size means less power is required for processing. In battery applications such as cell phones or MP3 players, reduced power consumption increases the useful lifetime. Additionally, smaller core footprint reduces die size of the SoC, thus positively affecting its cost.

## 2.13          SUMMARY

RISC microcontroller instruction set developments take many directions related to the application for which they are designed. Not constrained by legacy CISC architectures, innovative approaches can be taken. It is clear that RISC technology is the driving force in embedded microcontroller and processor design.

What is today considered the high-end processor will, by definition, become the mass market embedded core of the next generation of products. Engineers always find a way to use more processing power in their designs, and marketing managers always find new applications for them. And CEOs always demand greater profits for their company.

RISC, being new, was not limited by existing market constraints. A new group of semiconductor companies now had access to processor technology. They could develop a new group of microcontrollers based on RISC ISA.

With the increase in VLSI technology, it was possible to integrate the peripheral I/O functions with the processor core on a single chip. This was comparable to an Intel i8051, Motorola 68HC12, or Zilog Z8.

## QUESTIONS AND PROBLEMS

1. What was the driving force in the development of the microcontroller?
2. What does it mean to be called an electromechanical device?
3. What is the basis of the von Neumann architecture?
4. List the two most popular ISAs.

5. Define SSI.
6. Define the term "buss".
7. Why use a microcontroller on a PCB?
8. What is the partitioned design concept?
9. Diagram the basic signals of an RS-232C serial interface.
10. Describe a key difference between a microcontroller and an electronic calculator.
11. What was the original driving force behind microcontroller development?
12. What is decimal "173" in both binary and hex formats?
13. What is a key system design constraint of the microprocessor?
14. Why did the early microprocessor require multiple chips to operate?
15. List the five basic functions of a microcontroller.
16. What architecture is RISC based on?
17. What was the primary initial application for the RISC processor?
18. Define the concept of "second sourcing."
19. What is the "fabless" business model?
20. Embedded core technology is based on what key concept?
21. What are applications specific processors?

## SOURCES

C166S—A proven architecture takes on new shape. Technical presentation, Infineon Technologies, September 2002.

Dadda, L. 1991. The evolution of computer architectures. *IEEE:* 9–16.

*80C51FA/83C51FA Event-control CHMOS single-chip 8-bit* microcontroller data sheet. Intel Corporation.

8051 Microcontrollers hardware manual. 2005. ATMEL Corporation.

Godfrey, M. D. and D. F. Hendry. 1993. The computers as von Neumann planned it. *IEEE Annals of the History of Computing* 15 (no.1): 1993 11–21.

Miller, K. 2003. Development of the digital computer: Part 1. *IEEE Potentials:* 40–43.

von Neumann, J. 1945. First draft of a report on the EDVAC. 1993. *IEEE Annals of the History of Computing* 15 (4): 27–75.

# CHAPTER 3

# Embedded Microcontroller Technology

**OBJECTIVE: AN UNDERSTANDING OF THE INTERNAL ARCHITECTURAL DESIGN OF RISC AND CISC PROCESSORS**

The reader will learn:

1. Semiconductor technology development.
2. Concept of design-level abstraction.
3. Design performance measurement.
4. Software, hardware, and memory technologies.
5. Instruction set design concepts to microcode level.
6. Internal RISC processor design concepts.

## 3.0 INTEGRATED CIRCUITS

Since the invention of the transistor in 1949, the pace of semiconductor technology innovation has been the driving force in computer design. More than any other technology, putting more transistors into a smaller space has been the enabling technology leading to extraordinary developments in computing machines.

## 3.1 MOORE'S LAW

The market introduction of 256-bit dynamic random access memory (DRAM) composed of metal-oxide semiconductor (MOS) transistors (Figure 3.1) in 1969 is generally considered to have begun the semiconductor memory era. Many companies rapidly followed with

**FIGURE 3.1** CMOS (Complementory Metal-oxide Semiconductor) Transistor

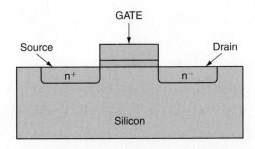

DRAM introductions of increasing density. By the end of the 1970s, 64-k DRAM were in volume production.

Gordon Moore, former chairman of the board of Intel Corporation, first quantified this in a mathematical formula. He postulated that the rate at which the number of transistors put on a single die would double was every 18 months. For microprocessors, this has meant higher clock frequencies. Figure 3.2 shows this relationship by plotting chip clock frequency against cell density.

**FIGURE 3.2**  Cell Density versus Clock Frequency (Reprinted by permission of Tensilica Incorporated.)

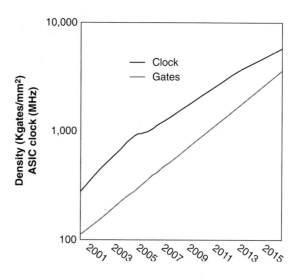

It is not simply shrinking the transistor cell size that accounts for this density increase. It is the entire collection of supporting technologies in the design and fabrication process. In addition to shrinking transistor size, fabrication technology enables larger die size. These two factors combine for the continually increasing density.

## 3.1.1  Microprocessor Performance

DRAM design and fabrication technology are considered the leading edge of the semiconductor revolution. Techniques developed for increasing DRAM density are subsequently used in microprocessor and microcontroller design. The number of transistors available to processor design enables increasingly powerful processors.

Moore's law can be applied to microprocessor design. The introduction of key microprocessors is plotted against transistor density in Figure 3.3. It is not simply the fact that

**FIGURE 3.3**  Microprocessor Transistor Count per Moore's Law (Reprinted by permission of Intel Corporation.)

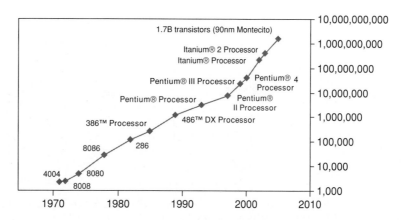

more transistors are available to design engineers; the additional transistor count enables more sophisticated processor architectures. New generations of designs are possible as the underlying suite of semiconductor technologies advances.

The consequence of Moore's law has resulted in spectacular increases in computing performance of microprocessors: from the 4.77-MHz crystal frequency of the first 8-bit PC to the performance of a 4-GHz 64-bit AMD Athalon. The performance of a $20,000,000 supercomputer of 1970 is now attained by a $500 personal computer—a price performance increase of a factor of 40,000.

### 3.1.2 Enabling Technologies

DRAM production can pioneer new transistor fabrication technologies that microprocessor design engineers can utilize in next-generation processors. In the same manner, microprocessor design technology leads the way for innovations in embedded microcontroller design (Figure 3.4). Thus, microcontrollers are derivative designs from their microprocessor parents.

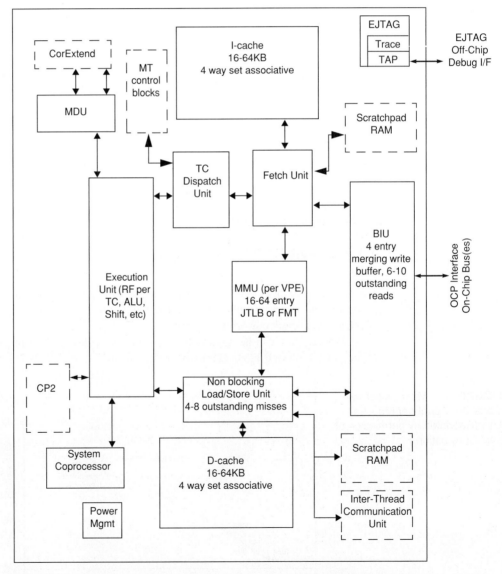

**FIGURE 3.4**   MIPS 34-Ke 32-Bit CPU Core Block Diagram (Reprinted by permission of MIPS Technologies Incorporated.)

This march of technologies led to current 64-bit microcontrollers. Supercomputer levels of performance can be incorporated into a single system on a chip (SoC). This allows design engineers to embed tremendous amounts of processor performance that is optimized to specific applications.

At any given point in time, engineering design is a balance of available technologies. Computer design engineers strive for the maximum performance given the constraints of the available technology. Its basic architecture is a result of this interplay of design and manufacturing technologies.

Von Neumann developed his design innovations from the technology available to engineers in the early 1940s. It was designed with slow vacuum tube circuit technology with data memory that was limited to a few words. The software program, as such, was hardwired.

The resulting von Neumann architecture was a natural consequence of the intersection of these three primary technologies. Alternative possible architectures were not pursued, not because they did not exist, but rather because the technology then available did not enable them. With the Princeton architecture selected, the goal became to optimize its performance.

The convergence of the computer triad of technologies (Figure 3.5) continues to be the enabling parameters for computer architects. Transistors have supplanted vacuum tubes, memory is measured in megabytes, and software is optimized. Designing for maximum computer system performance requires balancing these technologies. Emphasis on one technology over another will yield a less-than-optimal design.

**FIGURE 3.5** Three Primary Computer Technologies

### 3.1.3 Amdahl's Law

The challenge of balancing technologies to achieve maximum performance is well stated by Amdahl's law (Figure 3.6). Speedup due to an enhancement is proportional to the fraction of time that the enhancement can be used. This means that doubling the clock frequency of a processor will not correspondingly double throughput. The memory must also be able to provide instructions and data at that new speed. The software program must be structured in a way that supports uninterrupted instruction execution rates.

**FIGURE 3.6** Amdahl's Law

$$\text{ExTime}_{new} = \text{ExTime}_{old} \times (0.9^*1.0 + 1^*0.5)$$
$$= 0.95 \times \text{ExTime}_{old}$$

$$\text{Speedup}_{overall} = \frac{\text{ExTime}_{old}}{\text{ExTime}_{new}} = \frac{1}{0.95} = 1.053$$
$$<< 2$$

Technologies advance in their own timelines. As new innovations become available, design engineers incorporate them into their designs. However, outside technology lies the constraint of the marketplace. Economics plays a key, if not the overriding role, in what is actually designed. What exist as microcontroller design architectures today represent technology achievement within the complex constraints of market forces.

### 3.1.4 Technology Convergence

To that end, processor architecture can become frozen in time. Innovative architectural techniques may not be brought to the market. This can be seen with the case of the Intel processor architecture–based personal computer. With software essentially frozen on a fixed instruction set, only advances in logic and memory technologies could be pursued. The constrained advances in software technology resulted in less-than-optimal performance than otherwise could be achieved.

Applications that do not require a high degree of software compatibility free processor design engineers to pursue alternative architectures. Early hardware and memory technology of the 1980s developed to enable them. The Harvard architecture, an alternative to that of Princeton, was now enabled by this convergence of processor, memory, and software technologies.

## 3.2          DESIGN ABSTRACTION

The software instruction set is defined at the time of the processor design conception. It is not an output of the hardware design. It is what drives the hardware design. Engineers make trade-off decisions of what the three technologies are capable of contributing to maximize performance. The instruction set architecture (ISA) is a fundamental part of this optimization process.

Once the ISA is determined, processor design can proceed. The basic design goal becomes executing instructions at the fastest rate possible. The processor logic is designed around the instruction execution requirements. The memory structure is tailored to support feeding instructions and data to the processor at the maximum rate. Software programs are optimized to the fewest possible instructions for the application.

The concept of ISA is an abstraction of the computer system physical design. It encompasses a broad range of possible design implementations. It is more a description of design philosophy than specific design criteria. ISA also connotes a sense of the logic design of the processor as a system, not just a set of instructions. Figure 3.7 shows how the ISA is at the base of the layers of software and hardware abstraction.

**FIGURE 3.7**  Levels of Abstraction

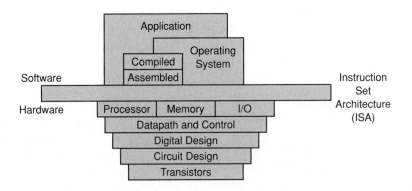

Within this concept, microprocessor architectures are categorized into two very broad ISAs: CISC and RISC. They are not mutually exclusive and have clear overlapping functionality. The core of the philosophy of design implementation is their differentiating characteristic.

### 3.2.1  Instruction Set Architectures

There are many significant instruction set architectures. A list of commonly defined ISAs is given in the table in Figure 3.8. An instruction set architecture is defined to overcome significant application-oriented engineering obstacles. Practical ISAs yield a sufficient price/performance enhancement to overcome their implementation costs.

**FIGURE 3.8**   Instruction Set
Architectures

| | |
|---|---|
| CISC | Complex instruction set architecture |
| RISC | Reduced instruction set architecture |
| VLIW | Very long instruction word |
| EPIC | Enhance parallel instruction computer |
| MISC | Multiple issue instruction set computer |
| DSP | Digital signal processing |
| GPU | Graphics processing |

At a given level of detail, processor architectures incorporate attributes of each other. From the top down, CISC and RISC implementations may have strong similarities. From the bottom up, they remain philosophically different. However, as the underlying processor logic, memory, and software technologies advance, the interplay between CISC and RISC becomes more advantageous. New designs with appropriate enabling technologies can incorporate design attributes of both ISAs. This, in turn, promises greater performance with these evolving "hybrid" architectures.

### 3.2.2  Processor Family Tree

Instruction set architecture definitions can be pictured as branching on a family tree. The times are based on their generally accepted introduction to the market in practical form. This chart also depicts, rather strikingly, the points in time when technologies converge to enable the new ISAs.

What can also be seen from the chart is that traditional CISC and RISC definitions are giving way to new ISAs. This reflects continuing development of processor logic, memory, and software technology. The graphs can be extrapolated. However, future commercial success is not so predictable. Determining when an ISA moves from theory to design is the stuff of which fortunes are made.

## 3.3   RISC AND CISC

At a very general level, CISC and RISC can be compared using their global design tendencies. The table in Figure 3.9 is a simple comparison table. It is not meant to be either exhaustive or detailed.

| CISC | RISC |
|---|---|
| Many instructions | Few instructions |
| Variable length instruction words | Fixed length instruction words |
| Many addressing modes | Few addressing modes |
| Memory-based data operations | Register-based data operations |
| Many special registers | Orthogonal register set |
| Instruction execution in many microcoded cycles | Single-cycle instructions |
| Assembly language driven | High-level language driven |

**FIGURE 3.9**   Comparing CISC and RISC

### 3.3.1 Processor Technology

In an idealized computer, the program executes as fast as the programmer has designed it. Nothing more complicated than that. The processor executes instructions that perform the required operations. Data processing is synchronized with the input and output. With all functional units of the system working in unison, maximum performance is achieved. In practical terms, the theoretical peak performance is a design goal.

The switching speed of transistors is fixed by their semiconductor fabrication technology. This in turn fixes the speed of the gate level processor logic. The graph in Figure 3.10 shows the relative transistor switching frequency versus the drawn feature size in microns. At deep submicron technology levels, switching frequency exceeds 10 GHz. This translates to gate delays under 100 picoseconds. At the same time, the transistor size is reduced, which means more logic gates are available for the design.

**FIGURE 3.10**  Transistor Switching Frequency

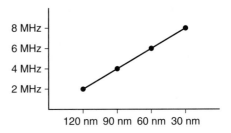

### 3.3.2 Performance Measurement

Performance measurement of a computer system at any level of abstraction is a relative process. It is like quantum measurement. The more exact the measurement of a given, the less exact it is. To judge which architectural design trade-offs to make, engineers need some form of performance measurement.

Simply put, the faster a program executes instructions, the more data it can process. This is a basic principal of data processing. The following equation is a basic measurement of design performance or benchmark. It is a determination of the efficiency of the processor design.

$$\frac{Time}{Program} =$$

$$Time_{Program} = Instructions_{Program} * Cycles_{Instruction} * Time_{Cycle}$$

The time per cycle is fixed by the processor design technology. It is considered to be the clocking frequency of the processor logic. Reducing the number of instructions in the program not only increases performance, but also simplifies the program. Programmer productivity and system performance are increased.

The time required to execute a program can be reduced in three different ways. Fewer instructions can be used to code the program. Fewer processors clocking cycles can be used to execute an instruction. Finally, the time per cycle can be reduced.

### 3.3.3 Program Instructions

Assume a program is written with the fewest possible instructions. This will fix the first term of the equation. The hardware will be fixed by the design technology. The only method left to improve performance (decrease the program execution time) is to reduce the number of instructions. This is the driving force of CISC design.

The three terms interact in a recursive manner. Adding instructions will correspondingly increase processor complexity. This may result in slower instruction execution cycle times. There is always a trade-off. Amdahl's law quantifies the performance value of the trade-off decision.

In real-world applications, much more than simple instruction and processor cycle time are the true measure of system-level performance. A suite of benchmark programs are used for processor evaluation. And even at that, there is no agreement on a single benchmark number.

### 3.3.4 Cost per Instruction

The cost per line of code is independent from the content. It is directly related to programmer efficiency. Assume a programmer generates about 40 lines of fully documented and debugged code per day. The program cost per instruction can be easily derived. It is the number of instructions divided by the programmer cost. For example, at $800 per programmer man-day, the cost per instruction is roughly $20, figuring a fully documented and tested 40 lines of code per day.

$$\text{Time}_{inst}\text{-Cost} = \text{Programmer Cost...Day/\#\_Instructions}$$
$$= \$800/40$$
$$\cong \$20$$

As we have seen, reducing the number of instructions increases program performance. At the same time, it also reduces the program's memory size. This means a cost savings in memory (Figure 3.11). Reducing the number of instructions will reduce program cost, increase performance, and reduce hardware (memory) cost. This is a powerful driving force in CISC design.

**FIGURE 3.11**   Program Size versus Cost

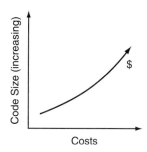

### 3.3.5 Microcoded Instructions

Increasing the number of instructions correspondingly increases the complexity of the processor logic. As per Amdahl's equation, there is a point of diminishing returns. An alternative method is required to process more instructions efficiently. The key enabling processor technology for this concept is microcoded instructions.

Within the processor, execution of machine instructions is partitioned into a series of steps. The instructions become small programs themselves that are executed with the processor core logic. Adding new instructions simply requires programming more microcode instructions.

Fewer instructions results in smaller code size. Expensive memory is more efficiently utilized. In turn, larger, more functional programs can be stored in the same address space. This effective technique is what drives the development of ISA extensions. A good example is the MMX$^{TM}$ feature of Intel Pentium processors. These instructions significantly increase graphics performance and simplify the programming software.

This is the creed of the CISC ISA processor design. Complex instructions, as opposed to "simple" ones are good. For the three converging technologies, this means faster execution, less memory, and lower programming costs.

Machine instructions are visible to the programmer; microcode is not. The microcoded instruction sequences are fixed by the processor designers. They are embedded within the core logic. In a sense, CISC processors contain embedded microcontrollers. For example, the Pentium processor could be thought of as an application-specific processor targeted at general-purpose computer applications.

From this simplified performance measurement, you can see the interplay of the three basic technologies. Hardware is designed to support the fastest execution of instructions with the least amount of program memory. This is true, not only at the high-level software programming, but at the microcode level as well.

## 3.4        MEMORY TECHNOLOGY

A key aspect of ISA development results from the disparity between accessing speed and size in the memory hierarchy. On a relative basis, memory with faster access time is expensive. To the degree that instructions can more efficiently use this memory, the greater the potential processor performance and the lower the system cost.

Stored program computers require memory for the instructions and data. Incorporating that memory as part of the processor logic is not practical or desirable. There is a finite limitation to the number of transistors that can be put on a die. Dedicating them to the processor logic is far more valuable.

There are, it is said, three fundamental requirements for memory: fast, faster, and fastest. Programmers might add large, larger, and largest. Regardless of speed and size, processors cannot execute instructions faster than they can be fetched from memory (refer to Figure 3.12). Nor can they process data faster than it can be fetched from and stored to memory. These are limiting factors in processor design.

As we know from Moore's law, memory doubles in size (per die) about every 18 months. Concurrently access times have been decreasing, as shown in the table in Figure 3.13. But they do not match the rate at which processor logic is clocked. Although fast, DRAM is still a magnitude of 10 slower than the processor.

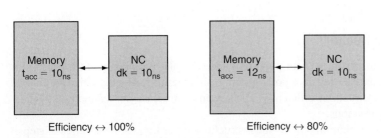

| DRAM Type | Frequency (MHz) |
|-----------|-----------------|
| PC-3200   | 400             |
| PC-5300   | 533             |
| PC-6400   | 800             |
| PC-8000   | 1000            |

**FIGURE 3.12**  Memory Processor Interface Speed

**FIGURE 3.13**  PC Memory Speeds

To achieve the computer's theoretical performance (CTP) requires that instructions and data be available to the processor at its instruction execution frequency. If either instruction or data are not available, the processor will delay execution, and the maximum performance will not be achieved. Direct execution of the instruction and data streams from main memory must effectively occur as close together as possible to match the processor speed.

### 3.4.1 Locality

Studies of program software flow show that programs execute in small groupings of instructions. Examples would be subroutines or repetition loops acting on a data set, as shown in Figure 3.14. The program code will execute a number of instructions and then branch to a new location. At the new address, the program will again execute a series of instructions. This procedure is called locality. It is in general true for both programs and data sets.

**FIGURE 3.14**  Program Locality

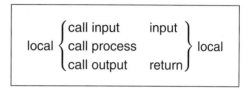

It is important to note that both the instructions and data must be made available to the processor. However, they have significantly different attributes. Although they are both thought of as streams, instructions must be executed in a predetermined sequence. Data, although exhibiting locality, is not inherently a sequential process.

Real-world applications range in their requirements for memory. Some applications require large complex programs to process relatively small amounts of data (word processing). At the other extreme are smaller programs that repetitively process a vast amount of data (Internet search engines).

With the single-memory von Neuman architecture of CISC, the challenge to fetch instructions and data from memory is effectively the same. Modified data must be written back to memory. The processor must store the data back to memory as fast as it is processed to make way for incoming data.

At any given point in time, the design engineer uses the currently available technology. As computer design technology has evolved over time, many techniques have been developed to keep the processor "busy." Many different methods determine what busy means. It is generally understood to mean keeping as many of the processor's functional units at work executing instructions as possible.

### 3.4.2 Memory Hierarchy

The memory hierarchy concept utilizes the principal of locality. A buffer, or cache memory, is inserted between the processor execution units and the main memory. The instruction fetch portion of the processor logic is designed to fetch and execute instructions from cache without delays. The data is moved from the cache to the register set. To the degree that it can be made available to the instructions, the processor can attain its CTP rating.

A general memory hierarchal structure has developed. In the abstract, four levels of hierarchy are defined: register, cache, main, and file storage (see Figure 3.15). The closer the level is to the

**FIGURE 3.15**  Memory Hierarchy

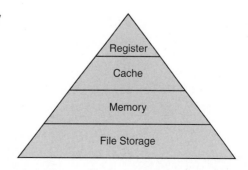

processor, the faster the access time required. Depending on the design goal of the application, the levels will vary in size and performance. For microcontroller design, one or all may be incorporated.

The speed of access across the levels of the hierarchy can vary widely. The greater the amount of storage capacity, the slower the access speed, as we have seen. Figure 3.16 shows the relative access time across a four-level hierarchy. File storage is electromechanically implemented (hard disk) and is orders of magnitude slower than semiconductor memory.

**FIGURE 3.16**  Memory Hierarchy Access Times

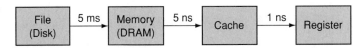

The memory hierarchy works because of locality. At the file level, this can be applied to complete programs within a complex application, such as Internet routing. Between levels of the hierarchy, additional caching (buffering) may be implemented to reduce the access latency times further. Hard disks and CD drives include cache memory to improve performance interface to main memory. As long as the processor receives the instructions and data as fast as required, the goal of the memory hierarchy is successful.

### 3.4.3 Cache Memory

As we have seen with the partitioning of memory into hierarchy the fastest memory is cache. As per Moore's law, devices for main memory continue to double in size. Two-gigabit memories will give way to 4-gigabit, then 8-gigabit as DRAM technology continues to evolve.

At the other extreme is the continuing progression of fast memory. Decreasing fabrication design rules increase the number of transistors that can be put on a single die. Additionally, more transistors can be used for on-chip cache memory. The larger cache size increases processor performance.

Putting 25 million transistors on a single die enables new design techniques. Not only registers but also cache memory can now be incorporated into the processor die. It is as if the processor itself now includes an instruction and data memory of its own.

### 3.4.4 L1 and L2 Cache

At this point, cache memory itself becomes partitioned. The fastest cache, the cache memory designed as part of the chip itself, becomes the L1 cache (Level 1). And the larger cache, on a separate die but within the same chip package (e.g., Pentium) becomes the L2 (Level 2) cache, as shown in Figure 3.17.

**FIGURE 3.17**  Processor with L1 and L2 Cache

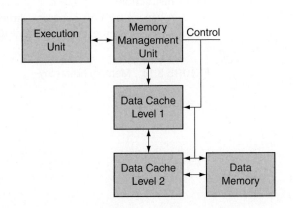

### 3.4.5 Data Registers

Memory incorporated within the processor logic is called a register file. It holds data words and can be read from and written to at the operating frequency of the processor. That is, the data in the registers is available as fast as the instructions can process it. As long as data can be moved between cache and the register set as the instructions require them, the processor will operate at the maximum performance.

Addressing the registers can be accomplished in one of two ways. They can be defined as unique addresses located within the processor. In this case, they will be implicitly coded within the instruction itself. Alternatively, they can be incorporated as part of the memory address space, as shown in Figure 3.18. Depending on the application, either approach may offer performance benefits.

**FIGURE 3.18**  Memory Mapped Registers

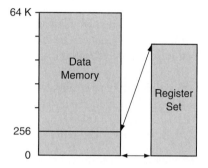

Studies have been made to determine the optimal number of registers. The number needs to be enough to adequately buffer cache memory, but not so many as to complicate the programming. There is a point of diminishing returns. This point determines register versus cache sizes.

### 3.4.6 Instruction Queues

Moving instructions into and out of the processor core logic is a key challenge facing processor design engineers. The goal is to make the program instructions available to the processor in the sequence defined by the software code. This is a distinctly different task from reading and writing data to memory.

Instructions cannot be simply moved into multiple memory locations within the processor logic and executed in a random order. There is no corresponding concept of registers for instructions. Instructions are buffered in a queue. Instructions within the queue are accessed sequentially by the processor, as shown in Figure 3.19.

**FIGURE 3.19**  Instruction Queue

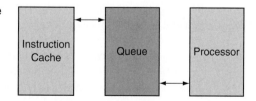

### 3.4.7 Branch Instructions

The crucial time when the sequential execution of instructions is broken occurs when a branching-type instruction is encountered. There are several types of branches, as shown in Figure 3.20. Whatever the type, it will result in a break of the sequential execution sequence and force the

**FIGURE 3.20** Branch Types in the MicroChip PIC 18F4520

| Type | Code |
|------|------|
| GOTO | Goto Skip |
| CONDITIONAL | If $x = 5$ then goto Skip |
| CALL (n) | Call Raster |
| RETURN | Return to (Call address $+1$) |
| RETFIE | Return from Interrupt Enable ($n + 1$) |
| BRA | Branch Skip |

processor to initiate a fetch from a new location. This will stall instruction execution until the new instruction is fetched.

With locality, the address of the branch instruction is not totally random. If the queue has sufficient memory, it can have the branch instructions already fetched. In this case, instruction execution can proceed without a stall. The processor will continue to operate at the full rated speed.

The two major instances when program instruction execution can be stalled are that an instruction is encountered out of sequence or data is not available in the register. Both problems can be resolved to a high degree through the memory hierarchy. This is based on the concept of locality.

### 3.4.8 Memory Latency

Moving a single instruction or data word from disk storage directly to the processor is a slow process. It is the sum of the access times for the cache, main memory, and the file storage device. The file storage device, being electro-mechanical, completely dominates the equation. A typical access time, or latency, through the hierarchy would be

$$\begin{aligned} T_{access} &= t_{cache} + t_{men} + t_{file} \\ &= 2\text{ ns} + 8\text{ ns} + 7\text{ ms} \\ &= 7.010\text{ ms} \end{aligned}$$

In most programs, less than 5% of the code is typically being executed at any given time. An example is a spreadsheet program, where only a very few functions are active at any given moment. As a result, only a very small portion of the spreadsheet's complex code needs to be memory resident.

In practical terms, instructions cannot be moved singly through the hierarchy. The average latency is too long. Memory is defined in terms of blocks of contiguous memory locations. Depending on the application, blocks may vary in sizes from 1 k to 4 k words. If an instruction is not resident in cache, then the block containing it is fetched from main memory.

Instruction latency is a critical definition. It is different for each application. If the required instructions are available as the program needs them, from the application point of view, they are always available. The processor will not recognize there is a memory hierarchy. From the application standpoint, there is one large and fast memory.

### 3.4.9 Cache Blocks

Data, besides being fetched, must also be stored. If a block of cached memory is modified, then it must be written back to the main memory. There is movement in both directions in a cached memory design. However, only blocks with modified data words need to be written back. This reduces cache block data movement to a minimum.

The importance of moving memory blocks between hierarchal levels cannot be over-stressed. In any processor-based design involving cache memories, this is a gating factor to system performance. For a microcontroller, the instruction and data streams must have a high degree of coherence. In real-time applications, the processor must be executing the proper instructions at the proper time to handle the event at the moment it occurs.

Managing a cache-based design requires incremental amounts of logic. This in turn increases the design complexity and ultimately the cost of the chip. Adding on-chip high-speed cache increases the die size with direct increases in fabrication costs. A "cacheless" microcontroller design will provide a lower cost point at the sacrifice of some degree of processor performance. In the PIC family, from Microchip, all on-chip memory is defined in terms of registers. This eliminates the cache problem altogether, as shown in Figure 3.21.

| Address | Name | Address | Name | Address | Name | Address | Name |
|---|---|---|---|---|---|---|---|
| FFFh | TOSU | FDFh | INDF2[1] | FBFh | CCPR1H | F9Fh | IPR1 |
| FFEh | TOSH | FDEh | POSTINC2[1] | FBEh | CCPR1L | F9Eh | PIR1 |
| FFDh | TOSL | FDDh | POSTDEC2[1] | FBDh | CCP1CON | F9Dh | PIE1 |
| FFCh | STKPTR | FDCh | PREINC2[1] | FBCh | CCPR2H | F9Ch | —[2] |
| FFBh | PCLATU | FDBh | PLUSW2[1] | FBBh | CCPR2L | F9Bh | OSCTUNE |
| FFAh | PCLATH | FDAh | FSR2H | FBAh | CCP2CON | F9Ah | —[2] |
| FF9h | PCL | FD9h | FSR2L | FB9h | —[2] | F99h | —[2] |
| FF8h | TBLPTRU | FD8h | STATUS | FB8h | BAUDCON | F98h | —[2] |
| FF7h | TBLPTRH | FD7h | TMR0H | FB7h | PWM1CON[3] | F97h | —[2] |
| FF6h | TBLPTRL | FD6h | TMR0L | FB6h | ECCP1AS[3] | F96h | TRISE[3] |
| FF5h | TABLAT | FD5h | T0CON | FB5h | CVRCON | F95h | TRISD[3] |
| FF4h | PRODH | FD4h | —[2] | FB4h | CMCON | F94h | TRISC |
| FF3h | PRODL | FD3h | OSCCON | FB3h | TMR3H | F93h | TRISB |
| FF2h | INTCON | FD2h | HLVDCON | FB2h | TMR3L | F92h | TRISA |
| FF1h | INTCON2 | FD1h | WDTCON | FB1h | T3CON | F91h | —[2] |
| FF0h | INTCON3 | FD0h | RCON | FB0h | SPBRGH | F90h | —[2] |
| FEFh | INDF0[1] | FCFh | TMR1H | FAFh | SPBRG | F8Fh | —[2] |
| FEEh | POSTINC0[1] | FCEh | TMR1L | FAEh | RCREG | F8Eh | —[2] |
| FEDh | POSTDEC0[1] | FCDh | T1CON | FADh | TXREG | F8Dh | LATE[3] |
| FECh | PREINC0[1] | FCCh | TMR2 | FACh | TXSTA | F8Ch | LATD[3] |
| FEBh | PLUSW0[1] | FCBh | PR2 | FABh | RCSTA | F8Bh | LATC |
| FEAh | FSR0H | FCAh | T2CON | FAAh | —[2] | F8Ah | LATB |
| FE9h | FSR0L | FC9h | SSPBUF | FA9h | EEADR | F89h | LATA |
| FE8h | WREG | FC8h | SSPADD | FA8h | EEDATA | F88h | —[2] |
| FE7h | INDF1[1] | FC7h | SSPSTAT | FA7h | EECON2[1] | F87h | —[2] |
| FE6h | POSTINC1[1] | FC6h | SSPCON1 | FA6h | EECON1 | F86h | —[2] |
| FE5h | POSTDEC1[1] | FC5h | SSPCON2 | FA5h | —[2] | F85h | —[2] |
| FE4h | PREINC1[1] | FC4h | ADRESH | FA4h | —[2] | F84h | PORTE[3] |
| FE3h | PLUSW1[1] | FC3h | ADRESL | FA3h | —[2] | F83h | PORTD[3] |
| FE2h | FSR1H | FC2h | ADCON0 | FA2h | IPR2 | F82h | PORTC |
| FE1h | FSR1L | FC1h | ADCON1 | FA1h | PIR2 | F81h | PORTB |
| FE0h | BSR | FC0h | ADCON2 | FA0h | PIE2 | F80h | PORTA |

**FIGURE 3.21** Register Map for the PIC18F4520 (Copyright 2004 Microchip Technology Incorporated. Reprinted with permission.)

Higher performance inevitably comes at higher cost. However, with higher performance comes greater functionality. The trick, of course, is to get the improvement in performance at the minimal cost. This most often is the difference between success and failure in the high-technology world.

## 3.5          INSTRUCTION PROCESSING

A processor by itself is nothing more than a complex design of logical functions. Assembler-level instructions are broken down to a more basic level called machine instructions. In general, it requires multiple machine instructions (microcode) to execute a single assembler instruction. Binary representations of logic 1s and 0s feed directly into the processor's gate-level logic (Figure 3.22).

**FIGURE 3.22**    Instruction Machine Code

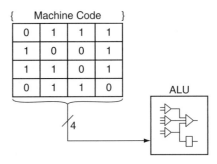

Writing even a short program of machine instructions is a tedious task. Each instruction is coded directly in its binary representation. A GOTO branch instruction needs the binary address of the branch address. It is difficult to read the program and an error-prone task to write it. And most important, it takes valuable software engineering resources to create a short program.

### 3.5.1  Symbolic Assembly

The increase in computer performance and memory size are enabling technologies for software developments. An assembler was developed to translate symbolic code to machine-level instructions. The pattern that produced the add function in the computer was simply named ADD.

Assembly language instructions are easier for the programmer to use. They are typically abbreviations of equivalent symbolic words. Addition becomes ADD, multiply becomes MUL, and so forth (Figure 3.23). The registers are given symbolic names such as R1, R2, and R3. Data variables are also named symbolically rather than having explicit binary addresses defined.

| Mnemonic, Operands | | Description | Cycles | 16-Bit Instruction Word | | | |
|---|---|---|---|---|---|---|---|
| | | | | **MSb** | | | **LSb** |
| **BIT-ORIENTED OPERATIONS** | | | | | | | |
| BCF | f, b, a | Bit Clear f | 1 | 1001 | bbba | ffff | ffff |
| BSF | f, b, a | Bit Set f | 1 | 1000 | bbba | ffff | ffff |
| BTFSC | f, b, a | Bit Test f, Skip if Clear | 1 (2 or 3) | 1011 | bbba | ffff | ffff |
| BTFSS | f, b, a | Bit Test f, Skip if Set | 1 (2 or 3) | 1010 | bbba | ffff | ffff |
| BTG | f, d, a | Bit Toggle f | 1 | 0111 | bbba | ffff | ffff |

**FIGURE 3.23**    Assembler Instructions from the PIC18F4520 (Copyright 2004 Microchip Technology Incorporated. Reprinted with permission.)

The assembler greatly increases the software programmer's ability to write code. Writing working code means not only generating instructions but also debugging them to a working program. The assembler increases computer programmer productivity dramatically. It is an enabling technology to increase the utility and performance of the computer as a system.

## 3.5.2 Program Compilers

Software is assembled line by line from the creativity of the programmer's mind. The higher the level of abstraction, the greater is the productivity, as measured by lines of production code. Compilers generate sequences of machine-level instructions from a single high-level statement. Programmers use a high-level language like C to generate correspondingly more code than one using assembly language. The software technology embodied in the C-compiler enables a significant degree of productivity in generating production code, as shown in the following C-code example.

```
void ee_byte(byte x)
{int i;
i=0x80;
do
  {
   SCL=0;
   TRIS1 =0;
   ee_delay();
   if (x&i) SDA=1;
   else SDA=0;
   ee_delay();
   SCL=1;
   ee_delay();
   i>>=1;
  }while(i!=0);
}
```

Using a compiler dramatically increases programmer productivity, but as we have seen, this comes at a price. Optimum performance is not achieved. In CSIC, the compiler generates larger code size and less-than-optimal performance. This is in part because of the difficulty in manipulating the special-purpose register set. In microcontroller design, all possible techniques must be utilized to maximize performance. Assembly-level programming is mandatory to optimize critical code.

## 3.5.3 Hard-Coded Instructions

Software development often exceeds the hardware design costs for a project. It can be shown that reducing the number of instructions results in increased performance. The optimum number is a relative definition. It is generally considered to be the minimum amount of code needed to achieve the maximum performance for that given function.

Programmers have fewer instructions to generate, thus reducing the cost of code. Program memory requirements are reduced, and expensive fast memory is conserved.

| 3.6 | PROGRAM DESIGN |
|-----|----------------|

CISC processors are designed to be general purpose. The instruction set is developed to handle the widest variety of applications reasonable. Responding to the need for increasing programmer productivity (reducing the cost of software), additional instructions are defined. These instructions, by definition, are application oriented. That is, they are specifically added to the processor logic to solve programming obstacles for applications such as data processing.

The demand for new instructions will continue as new applications are introduced. This creates a tension with the trade-off of more complex processor logic and the intrinsic value of a new instruction. Processors composed of hardwired logic functions become increasingly complex to design.

As we have seen, compilers increase programmer productivity at the expense of system performance. To get maximum performance, software programmers write critical routines in assembly code (Figure 3.24). This is true for microcontroller-based systems as well, if not more so. To achieve maximum performance, assembly language is the only way to go. This is a direct result of Amdahl's law.

**FIGURE 3.24**  High-Level Language Efficiency

> Assembly
> Instructions
> C-code 20 → 30
> Assembler → 20 → 33% improvement

### 3.6.1  Program Code Size Creep

Another key factor in microcontroller design is code size. With the increasing power of the hardware, more complex applications are possible. This means larger compiled programs for the increased functionality. However, fast memory continues to get larger and less expensive. The ratio of memory cost to software code size remains fairly constant. Memory availability always enables larger programs. Larger programs increase the demand for larger memory.

### 3.6.2  CISC Instruction Set

A CISC-based processor can have hundreds of possible instruction combinations. The Intel Pentium IV introduced in 1998 has more than 200 possible opcodes. Support of all these types of instructions requires increasingly complex microcode engines. The number of microcycles to execute complicated instructions limits the execution speed of the processor.

With the introduction of the L1 and L2 cache concepts (Figure 3.25), memory speeds closely matched those of the processor. This part of the performance equation was solved through semiconductor fabrication technology. This turns the focus to software technology. The increasingly complex instruction and register sets, coupled with data movement through the caches, were becoming the bottleneck.

**FIGURE 3.25**  Cache Memory Conflict

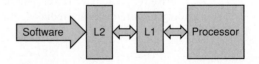

Another factor, the overriding factor, was also imposed from outside the technology triad. The marketplace was coalescing around the Intel ISA as used in the PC. The software, as defined by Windows, became a de facto standard for PCs. As a consequence, fundamental architectural changes to the ISA could not be made. Performance enhancement was limited to incremental changes. This in turn limited the increase in performance to less than the maximum allowed by the technology triad.

**3.7**            **UNIFIED INSTRUCTION SET**

High-level programming languages are continually becoming more efficient. The demand for programmer efficiency produces new generations of compiler technology. CISC compilers still have difficulty dealing with register sets that incorporate fixed function registers.

The CISC register set is intended to be manipulated by programmers at the assembly level. The complexity of the instruction reaches a point of diminishing returns. For Intel Pentium and AMD Athalon microprocessor design, the key factor in improving processor performance is through the underlying semiconductor fabrication technology.

Architectural enhancements are a derivation of the fabrication technology advancements. Through market forces, the software technology coalesced around a fixed number of instructions. Unlike earlier points in CISC development, new instructions were, for all practical purposes, forbidden.

### 3.7.1 Industry Standard Software

The instruction set architecture introduced by Intel when the i8086 became the de facto industry standard. The instruction set of future processors was fixed at that point in time. It was possible to make performance enhancements to the processor logic, but the basic instructions had to be executed in a manner that was compatible to legacy processors.

The instruction set can become the limiting design constraint for the processor engineers. Performance enhancements could continue with evolving software, memory, and semiconductor technologies, but within the scope of the existing instruction set. This fixed possible chip design enhancements that limited the potential for achieving maximum performance, something that had, until this point, been the overriding criteria for processor design.

New applications demand new software solutions. Consistent with the CISC approach, new instructions became unavoidable to meet the performance requirements of new applications. Computer graphics, for example, become limited by processor speed. Processor performance becomes limited by instruction execution rates. New instructions can be optimized for increased performance.

### 3.7.2 Instruction Set Extensions

The existing instruction set is maintained, and a limited number of new instructions are introduced. The new applications can take advantage of the additional instructions while the processor retains instruction set compatibility. The processor becomes upward compatible to new applications and backward compatible to existing legacy code.

The development extending the ISA is a key methodology to protect the large base of software programs from obsolescence. The processor design is not fully maximized to the performance of the new application. However, the performance improvement is a compromise. This compromise enables the new application while maintaining existing applications.

The combined cost of introducing a new processor design to the marketplace can be enormous. The aggregate investment of engineering effort on the part of the processor designers and the end users must be evaluated. New technology innovations in processor design bring attendant increases in performance and functionality. This achieves a greater price to performance level than existing processors. Without the cost benefit, no matter how elegant a design, it will not be introduced to the marketplace.

With the complete dominance of the Intel 808x–based architecture in the PC marketplace, alternative processor designs could simply not be brought to market. The PC constrained the balance of innovation with improvements in memory, software, and processor technologies. That is to say, the processor could achieve higher performance incorporating enabling technologies.

## 3.8          RISC INSTRUCTION SET ARCHITECTURE

Above all, RISC design is about simplicity, not just of the instruction set, but also simplicity across the three key technologies. The processor logic is simplified for greater speed. The memory is separated into program and data for faster access. Compiler-generated software is incorporated into the basic design.

### 3.8.1  Microcode

We have discussed three key points of program behavior first, software programs have a tendency to execute in small segments of code. This execution attribute is independent of the overall program structure. It is a characteristic of programs in general.

A second key aspect is that a small percentage of instructions are used repetitively during program execution. Basic operations such as data movement, Boolean, and arithmetic instructions tend to dominate the total instructions a program executes. Again, this tendency is independent of the software program itself. For different applications, the execution instruction set may vary, but it will be a subset of the complete instruction set.

The third level of locality is at the microcode level. As we have discussed previously, CISC processor design implements processor instructions as sequences of microcoded instructions. Of all the microcode instructions, it is also true that a small subset is used repetitively to implement the processor-level instructions.

### 3.8.2  Micro Instruction Cycles

A concept of microcode was developed by IBM for its System/360 line of computers introduced in the early 1960s. Microcode is a small software program stored in a memory as part of the processor. The microcode instructions are incorporated within the basic processor logic functions. A machine-level instruction became itself a series of microcode instructions.

Microinstructions control the processor's internal logic and execute the instructions as if they are programs themselves. A machine-level instruction's cycle times are defined as the number of microinstructions necessary to execute it. A simple addition (ADD) instruction might take only a few microinstructions. However, a multiply instruction (MUL) could take 32 cycles to execute.

```
MUL A,B        LOAD A
               LOAD B
               SHIFT A
               SHIFT A
               (Repeat 8 times)
```

Correspondingly more complex instruction types take more microcode cycles to execute. Examples are types designed for indirect and relative addressing modes. These types of instructions are only useful in a subset of applications. For any given application, only a portion of the instruction set may be used.

### 3.8.3  Application-Specific Instructions

To achieve maximum performance for a specialized application, only those specific instructions required had to be executed at the fastest rate possible. With the complex processor logic and cumbersome microcoded instructions, the absolute maximum theoretical performance could not be achieved within the capabilities of the hardware/software/memory technology triad.

Again, it is a convergence of Amdahl's and Moore's laws that hints at the solution. According to Amdahl, performance is increased in proportion to the time the enhancement is used. Moore's law implies that microprocessor logic will become faster and more complex.

Most assembly-level instructions will only require a subset of the microcode instructions for their execution. By reducing the number of microcode cycles, overall performance will be increased.

### 3.8.4 Single-Cycle Instructions

One of the most innovative architectural features of RISC is the concept of single-cycle instruction execution. This is accomplished based on the principals of Amdahl's law as applied to CISC instruction sets. RISC implements only the most frequently used microcode instructions of a typical CISC processor. In this sense, there is no microcode in an RISC processor. The RISC instruction set is in and of itself the microcode instruction set.

Applying the principal of locality, the RISC processor focuses on executing the "microcode" of a CISC processor. Other than branch instructions, all RISC instructions will execute in a single cycle. This reduces the second term of the equation to almost unity. (Superscalar designs can exceed one instruction per cycle.)

Using an optimizing C-compiler can reduce the first term further over CISC. And with simplified logic, the third term can be reduced compared to CISC. RISC clearly has the potential to significantly improve performance over CISC with the same semiconductor technology.

```
CISC
    Time-pgm = Inst-pgm *cycles-inst *time-cyc
RISC
    Time-pgm = Inst-pgm *1 *time-cyc
             = Inst-pgm *time-cyc
Note: Not all instructions will execute in a single cycle.
```

This equation can be applied to both CISC and RISC. The overriding goal of CISC design is to reduce the first term, thus reducing the number of instructions to be executed by the program. This also attacks the two issues of expensive memory and expensive programmers. Both of these economic goals change over time as their associated technologies advance. However, they were of paramount importance in development of CISC architecture.

## 3.9 PROCESSOR LOGIC

A processor is composed of multiple functional units. Each unit is associated with a signal propagation delay time. This can be defined in terms of the sum of the gate delays for the functional block. Figure 3.26 gives an example of multiple level logic with associated delays.

Advances in semiconductor fabrication technology decrease the transistor switching time according to the time range of Moore's law. Faster switching times decrease gate delays and increase the corresponding clocking speed. This is true for a synchronous processor design, whether RISC or CISC based.

RISC focuses on increasing performance at the processor logic level. By limiting the number of instructions, the processor's logic can be implemented with fewer gates. Fewer gates decrease signal propagation times. This will in turn allow for higher clocking frequencies.

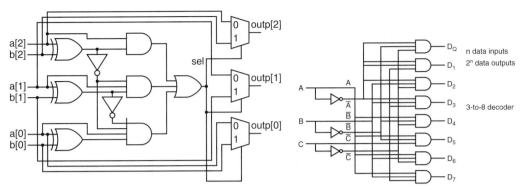

**FIGURE 3.26**   Logic Gate Delay Reduction

### 3.9.1 Synchronous Logic

In a synchronous processor design, each block is clocked at the machine cycle rate. An instruction execution cycle time is the sum of the clock cycles per block. The instruction execution rate can then be increased two ways. The gate delay within functional blocks can be reduced, or the number of overall clocking cycles can be reduced.

At its simplest form, a single cycle per clock is achieved if the processor logic is implemented in one block. This is accomplished in RISC by reducing the complexity of the logic associated with the instruction execution. The instruction set is defined so that all instructions can be executed through the processor logic in a single clock cycle.

### 3.9.2 Register Sets

The RISC architecture, although it has high theoretical performance, was not practical with the technology of the 1950s. Memory was slow and expensive, it required more of it, and programs were larger. A key to RISC philosophy is more efficient high-level language (HLL) technology. Technology did not yet exist to build the necessary optimizing compilers.

### 3.9.3 Orthogonal Registers

RISC has many desirable technology features. Instructions are designed to execute in a single cycle (except branches). They are the same word size. This greatly simplifies generating code for maximum performance. Another key factor is that the register set is general purpose. They are also interchangeable or orthogonal, as shown in Figure 3.27.

**FIGURE 3.27**   Orthogonal
Register Set Mapping

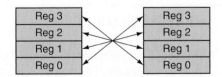

An orthogonal register set means the registers can be used interchangeably. The compiler does not have to worry about using specific registers. A compiler can be more efficient by using any register to store intermediate (temporary) data values.

### 3.9.4 Register Optimization

A CISC compiler stores intermediate data values into temporary memory locations. This increases the number of memory operations, which correspondingly increases the program execution time.

As per Amdahl's law, fewer instructions means greater performance. A RISC compiler can make greater optimizations in the use of register allocations.

HLL compilers are difficult to implement in CISC for a couple of significant reasons. The basic difficulty of optimizing a complex set of instructions is that they interact in complex ways. A further difficulty is the manipulation of a special-purpose register set. CISC compilers are designed more for efficiency of programming than maximally efficient code execution. Combining the software technology of a C-compiler with a fast processor produces higher performance than with CISC machines. Application orientation could be optimized by the compiler. With a general-purpose RISC processor, more optimal designs could be developed for specific applications.

### 3.9.5 Load/Store Data Operations

Instructions execute on data values in the register set. Data is moved, loaded, from memory to the registers. Conversely, modified register data is moved and stored back to memory (Figure 3.28). This is a key characteristic of RISC. Having data in the registers allows most instructions to execute in one cycle.

**FIGURE 3.28**   RISC Data Registers

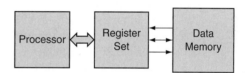

Orthogonal registers also affect the instruction execution rate. Instructions do not require special-purpose registers to be loaded in a certain sequence. Loads and stores to and from memory do not need to wait for specific registers to become available.

Once registers are loaded with data for an instruction, intermediate results can be stored in any available register. Subsequent instructions can operate on these data values. To the degree that the instruction sequence can be accomplished, manipulating the data within the register set, performance is maximized to the instruction cycle time.

## 3.10          PROCESSOR FUNCTIONAL PARTITIONING

As we have seen, at any given point in engineering space time, processor designs balance the available technologies to achieve the greatest target performance. Existing processors (CISC) were designed as a combination of more basic functional units. Partitioning the processor design allowed engineers, working within the constraints of the then-current technology, to build more complex processors than would be practical with a single monolithic block of logic.

The trade-off is a small reduction in potential performance but a large gain in design ability. Within microcontroller design, the trade-off between maximum obtainable performance and cost-effective design is a constant balancing act.

### 3.10.1 Instruction Pipelining

The functional units were controlled by the control logic. They are synchronized by clocking signals as an assembly line for instruction execution. Partial results of each stage of the instruction execution cycle are pumped like water through a pipeline. The instruction cycle time would be the total number of clocks per period to enter and then exit the pipeline.

Figure 3.29 shows a pipeline with water flowing at a fixed rate. If the pipeline diameter is increased, then more water can flow through at the same pressure. If the pressure is increased, then the water will flow faster. If the water pressure is fixed, then increasing the diameter of the pipeline will allow more water flow.

**FIGURE 3.29** Instruction Pipeline Analogy

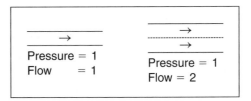

### 3.10.2 Execution Units

An important feature of RISC inherent in the architecture is the instruction pipeline. The RISC architecture organizes program and data into separate memories. This has significant implications. Within the program memory, instructions are stored in sequential memory locations, as shown in Figure 3.30.

This is not the case with CISC. Both the instructions and data are in the same memory space. Instructions can also vary in length, as shown in Figure 3.31. This means that instructions are not sequentially addressable on word boundaries. Calculating the next instruction $n + 1$ location can most often only be determined after the $n$th instruction had been fetched and decoded.

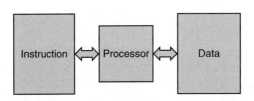

**FIGURE 3.30**   RISC Memory Architecture

| | |
|---|---|
| $n + 3$ | AND |
| $n + 2$ | ROT |
| $n + 1$ | STA |
| $n$ | MOV |
| $n - 1$ | SUB |

**FIGURE 3.31**   Sequential Instructions Stored in Memory

The constraint of memory speed differences in the hierarchy is a factor of technology. However, for fetching instructions from memory, RISC has a distinct advantage. Typically, the location of the next instruction address is known before the current instruction has finished execution. This is true except for branching type instructions.

### 3.10.3 Pipeline Stages

As an instruction progresses from stage to stage, only one of the functional units is operating at any given point in time. Increasing from a single processor logic unit to one composed of two stages can double the rate at which instructions are executed (refer to Figure 3.32). This limit cannot be reached in real-world designs because of incremental processor cycles generated for branch-type instructions.

**FIGURE 3.32**   Two-Stage Pipeline Instruction Execution

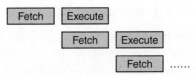

Partitioning a processor design leads to the natural design goal of keeping each functional block as busy as possible. To the degree that execution units can be kept busy, the more efficient the design is considered to be. RISC lends itself in a natural manner to that design goal: maximum performance with maximum efficiency.

This dramatic increase in performance makes pipelining so attractive. If the delay time of a functional block is 20 $n$s, a sequentially executing instruction would take 40 $n$s to execute. If both functional units can be kept busy simultaneously, the processor can complete a composite instruction execution in 20 $n$s, as shown in Figure 3.33.

**FIGURE 3.33** Effective Instruction Execution Cycles

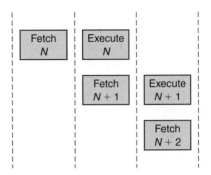

It will take the processor logic 40 $n$s to process the instruction; however, one-half of each instruction will be in execution phase, one in fetch, and the other in execute. A fetch and execute represents a full instruction cycle. In this manner, an instruction cycle becomes equivalent to a machine clock cycle for instructions requiring at most two machine clock cycles to execute. RISC instruction sets are comprised primarily of single instruction cycle operations.

Adding additional stages to the pipeline will give increased performance. The trade-off will be greater complexity, not only in the hardware but also in the software and memory. Again, all three technologies must be balanced to achieve the maximum performance within their constraints.

Within microcontroller design, the practical limit for RICS ISAs has converged in the range of seven stages. Simple microcontrollers might use two stages of fetch and execute, whereas higher-performance processors will break the execution phase into multiple phases.

The key point for utilizing pipeline technology is the performance increase that can be achieved within the same semiconductor technology. If the instruction set can be made to effectively execute in 20 $n$s, then the performance is doubled. In reality, this doubling effect (for a two-stage pipeline) is the theoretical maximum. Real-world programs contain branching instructions, which take additional cycles to execute. However, even if an 80% performance enhancement can be achieved with a relatively modest increase in processor design complexity, it is worth while.

The additional processor complexity from using logic into functional units will, on an absolute basis, reduce the theoretical maximum performance supported by the semiconductor technology (i.e., the basic switching rate of transistor and gate level logic). However, the economics of design ability and usability for the given degradation in performance is worth the price to pay.

### 3.10.4 Pipeline Throughput

If processor performance at a fixed clocking frequency can be increased by 200% with minimal effort, then the incremental complexity is worth the additional development and silicon penalty costs. More pipeline stages go against the RISC concept of minimal logic and single cycle instructions, but incremental pipeline stages offer tremendous performance rewards. For a 100 MHz clock frequency, a five stage pipeline would offer 400 MHz instruction throughput with an 80% pipeline efficiency.

The PIC 18F452x family of microcontrollers implement a simple two stage pipeline that effectively doubles their maximum performance. It is a simple overlap of the instruction fetch and execution phases. The instruction fetch and overlap with the execution of two sequential instructions results in the execution rate equivalent of one instruction per clock cycle. Branch-type instructions decrease the theoretical doubling of performance.

### 3.10.5  Sequential Execution

In a linear-processing processor, only one logic functional unit is active during the full instruction cycle. The processor function of fetching is not active during execution, and execution is not active during fetching. If a way can be found to utilize both functional units at the same time, then two instructions can be in partial execution at the same time.

This will occur for all instructions that execute in one cycle and are sequentially stored. The processor logic itself increments the program counter during execution of the $n$th instruction and automatically fetches the $n + 1$th instruction. This is not under programmer control.

The net effect is that one effective instruction is executed per cycle. This is the most important concept of pipeline design. Execution of a single instruction per cycle means partial execution of instructions such that in total, they represent a single instruction cycle.

### 3.10.6  Branch Execution

In the execution of branch instructions, the $n$th + 1 instruction will not be the correct next instruction (following $n$). Because the processor has already fetched the $n + 1$ instruction during the execution of the branch instruction at location $n$, it will be the wrong instruction (Figure 3.34). An additional machine cycle is required to fetch the branch address of $n + 4$. Once the new instruction is fetched, it will require an additional clock cycle to execute.

**FIGURE 3.34**  Branch Instruction Execution Delay

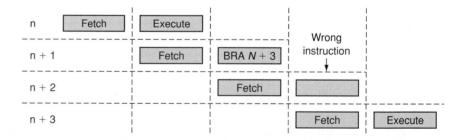

Branch instructions require two instruction cycles to execute because the new address cannot be calculated until after the fetch to the $n + 2$ address is performed. The fewer branch instructions that are executed, the closer the user is to achieving maximum performance. Consequently, one goal in programming the PIC, if performance is a consideration, is to reduce the number of branch instructions executed.

## 3.11     FIVE-STAGE PIPELINE

Five-stage pipelines are typical for multistage pipelined RISC architectures. This amount of pipelining offers a significant performance improvement with a tolerable amount of additional processor logic overhead. Figure 3.35 represents a typical five-stage pipeline.

**FIGURE 3.35** Five-Stage Pipeline

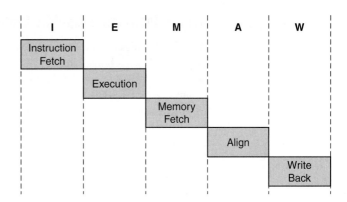

Microcontroller cost is fundamentally linked to the fabrications cost in an intimate way. The MicroChip PICs, with its simple core logic, means a minimal die size and directly lower manufacturing cost. The PIC is the most prominent success in 8-bit commercial microcontroller design: simple, elegant, and low cost.

The MIPS 4Kc implements a five-stage pipeline of instruction fetch, execute, memory fetch, align, and writeback. With processor logic that has a clock frequency of 400 MHz, the instruction processing rate would approach 2 GHz. Combined with the RISC ISA concept, this enables very high performance for applications such as Internet routers and switches. Figure 3.36 shows a detailed diagram of the five stages of the MIPS 4Kc family RISC processor unit.

**FIGURE 3.36** MIPS 4Kc Core Pipeline (Reprinted by permission of MIPS Technologies Incorporated.)

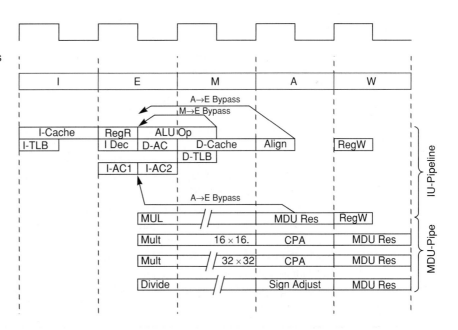

The principal of overlapping instruction execution is the same. For the MIPS 4Kc, this means overlapping the execution of the functional parts of the instruction cycle. If it takes five clock cycles to execute an instruction, then five instructions need to be in the processor unit in various stages of execution.

### 3.11.1 Instruction Pipeline Stalls

A key performance issue with a pipelined processor is what happens when the next instruction is not available. When a branch instruction is executed, the pipeline must restart with the new instruction address. This was shown in the two-stage PIC pipeline in Figure 3.34. In a five-stage pipe, this means five extra instruction cycles are lost. The pipeline needs to be flushed of any partially executed instructions and restarted from the new address. This can have a significant impact on performance.

If a branch instruction is taken only 5% of the time, it can degrade the potential performance increase by 20% (refer to Figure 3.37). Because of the significant performance hit from branching in multistage pipeline design, every effort is made in RISC to minimize branching instructions. This can be accomplished in both hardware design and software coding.

**FIGURE 3.37**    Pipeline Branch Penalty

$$100_{inst} *1_{inst/cycle} = 100_{inst\text{-}cycles}$$
$$(95_{inst} *1_{inst/cycle}) + (5_{inst} *5_{inst/cycle}) = 120_{inst\text{-}cycles}$$

The simplest solution is to eliminate branching instructions. By optimizing the software, many instruction branches can be eliminated. This is where the compiler technology interacts with processor hardware design. C and C++ compilers designed for RISC processors include optimizations to unroll loops.

### 3.11.2 Branch Predication Table

The second way is to prefetch the branch instruction target address. A branch lookup table can be implemented. As branches are encountered, the branching address instruction is also fetched. If the branch is taken, the instruction is already fetched and can be fed directly into the execution units. Thus, no cycles are lost, and the pipeline is kept full. This will increase the target object file size; however, it also increases the efficiency of the processor execution rate. This trade-off is not fixed and is application dependent.

### 3.11.3 Data Pipeline Stall

RISC is known as a load and store architecture. All data to be operated on is always moved from or to the data cache memory. In contrast to the simple two-stage pipeline of the Microchip PIC, the MIPS 5Kc implements a specific stage in the pipeline for the load/store data function. For the Microchip PIC18F4520, all data instructions operate with the data register files implicitly. In essence, the Microchip PIC 18F4520 has only a single data space addressable as registers implemented in on-chip RAM. Because of the simple two-stage design, no load/store stage is required in the pipeline.

With a more complex multistage pipeline, an execution step is dedicated for data memory. Depending on the instruction in partial execution, this will be a load or store operation. The data cache lies between the register set and the main memory. Pipeline efficiency can be gained by determining when load and store operations are required for the instructions currently in the pipeline, possibly saving stalls due to cache load/store operations.

## 3.12    SUMMARY

The processor logic and memory interface is designed to enhance the execution rate of the processor. The ISA becomes the driving force in the design of the hardware. Technological advances in software, hardware, and memory continue to be utilized to maximize performance.

However, they are utilized within the framework defined by how the instructions need to be executed. The architecture of the system evolves in a way consistent with the underlying architecture of the instruction set.

This provides a consistency for software program development that enables it to migrate forward to new generations of processors. It provides a consistent structure to support a market with a broad range of applications. The marketplace can coalesce around these key architectures. Software programs can be developed that can span families of processor designs. Technological advancements can continuously be incorporated to enhance performance as long as the integrity of the ISA is maintained.

At a point in time, when technologies converge to provide a significant advantage (to the marketplace) for performance to an application, a branch from the ISA can occur. This can happen by extending the ISA with additional specialized instructions or by changing the ISA itself.

Changing the ISA will result in the maximum performance because all three of the key technologies will be maximized for the solution. The processor design engineers will be free from the constraint to maintain any backward compatibility at a primitive level, the major limiting factor to processor innovation.

## QUESTIONS AND PROBLEMS

1. Define Moore's law.
2. What is the key product in developing new semiconductor technology?
3. What are "enabling" technologies?
4. Define Amdahl's law.
5. How is design abstraction used in system design?
6. List the primary differences between RISC and CISC.
7. Name three ways to reduce program execution time.
8. Define microcode.
9. What is program locality?
10. List the memory hierarchy of a RISC-based design.
11. What is cache memory used for?
12. What is the primary difference between assembler and compiler code?
13. What is the memory latency if the access time of the cache is 500 ps, the main memory is 2 $ns$, and the file system is 8 $ns$?
14. Give two examples of instruction set extension.
15. Describe microcode and how it is used.
16. What does a C-optimizing compiler do?
17. What does having orthogonal register set mean?
18. Why does instruction pipelining increase performance?
19. What does execution of a branch instruction do to the pipeline?

## SOURCES

Intel Web Site: http://www.intel.com/design/mcs51/manuals/272383.htm
Tensilica Web Site: http://www.tenselica.com/products/xtensa_LX.htm
MicroChip Data Books: http://www.microchip.com
MIPS Web Site: http://mips.com/products/overview/

# CHAPTER 4

# Microcontroller Functions

## OBJECTIVE: OVERVIEW OF MICROCONTROLLER DEVICE FUNCTIONALITY

The reader will learn:

1. Introduction to CMOS technology.
2. Different types of microcontroller memory.
3. Basic hardware features of microcontrollers.
4. Common microcontroller peripheral functions.

## 4.0  DEVICE FUNCTIONS

General-purpose single-chip microcontrollers share a basic set of device functions. These are developed from requirements of typical applications. Industry standard functions like $I^2C$ and USART are typical examples. Microcontrollers may also include application specific functions. The combination of the two gives the controller application focus with standards-based interfaces.

The number of functions incorporated into the chip depends on its intended application. Low-cost devices have fewer functions, whereas more expensive devices incorporate a more robust set. Determining the design of a microcontroller is a combination of market pricing and application-required functionality. There is an optimal range where pricing and functionality permit a design, as shown in Figure 4.1.

**FIGURE 4.1**  Microcontroller
Market Design Constraints

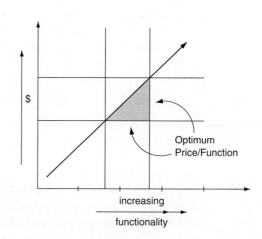

Optimum
Price/Function

increasing

functionality

## 4.1          TRANSISTOR TECHNOLOGY

Bipolar transistor technology preceded complementary metal-oxide semiconductors (CMOSs). It was used for standard small-scale integration (SSI) and medium-scale integration (MSI) logic devices. High-speed versions of these devices are designed using Schotky transistors. Requirements for minimum power consumption use low-power transistor design. A combination of the two is used for power reduction and increased speed over standard logic. Figure 4.2 lists the designations for a standard 2 input NAND (not and) logic function.

**FIGURE 4.2**   SSI NAND Logic Function

| Standard | Low Power | High Speed | Combination |
|----------|-----------|------------|-------------|
| 7400 | 7400L | 7400S | 7400LS |

### 4.1.1 CMOS Transistor

Many types of CMOS design technology have been developed. Embedded microcontrollers use different technologies based on the intended application. For battery-based microcontroller design, low power is the key criteria. For high-performance designs, speed becomes the design goal. Figure 4.3 is a list of CMOS technologies and typical applications.

**FIGURE 4.3**   CMOS Technology Types

| Technology | Application |
|------------|-------------|
| CMOS | Standard |
| CHMOS | Low power |
| Bi-CMOS | Mixed Signal |
| XCMOS | Flash Memory |

CMOS transistors inherently draw less current than bipolar transistors. This is due to their capacitive nature. Figure 4.4 shows a schematic of a CMOS transistor. It has a gate, source, and drain. $V_{cc}$ is replaced by $V_{dd}$, and ground is referred to as $V_{ss}$.

**FIGURE 4.4**   CMOS Schematic

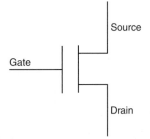

Figure 4.5 shows an inverter using a pair of complementary MOS transistors (CMOS). Both transistors will be "off" in the static situation. This means essentially no current flow. When the input voltage $V_{in}$ is switched, current will flow during the short time of the voltage change.

**FIGURE 4.5**   CMOS Inverter
Schematic

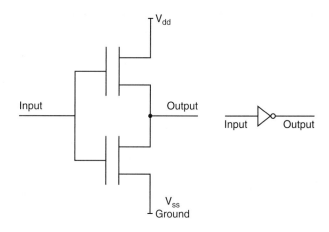

## 4.1.2  CMOS Power Consumption

Low-power consumption is a necessity for microcontroller design with high transistor counts. With bipolar technology, some portion of all transistors will be in the "on" (logic 1) state, which means they will be drawing current. This is not true with CMOS. Essentially, there is no current draw, except when switching logic states.

CMOS device power consumption becomes directly related to the switching frequency. This effect can be seen in personal computer design. The higher the clock rate of the microprocessor, the more elaborate the heat sink required. Graphics display microcontrollers, with their associated high device count and corresponding performance range, require heatsinks. Depending on complexity, the system chip on the motherboard often has some form of heatsink.

A limiting factor to performance in embedded microcontroller design can be power consumption. Adding mechanical devices of any type, such as heatsinks, to a microcontroller-based PCB is often prohibited by the system design requirements. Particularly when physical access to a microcontroller card is limited, mechanical parts should be avoided.

## 4.1.3  Packaging

The most common packaging material used for microcontrollers is plastic. It has become the de facto industry standard. The number of pins of a package is also standardized to reduce packaging costs. Customizing a specific package type for a project would be prohibitively expense.

Commercial microcontrollers are intended for high-volume production. Plastic is a low-cost material specifically designed for this purpose. Pin count of a package depends on the application requirements. More I/O functions require more pins. The least costly packaging is one with a low pin count in plastic.

**FIGURE 4.6**   (a) 40 Pin PDIP;
(b) 44 Pin TQFP (Copyright 2004
Microchip Technology Incorporated. Reprinted with permission.)

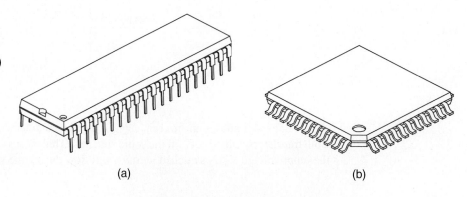

(a)                                             (b)

**FIGURE 4.7**   Package Types

| Package Type | |
|---|---|
| PDIP | Plastic dual inline package |
| SOIC | Small outline inline package |
| TSOP | Thin small outline package |
| PQFP | Plastic quad flat pack |
| PLCC | Plastic leaded chip carrier |
| TQFP | Plastic thin quad flat pack |

System-on-a-chip (SoC)-based microcontrollers can be packaged in either plastic or ceramic (Figure 4.7). For high-volume applications the same decision criteria exist as for commercial parts. Most high-volume SoC designs use a type of the quad flat pack configuration. It also allows for die bonding on all four sides of the die. This has significant advantages for maintaining chip voltage and balancing I/O port currents.

### 4.1.4  Operating Temperature Range

The three common temperature ranges defined for proper functioning of microcontrollers are listed in Figure 4.8. Within these ranges, a microcontroller is "guaranteed" to operate at the published specifications. Importantly, the part will function without a heatsink.

**FIGURE 4.8**   Standard Temperature Ranges

| Temperature Range (centigrade) | Designation |
|---|---|
| 0 to +70 | Commercial |
| −45 to +85 | Industrial |
| −55 to +125 | Military |

## 4.2          MEMORY TECHNOLOGIES

Four major types of memory technology are shown in Figure 4.9: read-write (RWM), nonvolatile read-write (NVRWM and FLASH), and read-only (ROM). RWM and NVRWM are implemented with different circuit design technology, depending on application. ROM is always based on mask-programmed fabrication technology.

**FIGURE 4.9**   Memory Types

| RWM | RWM | NVRWM | ROM |
|---|---|---|---|
| Random access | Non-random access | EPROM<br>E2PROM<br>Flash | Mask-programmed |
| SRAM<br>DRAM | FIFO<br>LIFO<br>Shift register<br>CAM | | |

### 4.2.1 DRAM

By its name, dynamic random-access memory (DRAM) will lose storage bits if not refreshed. Bits stored as logic 1s will, over time, be read as 0s. This means they must be constantly refreshed to maintain the logic 1 state. DRAM cells are designed with single transistors, often in a vertical arrangement. This advanced technology makes it compact geometrically, hence the term high-density DRAM.

DRAM is not used in microcontroller design except in specialized custom applications. Although it is high density, its underlying fabrication technology is fundamentally analog based. Combining DRAM and digital microcontroller logic requires a very complex fabrication process. In practical applications, DRAM and microcontrollers are not implemented on the same die.

### 4.2.2 SRAM

A static random-access memory (SRAM) cell will hold its charge without being refreshed. Essentially it is a form of flip-flop. A typical static RAM cell design with six transistors is shown in Figure 4.10. This makes SRAM much larger geometrically than a corresponding DRAM.

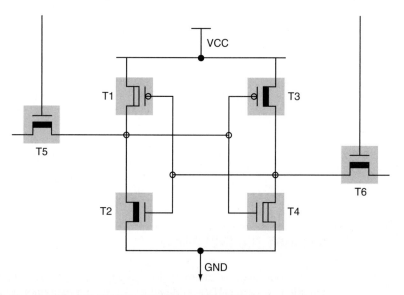

**FIGURE 4.10**   Standard CMOS Six-Transistor SRAM Cell

In the RISC-based PIC18F4520, 1536 bytes of SRAM are implemented. The basic i8051 has 256 bytes of scratchpad memory. For embedded SoC designs targeted to high-performance applications, 16 k bytes of SRAM is not uncommon. As per Moore's law, new processor designs are enabled to incorporate increasing amounts of SRAM memory.

There are four major types of SRAM. Each type is used for a special purpose. LIFO (Last In First Out) is used for the address stack, as shown in Figure 4.11, for an 8-bit PIC microcontroller. Instruction return addresses are "pushed" on and "popped" off the stack. Content addressable memory is used to translate page addresses to physical chip addresses. FIFO (first in first out) RAM is used to implement instruction buffers with program memory. Shift registers are used to serialize data in processor logic functions. The underlying technology for these different implementations remains the basic six-transistor SRAM bit cell.

**FIGURE 4.11**  LIFO Stack
Instruction Queue

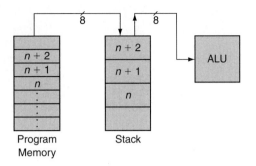

### 4.2.3 NVRWM

Applications that require data information to be fixed through power cycles use electrically erasable programmable read-only memory (EEPROM)-type memory. Configuration data for a television is an example. The receiver options, such as channel select, are programmed by the viewer. When the television plug is pulled, the appropriate channels remain selected.

Microcontrollers often include both types of NVRWM. FLASH memory is used to store the program. EEPROM is used for nonvolatile data memory. The name FLASH refers to the ability to quickly program the memory words by having them organized into blocks. New processors, like the Microchip PIC18F4520, support FLASH programming by the instruction set. This can be used to remotely update a program.

Three key parameters are associated with FLASH and EEPROM, in addition to $t_{acc}$, that make them unique. Two relate to the time it takes to program and erase the memory words. The third is the total number of erase/write cycles that can be performed. Figure 4.12 gives the EEPROM/FLASH technical parameters for the PIC18F4520 microcontroller.

**FIGURE 4.12**  PIC 18F4520 E/W
EEPROM/FLASH Specifications

| Data EEPROM | Minimum | Typical |
|---|---|---|
| Byte endurance | 100 K | 1 M |
| Number of total erase/ write cycles before refresh | 1 M | 10 M |
| Program flash memory Cell endurance | 10 K | 100 K |

Programming time is particularly important during the product design process. The software must be downloaded to the microcontroller for debugging. The faster the code can be programmed, the sooner the engineer can continue working. Time is money.

A second key aspect is the number of total program/erase cycles. EEPROM technology limits the number of times a cell can be programmed. It is always erased before being written, so it is referred to in program/erase cycles; 100,000 program/erase cycles is a large number. Over a 10-year product life cycle, where data is programmed once per hour, 43,810 cycles will be required.

### 4.2.4 EEPROM

The EEPROM cell is based on the concept of a floating gate (Figure 4.13). During programming, charge is increased on the floating gate to the threshold for a logic 1. This is done by raising

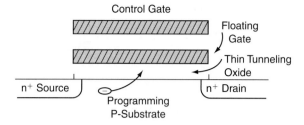

**FIGURE 4.13** EEPROM Floating Gate

the programming voltage above 5 volts. This causes charge migration across the thin tunneling oxide to the floating gate. When the chip power is subsequently turned off, the gate charge remains.

During the erase cycle, the charge is dissipated to the threshold of a logic zero. This is a unique aspect of EEPROM design. The cell is erased to a logic zero state before being programmed to a 1. This ensures that the floating gate will have a controlled charge level.

### 4.2.5 FLASH Technology

FLASH memory is combination of EPROM and EEPROM technology with a fast programming capability. Multiple number memory locations are programmed to sequential address locations during a single cycle. The PIC18F4520, for example, uses 64 bytes as the block size for erase/write cycles. The larger the block size, the faster the programming time over EEPROM.

### 4.2.6 ROM

Read-only memory (ROM) is the densest memory used for program store. It is a single transistor cell design. The key difference from DRAM is that the memory bits are permanently programmed to a logic 1 or 0 state. This is done during manufacturing. ROM is said to be mask programmable.

## 4.3　HARDWARE FEATURES

Commercial microcontrollers (COTS) are designed to be as general as possible to an application. The ability for the microcontroller to be more specifically configured to a given application increases its market potential. A microcontroller design that is targeted to low-cost applications, for example, may have a simple crystal oscillator as well as an internally generated RC clock.

### 4.3.1 Configuration Word

A certain number of options are hardwired. These options define fixed characteristics of the device as it is powered up. These hardware functions are enabled (or disabled) at the time the microcontroller software is programmed. They are deliberately fixed in hardware so the software will not be able to change them. For example, you would not want to inadvertently disable the chip's oscillator.

| File Name | | Bit 7 | Bit 6 | Bit 5 | Bit 4 | Bit 3 | Bit 2 | Bit 1 | Bit 0 | Default/ Unprogrammed Value |
|---|---|---|---|---|---|---|---|---|---|---|
| 300001h | CONFIG1H | IESO | FCMEN | — | — | FOSC3 | FOSC2 | FOSC1 | FOSC0 | 00-- 0111 |
| 300002h | CONFIG2L | — | — | — | BORV1 | BORV0 | BOREN1 | BOREN0 | $\overline{\text{PWRTEN}}$ | ---1 1111 |
| 300003h | CONFIG2H | — | — | — | WDTPS3 | WDTPS2 | WDTPS1 | WDTPS0 | WDTEN | ---1 1111 |
| 300005h | CONFIG3H | MCLRE | — | — | — | — | LPT1OSC | PBADEN | CCP2MX | 1--- -011 |
| 300006h | CONFIG4L | $\overline{\text{DEBUG}}$ | XINST | — | — | — | LVP | — | STVREN | 10-- -1-1 |
| 300008h | CONFIG5L | — | — | — | — | CP3[1] | CP2[1] | CP1 | CP0 | ---- 1111 |
| 300009h | CONFIG5H | CPD | CPB | — | — | — | — | — | — | 11-- ---- |
| 30000Ah | CONFIG6L | — | — | — | — | WRT3[1] | WRT2[1] | WRT1 | WRT0 | ---- 1111 |
| 30000Bh | CONFIG6H | WRTD | WRTB | WRTC | — | — | — | — | — | 111- ---- |
| 30000Ch | CONFIG7L | — | — | — | — | EBTR3[1] | EBTR2[1] | EBTR1 | EBTR0 | ---- 1111 |
| 30000Dh | CONFIG7H | — | EBTRB | — | — | — | — | — | — | -1-- ---- |
| 3FFFFEh | DEVID1[1] | DEV2 | DEV1 | DEV0 | REV4 | REV3 | REV2 | REV1 | REV0 | xxxx xxxx[2] |
| 3FFFFFh | DEVID2[1] | DEV10 | DEV9 | DEV8 | DEV7 | DEV6 | DEV5 | DEV4 | DEV3 | 0000 1100 |

**FIGURE 4.14**   Configuration Word for PIC18F4520 (Copyright 2004 Microchip Technology Incorporated. Reprinted with permission.)

## 4.3.2 Oscillator Types

Several different types of external clock generation are normally available to the engineer and are chosen depending on the application. The table in Figure 4.15 lists types for the Microchip PIC18F4520 family. Generally speaking, the greater the accuracy timing that is required, the more will be the cost in external components.

| Designation | Type |
|---|---|
| LP | Low-power crystal |
| XT | Crystal/resonator |
| HS | High-speed crystal/resonator |
| HSPLL | High-speed crystal/resonator with PLL enabled |
| RC | External resistor/capacitor with Fosc/4 output on RA6 |
| RCIO | External resistor/capacitor with I/O on RA6 |
| INTIO1 | Internal oscillator with Fosc/4 output on RA6 and I/O on RA7 |
| INTO2 | Internal oscillator with I/O on RA6 and RA7 |
| EC | External clock with Fosc/4 output |
| ECIO | External clock with I/O on RA6 |

**FIGURE 4.15**   PIC18F4520 Family Oscillator Types

Crystal type oscillators provide the most accuracy. They require capacitors on the leads to stabilize the oscillating frequency. Figure 4.16 gives the range of values for various crystal frequencies.

A resistor capacitor (RC)-based oscillator is an alternative to the use of a crystal. With just two passive components, it is less costly than a crystal. RC-based clocking is limited to applications that do not require a high degree of accuracy. Figure 4.17 shows the RC connection for the PIC18F4520.

| Osc Type | Crystal Freq | Typical Capacitor Values Tested: | |
| --- | --- | --- | --- |
| | | C1 | C2 |
| LP | 32 kHz | 30 pF | 30 pF |
| XT | 1 MHz | 15 pF | 15 pF |
| | 4 MHz | 15 pF | 15 pF |
| HS | 4 MHz | 15 pF | 15 pF |
| | 10 MHz | 15 pF | 15 pF |
| | 20 MHz | 15 pF | 15 pF |
| | 25 MHz | 0 pF | 5 pF |
| | 25 MHz | 15 pF | 15 pF |

**Capacitor values are for design guidance only.**

**FIGURE 4.16**   PIC18F4520 Crystal Capacitor Range Chart (Copyright 2004 Microchip Technology Incorporated. Reprinted with permission.)

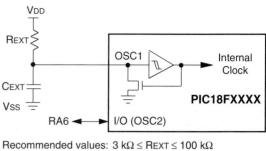

Recommended values: $3\ \text{k}\Omega \le \text{R}_{EXT} \le 100\ \text{k}\Omega$
$\text{C}_{EXT} > 20\ \text{pF}$

**FIGURE 4.17**   RC Oscillator (Copyright 2004 Microchip Technology Incorporated. Reprinted with permission.)

The clocking frequency is a function of the resistance, capacitance, voltage, and operating temperature. It will also vary from part to part due to manufacturing tolerances. A typical application would be an electronic-based child's toy in which a high degree of clock accuracy would not be required.

### 4.3.3  Reset

Microcontrollers reset themselves at power on time to a fixed initialization state. The configuration word will be read, default parameter values set, and the first instruction fetched for execution. Depending on the microcontroller design, the memory may be cleared and peripheral functions initialized.

The power on reset function detects the voltage rise on $V_{dd}$. When a specified voltage threshold is reached, the microcontroller logic will synchronize to the external clock source. After a fixed period of time, when the external voltage is stabilized at $V_{dd}$, an internal reset pulse will be asserted.

Most microcontrollers also support the concept of soft reset as an instruction. When executed, it will reset the program counter without reinitializing the microcontroller to the power on state. This restarts the program without modifying the control register file contents.

A soft reset is also used in multichip applications to synchronize the system after power-on. This can be done through an I/O port pin, for example. This will ensure the other system logic synchronizes with the master microcontroller.

### 4.3.4  Standby Modes

For low-power applications, such as cell phones, maximizing battery life is of paramount importance. Microcontrollers incorporate a sleep mode that will power off all nonessential logic. This is accomplished by reducing the clock frequency to zero or D.C. Only the minimum amount of circuitry needed to restart the microcontroller will actively draw power.

Executing a sleep instruction will cause the microcontroller to enter a power-down sequence. The instruction $n + 1$ will be pushed on the stack. The sleep instruction will be the last instruction executed before the program counter is halted.

Several events can cause the microcontroller to wake up. Typically they would be a chip reset signal, special timer expiration, or interrupt detection. Using a timer to force a wake-up signal ensures that the circuit will become active if it fails to receive an interrupt. This allows the program to examine the interrupts to determine if there is an error condition.

A cell phone microcontroller monitors, for example, device activity. It controls the system clock, and if there is no activity after a fixed period of time, it will initiate a power-down sequence. This will put the cell phone into standby mode. When activity is detected, it will power up the circuitry to initiate the proper response.

### 4.3.5 Low-Power Consumption

Reducing the overall power consumption of a microcontroller, independent of any sleep modes, can by done by reducing the supply voltage $V_{dd}$ (Figure 4.18). Transistor-transistor logic (TTL) traditionally has $+/-5\%$ on $V_{cc}$ for a voltage range of 4.5 to 5.5 volts. Voltage ranges for commercial CMOS microcontrollers often vary over a 4-volt range, from 1.5 to 5.5 volts. Lower voltages are possible for extremely low-power application requirements.

**FIGURE 4.18**  Microcontroller Operating Voltage Range

| Part | Range (Volts) |
|---|---|
| Zilog Z8 | 2.7 to 3.6 |
| Freescale 68HC11 | 3.0 to 5.5 |
| PIC18F4520 | 2.0 to 5.5 |

### 4.3.6 Watchdog Timer

The watchdog timer (WDT) function plays a critical role in robust interrupt-based system design. Microcontroller-based designs by definition are intended for applications that are self-contained. They are embedded as part of the application circuitry on a PCB (printed circuit board). It is important for the overall reliability of the module that they function properly.

Microcontrollers are also programmed devices. They are susceptible to software bugs, which may occur from unpredictable operating conditions. Comprehensive testing for fault conditions, no matter how thorough, cannot cover all possible combinations of real-world conditions.

An unpredictable single event or transient fault may occur that causes an error in the instruction execution sequence. The microcontroller may get stuck in an infinite loop or begin operating erratically. This in turn could require a complete power-on reset sequence.

A WDT function is designed to prevent this situation. It does so by creating a hard-coded interrupt to the microcontroller if the WDT is not reset before it expires. This forces the program counter to an interrupt address where the instruction sequence will begin executing properly.

The WDT timer is outside the interrupt enable tree. There is no corresponding enable bit for it. For the Microchip PIC18F4520 family, WDT is enabled or disabled at the time of chip programming. This prevents the software from disabling it due to faulty execution.

### 4.3.7 In-Circuit Programming

Microcontrollers are soldered to the PCB along with the other devices of the module. If an error is determined in the code, it cannot easily, if at all, be changed. The entire module would need to be replaced. In-circuit programming allows the microcontroller software to be updated without removing the PCB module from the system. This is a very valuable feature in keeping product costs down.

Special-functions registers control the programming sequence. The updated code is transferred to the data memory. This could be through the USART peripheral function, for example.

The microcontroller is then put into a program mode. A sequence of instructions is executed to program the FLASH memory with the updated software. For the Microchip PIC18F4520 family, special control registers associated with programming the FLASH memory are used. Four table control registers are used to define the instruction address in data memory tables that are to be programmed. Two control registers (EECON1, EECON2) then control the write phase of the programming sequence.

## 4.4          DATA INPUT/OUTPUT

Microcontrollers incorporate flexible methods for their input/output (I/O) capabilities. Many microcontrollers provide serial bit manipulation, including the standard parallel I/O function. Often both I/O mapped and memory mapped capabilities are incorporated.

### 4.4.1 Parallel I/O

The most basic I/O function for a microcontroller is the parallel port. It is 1 byte wide and, in most microcontrollers, each bit is selectable as either input or output. This increases its flexibility by allowing the program to treat them as either standard byte interfaces or bitwise signal control lines. Figure 4.19 shows a block diagram of the ubiquitous 8051 from NXP with the I/O ports.

By contrast, the table in Figure 4.20 lists the I/O bits of the ZiLOG Z8. The smaller package types have correspondingly fewer port pins available and cost less.

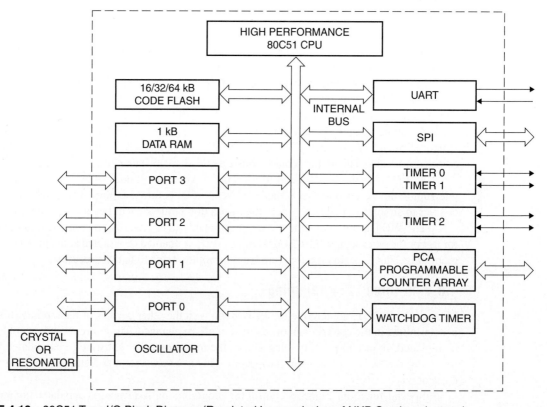

**FIGURE 4.19**  80C51 Type I/O Block Diagram (Reprinted by permission of NXP Semiconductors.)

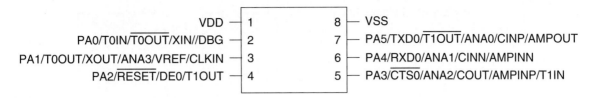

**FIGURE 4.20**   8-Pin package I/O Pin Designations (Reprinted by permission of ZiLOG Corporation.)

### 4.4.2  Tri-State Bit I/O

Circuit design techniques for bit I/O utilize the tri-state concept. An example of a tri-state capable I/O pin is show in Figure 4.21. For output, the data bit is clocked to the output D-latch. On input, the output drive amplifier is put into a high impedance state. This allows the input bit to be clocked by the read input D-latch.

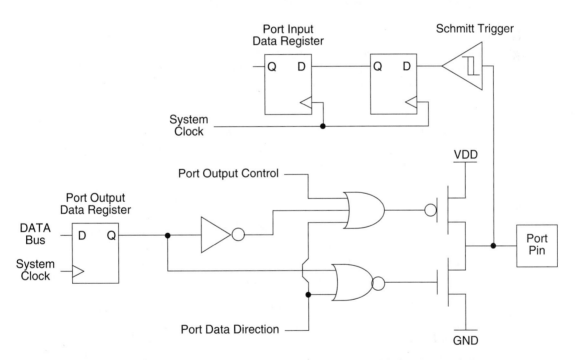

**FIGURE 4.21**   ZiLOG Z8 TRI-STATE I/O Pin (Reprinted by permission of ZiLOG Corporation.)

### 4.4.3  Memory Mapped I/O

In memory-mapped I/O (Figure 4.22), each port is addressable as if it were located in memory. This can also be done treating I/O locations as part of the register memory address space, as was done in the Microchip PIC family. Instructions implicitly define the address for PORT operations. For an 8-bit I/O port, it would be read as a complete byte. Bit manipulation instructions could then be executed for individual bits in the I/O byte. A write to the I/O port would happen correspondingly as the byte is written back to memory.

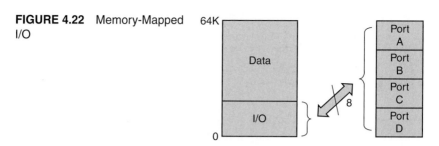

**FIGURE 4.22** Memory-Mapped I/O

A key advantage to memory-mapped I/O is more flexibility in programming. All memory access type instructions can be utilized with the I/O features. This provides for a more powerful I/O capability.

## 4.5    SYNCHRONOUS SERIAL COMMUNICATION

The simplest form of communication between devices is bit serial. On a printed circuit board, it takes only a single pin of two connected devices ($V_{dd}$ and $V_{ss}$ are common). Many microcontroller-based designs contain multiple intelligent peripherals. Communication between them is more complex than simply connecting pins. SPI and $I^2C$ have been developed to meet this requirement.

The simplest technique to connect two devices is bit-banging. One device, the master, transmits bits while the slave device continually checks for them. The roles can be reversed. A small amount of code can process the bits as they are transferred.

For a simple design, this can be an easy way to implement communication between microcontrollers. However, if more devices are introduced, it can be become complex and error prone. There is a lot to be said for keeping things simple. Bit-serial synchronous communication resolves these problems.

Many applications utilize multiple microcontrollers or intelligent peripherals. Devices such as display microcontrollers, A/D converters, and external EEPROM need a fast method to communicate. Parallel buses are not cost effective, physically practical, nor necessary in many microcontroller-based designs. A high-speed interdevice serial interface is a more practical solution.

Two synchronous-bit serial interfaces are defined as industry standards: serial peripheral interface (SPI) and inter integrated circuit ($I^2C$). Each provides the engineer with different design options. The interfaces are suited for multichip designs that can support the cost of their "overhead."

**FIGURE 4.23** Board Level Serial Communication

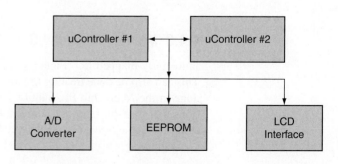

By definition, using SPI or I$^2$C implies the design and production costs are budgeted. The value of these functions lies in their ability to support complex microcontroller-based designs. They allow the design to be partitioned into manageable functional blocks.

## QUESTIONS AND PROBLEMS

1. Why do CMOS transistors use less power than bipolar?
2. List the three typical temperature ranges for microcontroller devices.
3. Describe the basic difference between DRAM and SRAM.
4. Why use an RC-based oscillator?
5. How does a "soft reset" differ from a "hard reset"?
6. What does the WDT provide?
7. What is a key advantage of in-circuit programming?
8. What are the three states of "tri-state" logic?
9. Describe a key advantage of memory mapped I/O.
10. What is meant by synchronous serial communication?

## SOURCES

*Microchip PIC18F2420/2520/4420/4520 data sheet*. 2004. 28/40/44-Pin enhanced flash micro-controllers with 10-bit A/D and nanoWatt technology. Microchip Technology.
*M68HC11E family data sheet*. 2005. Freescale Semiconductor.
*NXP (Philips), P89V51RB2/RC2/RD2 8-bit 80C51 Product Manual*. 2004. NXP Corporation.
*Z8 Encore! XP F08xA series high-performance microcontrollers with eXtended peripherals data sheet*. ZiLOG, www.zilog.com.

# CHAPTER 5

# Program Design

## OBJECTIVE: UNDERSTANDING SOFTWARE PROGRAMMING TECHNIQUES FOR MICROCONTROLLER-BASED DESIGN

The reader will learn:

1. Basic understanding of polling concept.
2. Basic understanding of interrupt concept.
3. Introduction to real-time operating system concepts.

## 5.0 PROGRAM DESIGN

Two primary styles of software programming are used in microcontroller-based design. Polling is a basic method in which the program repetitively "asks" a peripheral function if it requires service. In an interrupt-driven design, the peripheral function "tells" the program that it requires servicing (Figure 5.1). Polling is considered to be less complex and easier to program.

**FIGURE 5.1** Polling versus Interrupt

Polling: Are you full?          Interrupt: I am full!

With polling, the program always knows when it is communicating with the peripheral. The requirement for communication is synchronized in time between the two (Figure 5.2). The program is in complete control over the peripheral interface communication process.

**FIGURE 5.2** Polling Synchronization

need service?   No
need service?   No
need service?   Yes $\Rightarrow$ service routine

## 5.1    POLLING PROGRAM

Polling is simply the program asking a peripheral function for its status. If something needs to be done, the program can do what is required. This can be a data word transfer, update of a variable, or setting a status bit. An example would be a simple application to read a keypad, display the numbers, and print them. Figure 5.3 shows a typical polling arrangement.

What can be seen from the time flow of the program is that most of the processing time is spent waiting, such as for a keypad depression, a liquid crystal display (LCD) character to display, or a printed receipt to be completed. This is an important characteristic of the polling technique. The microcontroller spends most of its time waiting for something to happen.

### 5.1.1  Program Flow

Peripherals set a flag bit when they are ready for servicing. The program tests repeatedly for this bit to be set. When the test condition is met, it branches for the data transfer. Figure 5.4 shows the flowchart to test for the status of the keyboard, LCD, and printer. It is simple to implement and reliable for basic application design.

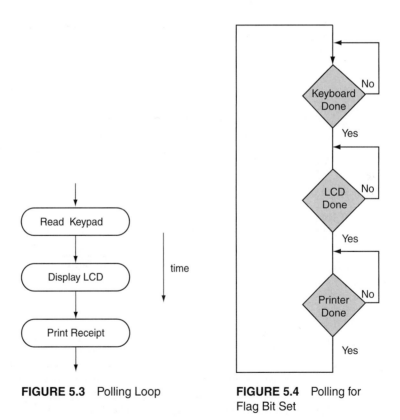

FIGURE 5.3   Polling Loop

FIGURE 5.4   Polling for Flag Bit Set

The instruction sequence for the PIC18F4520 is shown in Figure 5.5. Each device status bit is checked and a branch taken for servicing. If there is no service request, the program continues to loop. Notice that the program spends most of its time looping and waiting for status flags to be set.

**FIGURE 5.5**  Polling Instruction
Sequence

| Main-Loop | btfsz | Kypd, Rdy |
|-----------|-------|-----------|
|           | goto  | Kypd_service |
| LCD       | btfsz | LCD,IRdy |
|           | goto  | LCD_service |
| Printer   | btfsz | Printer, Rdy |
|           | goto  | Printer_service |
|           | goto  | Main-Loop |

## 5.1.2  Program Timing

Timing is a critical issue with polling. How long it takes to process service requests determines the structure of the program (Figure 5.6). Ultimately, it determines whether the polling technique can be used at all. A fixed amount of time is required to process the keypad, LCD, and printer requests. During the time the program is processing a service request it will not be available for another task.

**FIGURE 5.6**  Polling Loop Delay
Time

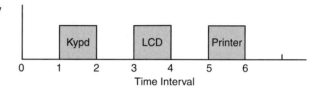

What happens if it takes longer to process the keypad request? If the keypad service processing time exceeds 20 us, the LCD character display will be delayed, as shown in Figure 5.7. However, the printer is expecting the next character to be processed. The extended processing time of the keypad routine will cause a printer error. The character will not be printed.

**FIGURE 5.7**  Overlapped
Processing Times

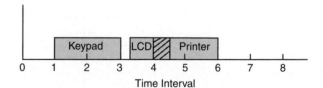

If something were to happen to the printer, its service time would be extended. The next key depression could be missed. When using polling, it is critical to have complete control over the program service routine timing. Extending time in a service routine can cause errors in other routines.

As we have seen, most of the microcontroller time in a polling arrangement is spent waiting for something to happen. As more input/output (I/O) service routines are incorporated in a design, timing becomes more crucial. Any deviation can create a ripple effect causing multiple errors.

## 5.1.3  Sequential Tasks

Polling can be compared to juggling multiple balls. The juggler's hands pass balls back and forth while the other balls are in the air. This is shown in Figure 5.8.

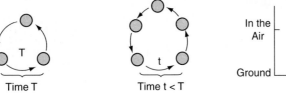

FIGURE 5.8    Juggling Balls

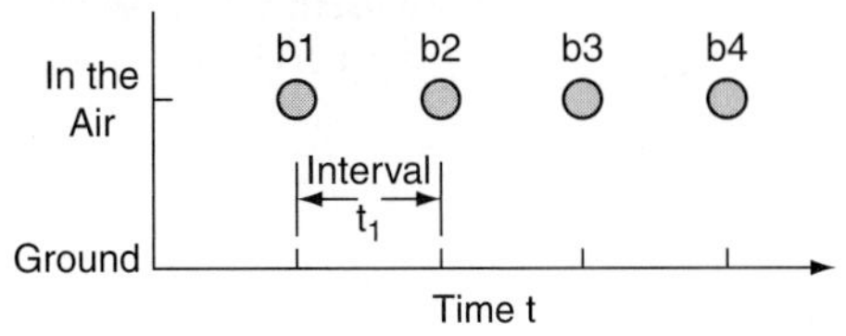

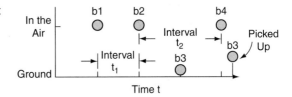

FIGURE 5.9    Time Interval of the Juggled Balls

We can show the time graphically by plotting the time on the *x*-axis and whether the ball is in the air (out of the hand) on the *y*-axis (Figure 5.9). For normal juggling, the time between tosses will be consistent. As the juggler increases the number of balls, the time between tosses decreases. At a certain point, there is not enough time to pass a ball before the next one arrives. This time represents the minimum task time for a polling loop.

What happens if the juggler drops a ball? It increases the task time to process the task for that ball, as shown in Figure 5.10. We can see that time interval $t_2$ exceeds the standard interval of $t_1$. Because of this, the ball b4 will be processed out of sequence, which is the same as a software failure. For example, if a valve is not turned off at the proper time, it may cause flooding.

FIGURE 5.10    Missed Time Slot

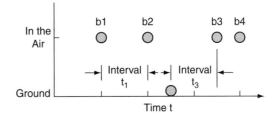

Another possibility is that after the ball is dropped, the juggler is able to pick it up and pass to the other hand, before the next ball arrives, as shown in Figure 5.11. If this is the case, then from the external perspective, nothing incorrect will have happened. It will be in sequence. For software, the task associated with b3 will be processed in time to keep the tasks in sequence, and the program will work properly.

FIGURE 5.11    Missed Time Slot Made Up

## 5.1.4  Task Timing

In real-world design, the process times for tasks are more likely to be variable. We live in an analog world. With polling, there is a minimum (worst-case) time during which a peripheral must receive attention. This translates into the maximum (worst-case) amount of time the microcontroller can spend doing any other or, more important, all other tasks. If the minimum time between servicing is exceeded for any reason, an error condition will occur (see Figure 5.12).

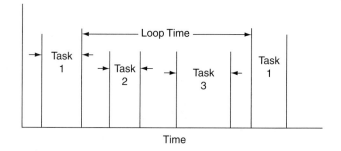

**FIGURE 5.12**   Minimum Task
Service Time

### 5.1.5 Multiple Sequential Tasks

A security card scanner application is shown in Figure 5.13. It loops until a card is inserted and then scans the bar code. A message of either acceptance or rejection is put on the display. The keypad is also scanned for any key depression. Once input is complete, it is processed and sent to a computer, which releases the door latch.

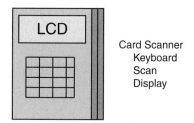

**FIGURE 5.13**   Security Badge
Reader

Card Scanner
Keyboard
Scan
Display

For this example, we will assign processing times to each of four operations. Scanning the keyboard will be $pt_1$, displaying a character on the LCD will be $pt_2$, scanning the card will be $pt_3$, and transmission of data to the computer will be $p_{t4}$. The total processing time is the sum of the individual service routines.

$$P_{total} = pt_1 + pt_2 + pt_3 + pt_4$$

From the equation, we can see that the minimum time between servicing the scanner peripheral function is

$$P_{m3} = P_{total} - P_{t1} - P_{t2} - P_{t4}$$

What happens if a second card is scanned quickly after the previous one? If a second card is scanned in less time than $P_{m3}$, it will not be read. It is a situation similar to the juggler dropping the ball.

This absolute time constraint is the limiting factor in using the polling program technique. The minimum time between service episodes of a peripheral must always be less than the total processing time for all program activities. Because it is difficult to predict asynchronous behavior, polling is best limited to applications with controlled I/O peripherals.

## 5.2          INTERRUPTS

The juggler is tossing four balls in the air. As he tosses one up, he waits for the next one to drop. Again, most of his time is spent waiting for a ball to drop into his hand. If he is better, he can juggle and chew gum at the same time (Figure 5.14). He is concurrently executing tasks.

**FIGURE 5.14** Multiplexing Tasks

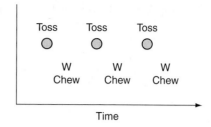

This is the essence of interrupts, seemingly being able to do more than one task at a time. Microcontrollers execute their programs so quickly that much of their time is spent waiting for something to happen. If that time can be used for other productive tasks, the performance of the system will be increased without increasing the speed of the circuitry.

## 5.2.1 Asynchronous Timing

An interrupt causes an immediate response by the microcontroller. It will transfer control to an interrupt service routine (ISR). The interrupt service code will process the event and then return execution to the main program.

The total of the timing of program execution is the foundation on which interrupts are based. It relies on the fact that something quick can happen while something slower is being processed. If, like the juggler, the microcontroller can handle an interrupt fast enough, the balls will not realize one has been dropped. They will continue to cycle in the proper order.

With interrupts, the program code does not have to be constrained to a fixed linear looping flow. The program can be written in a much more flexible manner to respond to events, as shown Figure 5.15. The program becomes event driven by responding to the events as they happen. Interrupt-based software design is referred to as real-time programming.

**FIGURE 5.15** Interrupt Program Flow

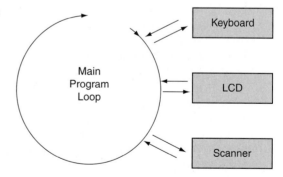

## 5.2.2 Interrupt Enable

By definition, interrupts occur asynchronously to program execution. For this reason it is important that they are allowed or enabled under program control. At minimum, the software needs to be aware that an interrupt might happen. If the software is doing something critical, an interrupt could cause an error.

Each interrupt has an associated enable bit, which is one input to a logic *and* gate. The enable bits are located in an associated bit mask for that function, as shown in Figure 5.16.

**FIGURE 5.16**   Interrupt Enable

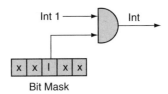

Most commercial microcontrollers have multiple possible interrupt sources. The on-chip peripheral functions can generate interrupts. There will be interrupts from external sources. Other interrupts may be provided to support software code debugging.

An enable bit will be associated with each interrupt. In addition, a global interrupt enable bit is often incorporated. Setting this bit to a logic zero will disable all interrupts. For microcontrollers with many interrupt sources, there may be additional peripheral enable bits, such as the Microchip PIC18F4520 family.

Interrupts that do not have an associated enable bit are referred to as nonmaskable, meaning they cannot be turned off by the software. No instruction can disable them. If an execution error disables all interrupts, the microcontroller would go into an infinite loop. The program is then forced out of the infinite loop and back into execution mode.

### 5.2.3 Machine State

When the interrupt occurs, the microcontroller will be in a specific logical state. This can vary by microcontroller type. Registers and status words will contain data specific to the current process. Part of servicing an interrupt is to save the critical processor values before executing the interrupt code. At the end of the service routine, the machine state can be restored.

Saving the machine state is the most important task to execute when an interrupt occurs. When the machine state is restored, the instruction sequence can be restarted, as if there had been no interruption. This requirement defines the machine state.

At the minimum, all values associated with execution of the current instruction must be saved. This includes the specific registers being operated on and any status flags that may be affected. These values are in transition until the instruction execution is complete.

The values that will be directly modified by any subsequent instruction must be saved at the time of the interrupt. Once the interrupt occurs, the processor hardware will change the program counter to the new instruction location. The first sequence of instructions must save critical values before executing instructions that could modify them.

Saving the machine state may not be necessary in all situations. However, it is good programming practice to do so. Except in high-performance applications, executing a small number of save-type instructions does not appreciably affect performance. For example, in the Microchip PIC18F4520, only the STATUS, W, and BSR registers need to be saved.

### 5.2.4 Latency

The time it takes from detecting the interrupt until the proper software routine begins to execute is latency. Latency is considered the total time from receipt of the interrupt until the first instruction of the interrupt service routine is executed (Figure 5.17). The lower the latency, the faster will be the response to the interrupt. At minimum, the latency must be less than the maximum service time for the interrupting function.

**FIGURE 5.17**   Interrupt Latency
Timing

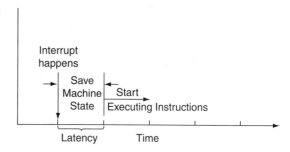

## 5.2.5  Context Switch

Another term used to characterize interrupt response is *context switch time*. This is the length of time needed to transfer control between software routines, including the time to save and restore the machine state. In very high performance microcontroller applications like gigabit switches, this is a limiting parameter on system performance. One large set of registers must be saved in addition to status bits. One solution is to provide multiple registers in hardware so the processor simply needs to remap the register addressing, eliminating the save and restore operations. This is shown with a register bank address pointer in Figure 5.18.

**FIGURE 5.18**   Register Banks

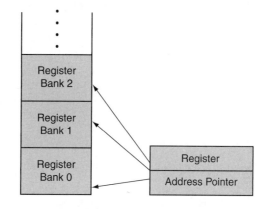

## 5.2.6  Interrupt Vector

Program control is passed to the interrupt service routine (Figure 5.19) through three methods:

> A single interrupt vector
> A table of interrupt vectors
> Programmable interrupt vectors

**FIGURE 5.19**   Interrupt Vectors

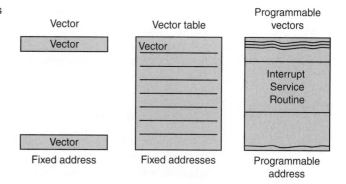

The Microchip PIC18F4520 has only two interrupt vectors. The high-priority interrupt is at location 0x0008, and the low-priority is at 0x0018. When an interrupt occurs, the next sequential address, $n + 1$, is pushed on the program address stack. A GOTO instruction at the vector location branches to the service routine (Figure 5.20). At the end of the service routine, a RETFIE (return) instruction pops the stack and returns execution to location $n + 1$.

**FIGURE 5.20**   Vector Interrupt

```
ORG      Ox0008              ; High Vector
GOTO     Int-Service-High    ; To Service Routine
         .
         .
         .
```

The Microchip PIC18F4520 has many possible interrupt sources. The first task of the interrupt service routine is to determine the source of the interrupt. It then takes the branch to the appropriate routine. This takes time. If many of the interrupts are in use, the time to respond to any given interrupt may be a problem. This is a performance limitation and could be a determining decision on microcontroller selection.

When an interrupt occurs, the microcontroller hardware will load the program counter with the address of the interrupt service routine. This address can be at a fixed memory address for all interrupts, as in the Microchip PIC18F4520, or, for faster response, a table of addresses that point to corresponding service routines called the interrupt vector table.

As with most microcontrollers, the ZiLOG eZ8 uses an interrupt vector table (Figure 5.21). The table contains the addresses for 20 interrupt service routines. For example, a TIMER0 interrupt will vector to location 0x000C.

Unlike the single-interrupt vector, this technique provides for faster response. Any interrupt can be programmed to one of three levels of priority. This significantly enhances the ability to control response time (latency) for a specific function.

### 5.2.7 Nested Interrupts

In a system with multiple interrupts, it is possible that a new interrupt will occur while the program is executing an ISR. In this situation, the interrupts are considered to be nested (Figure 5.22). It is a design decision as to the number or levels at which interrupts can occur.

When multiple interrupts can occur, an address stack is often used to store the return address. As each interrupt occurs, it will push the return address on the stack. The number of nested interrupts is limited by the depth of the stack. In a stack-based architecture, subroutine call instructions are also used. Instructions push or pop return addresses on the stack.

The stack is a LIFO (last in first out) (random-access memory) RAM as opposed to a buffer, which is a FIFO (first in first out) RAM, as shown in Figure 5.23. The instructions push addresses to the top of the stack. The next return executed will pop the stack and transfer program execution. The programmer must be sure to match pushes and pops to avoid stack overflow or underflow conditions.

### 5.2.8 Critical Code

It may not be desirable to allow an interrupt. When a process is in critical code, an interrupt may have undesirable consequences. A medical instrument would not want unpredictable interrupts when controlling heartbeats.

| Interrupt Priority | Program Memory Vector Address | Interrupt or Trap Source |
|---|---|---|
| Highest | 0002H | Reset (not an interrupt) |
| | 0004H | Watchdog Timer (see Watchdog Timer chapter) |
| | 003AH | Primary Oscillator Fail Trap (not an interrupt) |
| | 003CH | Watchdog Oscillator Fail Trap (not an interrupt) |
| | 0006H | Illegal Instruction Trap (not an interrupt) |
| | 0008H | Reserved |
| | 000AH | Timer 1 |
| | 000CH | Timer 0 |
| | 000EH | UART 0 receiver |
| | 0010H | UART 0 transmitter |
| | 0012H | Reserved |
| | 0014H | Reserved |
| | 0016H | ADC |
| | 0018H | Port A7, selectable rising or falling input edge or LVD (see Reset, Stop Mode Recovery and Low-Voltage Detection on page 21) |
| | 001AH | Port A6, selectable rising or falling input edge or Comparator Output |
| | 001CH | Port A5, selectable rising or falling input edge |
| | 001EH | Port A4, selectable rising or falling input edge |
| | 0020H | Port A3 or Port D3, selectable rising or falling input edge |
| | 0022H | Port A2 or Port D2, selectable rising or falling input edge |
| | 0024H | Port A1, selectable rising or falling input edge |
| | 0026H | Port A0, selectable rising or falling input edge |
| | 0028H | Reserved |
| | 002AH | Reserved |
| | 002CH | Reserved |
| | 002EH | Reserved |
| | 0030H | Port C3, both input edges |

**FIGURE 5.21**  ZiLOG eZ8 Interrupt Vector Table (Reprinted by permission of ZiLOG Corporation.)

**FIGURE 5.22**  Nested Interrupts

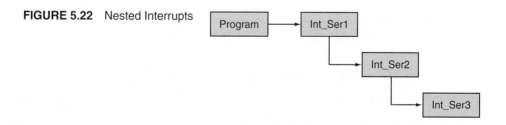

It is possible to poll the interrupt flag. A timer can be used in a delay mode. Because the program is waiting for the timer to go off, which it will by interrupt, the program can simply monitor the timer interrupt flag. This saves program code in not implementing an interrupt service routine.

**FIGURE 5.23**    LIFO and FIFO
Stacks

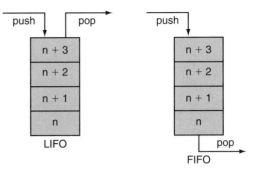

### 5.2.9  Interrupt Service Routine

The example program code shows in Figure 5.24 how the W, STATUS, and BSR registers of the Microchip PIC18F4520 are saved. First, they are put into temporary memory locations. After they have been saved, the interrupt service routine can proceed. At the end of the routine, W_TEMP, STATUS_TEMP, and BSR_TEMP are restored. This has the effect of preserving the machine state at the time of the interrupt.

```
Int_service
      MOVFF W_TEMP                    ; copy w to temp register
      MOVFF STATUS, STATUS_TEMP      ; save status register
      MOVFF BSR, BSR_TEMP            ; BSR located anywhere
      clrf  PORTB              ; clear    PORTB to show interrupt
      movlw h'20'             ; Green, RD<5>
      movwf PORTD             ;

;
      MOVFF BSR_TEMP, BSR           ; Restore BSR
      MOVF  W_TEMP, W               ; Restore WREG
      MOVFF STATUS_TEMP,STATUS      ; Restore STATUS
;
;     don't forget to reset the interrupt flag for RB0/INT
;
      bcf   INTCON,INTF       ; set INTF to zero
      retfie                  ; return with interrupts enabled
```

**FIGURE 5.24**    Example of an Interrupt Service Routine

## 5.3    REAL-TIME OPERATING SYSTEM

A real-time operating system (RTOS) is an operating system for real-time applications. Like Microsoft's Windows, it provides the underlying structure for the application software program and the standardized interfaces, such as terminal and printer communication. As with Windows, it greatly simplifies the software effort to write the program by letting the programmers focus on the application code. Unlike Windows, it supports program execution in real time.

In a simple application with a few interrupt sources, the main program may simply loop on a single instruction (Figure 5.25). When an interrupt occurs, the program will transfer control to the location of the interrupt vector and begin executing instructions from there. At the completion of the service routine, the program returns to the calling routine.

**FIGURE 5.25** Single-Instruction
Interrupt Loop

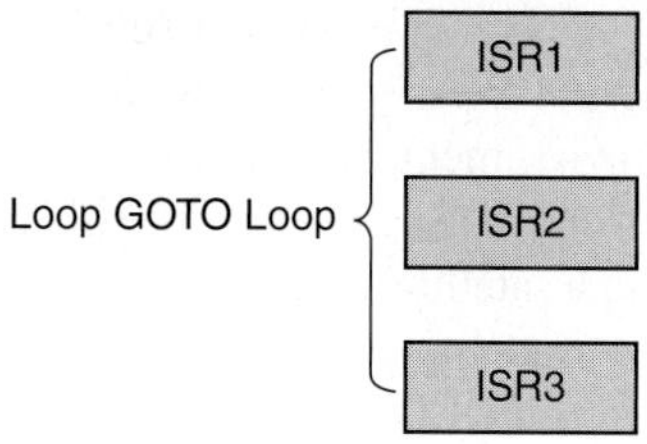

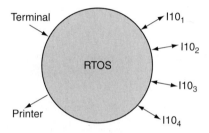

## 5.4 EVENT-DRIVEN SYSTEM

An RTOS continually executes code waiting for an interrupt to occur. By definition, the application's complexity must be sufficient to require an RTOS. A satellite control system is a good example. The RTOS is designed to expect interrupts to occur and is prepared to act on them (in real time) (Figure 5.26).

**FIGURE 5.26** RTOS System
Interrupts

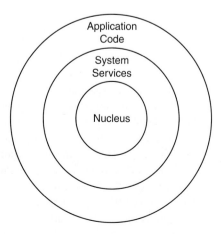

More complex applications, with a robust interrupt structure, require more than a simple loop. Common tasks, such as output to a printer or input from a terminal, can be done by the RTOS. The RTOS provides systemwide services.

## 5.5 NUCLEUS

The heart of the RTOS is referred to as the nucleus or kernel. It is the minimally required program code for the system to function. Application and system code is layered around the nucleus, as shown in Figure 5.27.

**FIGURE 5.27** RTOS Layers

The kernel must be memory resident at the lowest level of the memory hierarchy. It needs to interact directly with the processor logic to handle interrupt requests. In applications where minimizing memory is a key goal, the smallest size kernel is preferred.

An RTOS simplifies the software programming task. In a complex application such as controlling a satellite, a commercial RTOS such as $V_x$Works from WindRiver provides a proven operating system platform. Basic I/O routines are standardized. For example, communication with a PC over the COM1 serial link is done via a USART.

## 5.6    SYSTEM LAYERING

For applications that have specific I/O functions with critical timing constraints, specialized interfaces directly to and from the kernel are implemented (Figure 5.28). Exposing a kernel to the application code increases the risk of failure. It must be done in a very controlled manner.

**FIGURE 5.28** ThreadX Kernel (Reprinted by permission of Green Hills Software, Incorporated.)

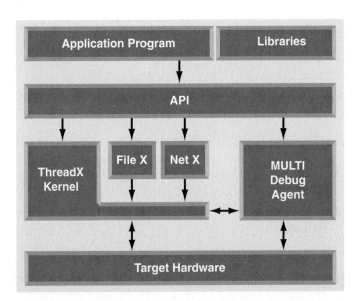

## 5.7    RISK

Microcontroller-based applications operating in real time always face the risk of failure. Some combination of unpredictable asynchronous events could cause a system to latch up. Fault recovery is a key aspect of system design. A robust microcontroller design is fault tolerant. It can withstand the unpredictable failure, recover, and continue to function. For physically remote systems, such as satellites, this is a major challenge for the design engineers.

A good design decision is to use a standardized RTOS with proven reliability, such as Wind River's VxWorks. Using an RTOS is a preferable method for designing more complex systems. However, not all applications require a commercial RTOS. For small applications, using a commercial real-time kernel can not only save time, but also increase code reliability.

Interrupts are a powerful programming tool. They can dramatically increase the perceived performance of a microcontroller design. They do carry a cost. The system must be carefully designed to avoid unpredictable fault conditions. Robust design techniques and conservative design rules enable the engineer to create complex customized real-time applications within a reasonable budget.

## QUESTIONS AND PROBLEMS

1. Describe the main differences between polling and interrupt-based designs.
2. What is the primary function of a polling program?
3. In an interrupt-driven design, who sets the flag bit? Who clears it?
4. Why is timing a key issue with polling?
5. What happens if a time slot is missed?
6. Describe worst-case timing design.
7. In a design, if the $P_{total}$ = 15 ms, $P_1$ = 4 ms, $P_3$ = 3 ms, and $P_4$ = 5 ms, what is the maximum time allowed for $P_2$?
8. What are the basic advantages of interrupts?
9. Describe what is meant by event.
10. Why do you need an interrupt enable bit?
11. Describe and give an example of an interrupt mask.
12. What is the machine state?
13. What does latency mean in relation to an interrupt-driven design?
14. Give an example of an interrupt vector.
15. Show how it is possible for one interrupt to interrupt another.
16. Give an example of an ISR.
17. What three registers define the machine state in the PIC18F4520?
18. What is meant by layered code in an RTOS-based design?

## SOURCES

Joseph, M. 2001. *Real-time systems specification, verification and analysis*. Pune, India: Tata Research Development & Design Centre.

*MicroChip PIC 18F4520 data sheet*. July 2007. Microchip Technology.

Moore, R. 2001. *How to use real-time multitasking kernels in embedded systems*. Costa Mesa, CA: Micro Digital Associates, Inc.

*Thread X data sheet*. 2008. Green Hills Software, Inc.

*Why is a different operating system needed?* Symbian White Paper, Symbian. October 2003.

*ZiLOG Z8 Encore! XP F08xA Series with Extended Peripherals*. Product specification, ZiLOG. June 2006.

# CHAPTER 6

# Hardware/Software Debug

**OBJECTIVE: TO GAIN AN UNDERSTANDING OF THE HARDWARE/SOFTWARE DEBUG PROCESS FOR COTS AND SOC MICROCONTROLLER DESIGN**

The reader will learn:

1. Concept of IDE and ICE.
2. Real-time debugging.
3. Steps in the debug process.
4. COTS and SoC debug tools.

## 6.0 HARDWARE/SOFTWARE DEBUG

Embedded microcontroller design incorporates three major steps to achieve a working product. The software and hardware must be debugged, and more important, the integration of the software with the hardware must be debugged (Figure 6.1). Tool suites or tool chains have been developed for each aspect of the debug cycle.

**FIGURE 6.1** Hardware and Software Debug

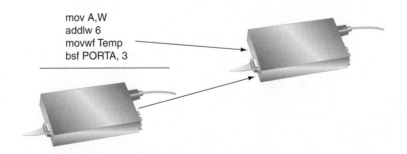

```
mov A,W
addlw 6
movwf Temp
bsf PORTA, 3
```

The time spent in debug will exceed the initial software or hardware design time. It is an inherent part of the design process to debug the design. For commercial off-the-shelf (COTS)-based applications, this is a well-defined iterative process. For embedded microcontrollers, this process carries much higher importance. The cost of getting the first working chip, first silicon, is hundreds of thousands of dollars or more.

## 6.1 COTS CONTROLLER TOOLS

New tools are constantly being introduced to make the debug process more productive. For COTS microcontrollers, where architectures are well defined (i8051, HCS12, PIC18F4520), individual tools are widely available in the established market. Some innovation is done at the integrated development environment (IDE) level, but it is mostly a question of how robust the software is.

The more flexible an IDE is to support families of chips from multiple vendors, the more popular it is in the market (Figure 6.2). With COTS microcontroller design, where chip costs are well defined and up-front design costs are relatively low, the capabilities of the IDE are of paramount importance. An extended version of the IDE is the in-circuit development environment (ICD). This extends the hardware/software integration to include real-time debug of the target board. This is particularly valuable in locating transient bugs.

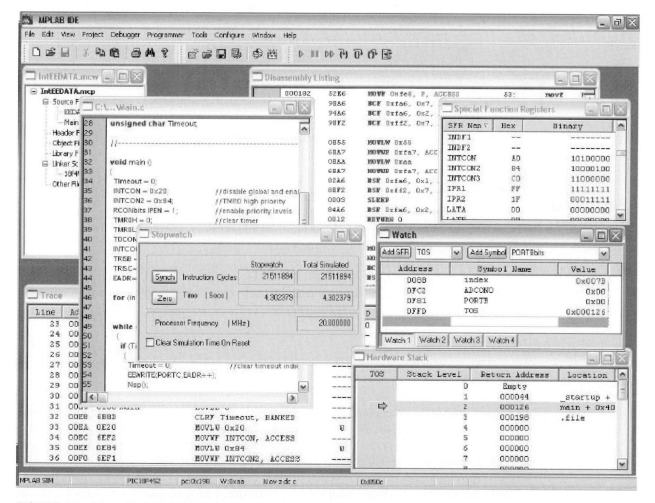

**FIGURE 6.2** Microchip MPLAB IDE (Copyright 2004 Microchip Technology Incorporated. Reprinted with permission.)

## 6.2          EMBEDDED CONTROLLER TOOLS

Embedded microcontrollers present significant technical difficulties for hardware and software debug. The microcontroller core is not directly accessible because it is inside the physical integrated circuit. Special methods have been created to access the core. Both ARM and MIPS have defined debug modules that are embedded with the core on the chip. They provide access to the internal core status (Figure 6.3).

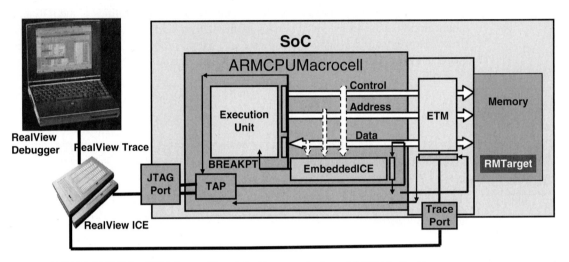

**FIGURE 6.3**    ARM Multi-ICE for ARM Cores (Reprinted by permission of ARM Limited.)

At the higher level of software debug, embedded core tool suites have similar design to those for COTS, but are more complex. By definition, an embedded microcontroller design is magnitudes of orders more expensive than a COTS-based design. With such high economic risk, the importance and cost of the tool chain is high.

A simple embedded core application, such as a security badge reader, still requires the up-front cost of first silicon. A COTS-type software development tool suite may be more than sufficient for program development. The debug process requirements are significantly more stringent.

## 6.3          FIRST SILICON

For an embedded microcontroller-based project, the silicon must work the first time. Not only is it expensive to fabricate the chip, but it can also take many months for fabrication. The cost to the project includes time lost waiting for the chip to return from fabrication. This is compounded by the fact that the product will be delayed for introduction to the market.

With such high stakes for getting it right the first time, embedded microcontroller tool suites are significantly more complex and expensive than their COTS counterparts. Unlike COTS, which has low costs and repetitive development cycles, the embedded microcontroller cycle is long and must be correct the first time.

## 6.4          BROAD-LEVEL PROBES

Debug strategies for embedded microcontrollers are determined by their architecture. The processor is embedded with peripheral functions in the chip. This makes the debug effort more difficult. Figure 6.4 shows the ICD TRACE setup connections from the PC through the debug unit(s), which incorporate the target microcontroller.

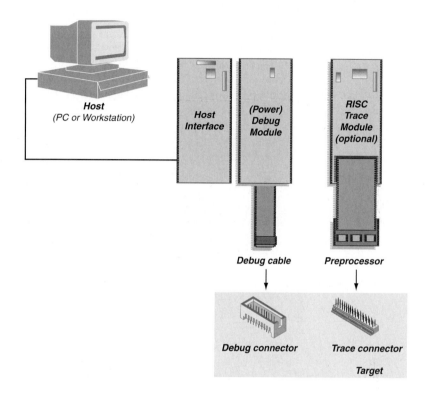

**FIGURE 6.4** ICD Controller Internal Buses (Reprinted by permission of Lauterbach Datentechnich, Gmbh.)

Onboard probes have access to the processor buses, including address, control, and data. As such, they perform logic analyzer type functions. This is limited to the bus clocking speeds of the processor. A key point is that the probe is nonintrusive. That is to say, it does not alter the execution of the software program. It watches what is happening on the buses and can be triggered on by events defined by the debug software. These conditions can be instructions, memory access, or data values. Key features of the Microchip MPLAB ICD 2 are as follows:

- USB (full-speed 2 M bits/s) & RS-232 interface to host PC
- Real-time background debugging
- MPLAB IDE GUI (free copy included)
- Built-in overvoltage/short-circuit monitor
- Firmware upgradeable from PC
- Totally enclosed
- Supports low voltage to 2.0 volts. (2.0 to 6.0 range)
- Diagnostic light emitting diodes (LEDs) (Power, Busy, Error)
- Reading/writing memory space and EEDATA areas of target microcontroller
- Programs configuration bits
- Erase of program memory space with verification
- Peripheral freeze-on-halt stops timers at breakpoints

## 6.5     DEBUG PROCESS STEPS

There are five basic steps to reaching a debugged and working design:

Software entry
Assembly/compilation
Program build
Software simulation
Software/hardware verification.

Figure 6.5 shows the working screen of the Lauterbauch TRACE32 with the ICD option. The TRACE32 combines all five tools for debugging a design in a single unified package. This increases engineer productivity because the tools have a common user interface. Figure 6.5 shows the debugger with the high-level language (HLL) and debug windows open.

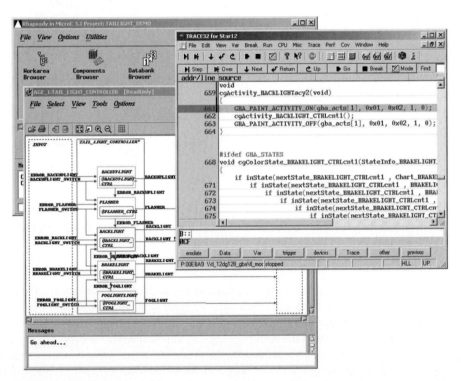

**FIGURE 6.5**   Lauterbach TRACE32 Debug Suite (Reprinted by permission of Lauterbach Datentechnich, Gmbh.)

For COTS-based applications, the availability of an IDE can be the overriding factor in choosing the microcontroller. At the level where the volume chip costs become nearly identical, the cost of product development can be an important issue. A good IDE can save programmer time, increase debug efficiency, and provide downstream product support.

### 6.5.1  Software Editor

Program entry is accomplished using an editing tool. This can be a simple text editor like Notebook. There are also more complex editors that have special features specifically for code writing.

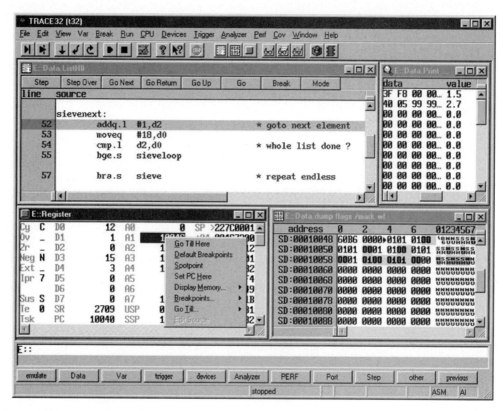

**FIGURE 6.6** Lauterbach Assembler Editor Mode (Reprinted by permission of Lauterbach Datentechnich, Gmbh.)

They can check the syntax of the statement as each line is entered. The output will be in a source code file (.src) for C Code and an (.asm) file for assembly language (Figure 6.6).

## 6.5.2 Compilation

The next step is compiling the program. This is an iterative process repeated until an error-free compilation is reached. If it resides on a computer, other than the one being debugged, it is referred to as a cross compiler or cross assembler. The compiler will generate the executable file and corresponding listing of the instruction sequence (Figure 6.7).

**FIGURE 6.7** Green Hills Optimizing C/C$^{++}$ Compiler (Reprinted by permission of Green Hills Software, Incorporated.)

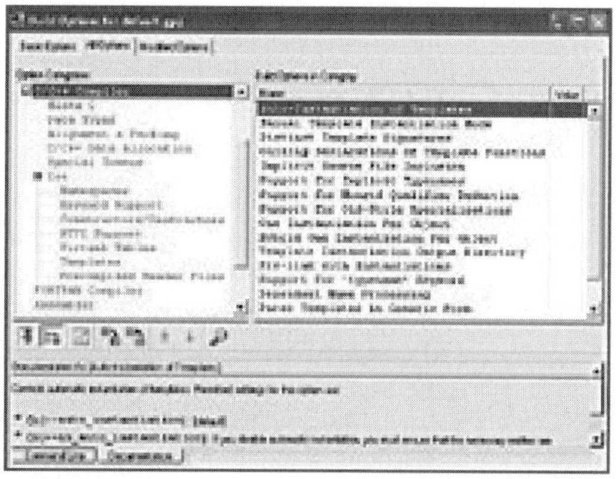

When compiling a C-program, the debug switch is set. This will generate additional information that is used by the simulator program to track instruction execution. The corresponding assembly-level instructions are correlated to the C instructions.

The debug information provides the connection between the probe and software. The instruction execution taking place in the processor can be back annotated to the C instructions. This provides the closure needed to trace the processor instruction flow.

### 6.5.3 Program Build

This is a step to incorporate software modules into a single program image. In addition to the assembler/compiler, most IDEs will include program modules to simplify the program-generation process. MPLAB includes files that contain the list of special files and associated bits for each processor. This simplifies the programming task.

For larger programs, it is recommended practice to divide the program into functional modules. This simplifies the debug process. It also provides a mechanism for multiple programmers to work in parallel, saving time for software development (Figure 6.8).

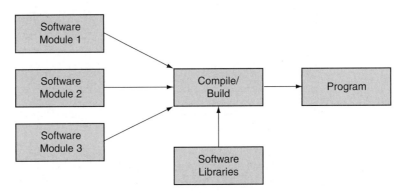

**FIGURE 6.8**   Build Process

A build step is added after the compilation process. Input to linker is the program modules and include files. More robust programming may also incorporate system library files. The output of the build step will be the executable file. This is the file that will be programmed into the microcontroller memory (Figure 6.9).

### 6.5.4 Simulator

When a compiled error-free program is completed, the next step is to debug for logical errors. This is accomplished through the software simulator. A simulator is a program that supports execution of the target processor's instructions.

A simulator debugs the logic of the code to the degree it can be accomplished without residing in the microcontroller memory. A key aspect of the simulator is to support input stimulus. If a keypad depression is being scanned by an input port, it is possible to create a data word with the necessary patterns of 1s and 0s that simulate the input if a key is depressed.

This is a key feature of simulation, the ability to test the software with external test vectors. These vectors are in the form of a hex table that, for example, can be implemented in Excel and

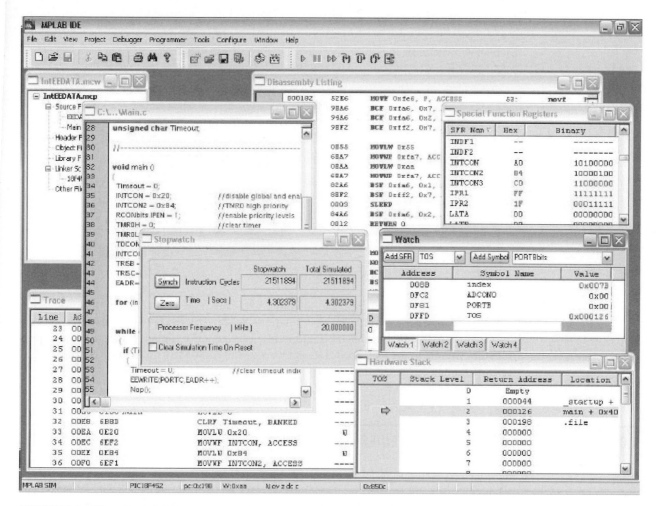

**FIGURE 6.9** Microchip MPLAB Hex Listing (Copyright 2004 Microchip Technology Incorporated. Reprinted with permission.)

**FIGURE 6.10** Test Vectors

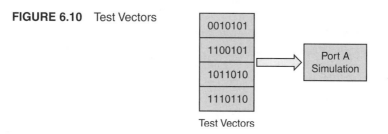

Test Vectors

used as input to a port (Figure 6.10). More powerfully, they can be files that test all input sources. This allows a comprehensive testing of the software code in a simulation that executes the program.

### 6.5.5 In-Circuit Emulation

The final step is to test the software with the microcontroller in the target application. With COTS microcontrollers, it is possible to directly connect the IDE to their socket in the printed circuit board (PCB). This is in-circuit emulation (ICE) (Figure 6.11). The microcontroller chip is plugged into the probe socket pod, which is plugged into the microcontroller socket on the PCB.

**FIGURE 6.11** ARM RealView
ICE and TRACE Modules
(Reprinted by permission of
ARM Limited.)

ICE has electrical and physical limitations. It cannot be used with a core embedded in an SoC. COTS microcontrollers have limitations as well. As the oscillator frequency of the microcontroller increases, it becomes more difficult to control timing. Bus signals must pass from the ICE pod to the PC-based IDE and back again. A 20-MHz oscillator frequency is fast for an ICE unit.

## 6.6    SoC DEBUG STRATEGIES

Embedded cores in system-on-a-chip (SoC) designs cannot have a physical probe head attached. In addition, they are isolated by the complementary metal-oxide semiconductor (CMOS) circuitry from outside probing. They truly are embedded, isolated both physically and electronically from the outside world. This presents an insurmountable obstacle to using traditional COTS microcontroller ICE debug strategies.

With SoC design, working silicon on first pass is a holy grail. The costs are so high that successful first silicon is a must. Many debug solutions on the market meet this need. In general, the method to use correlates to the design cost of the project. The greater the economic impact of first silicon failure, then the greater the up-front investment in tools should be (Figure 6.12).

**FIGURE 6.12** Debug Tool Costs

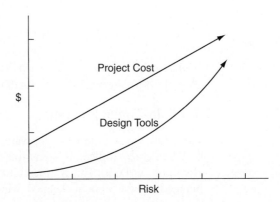

## 6.6.1 SoC Software Debug

Several alternative software-oriented strategies can accomplish microcontroller debugging. As we have seen with the ICD, one method is to simulate the software, program the microcontroller, and give it a try. This is fine for applications with a small amount of code and physically stable hardware, but not practical for larger designs.

Another alternative for program simulation is the copious use of the C-Code PRINTF statement (Figure 6.13). PRINTF statements are planted like markers on a trail. If the software sequence continues along the premarked path, the PRINTF will put a message on the terminal, "I am here."

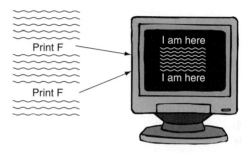

**FIGURE 6.13**   Use of PRINTF

The problem with PRINTF is what happens if you don't make it from marker 1 to marker 2. In this situation, you would add PRINTF statements closer to where you thought the incorrect instruction execution occurred. Through trial and error, you would narrow down the problem section of code. This is fine for well-behaved programs, but for complex code in real-time applications, this system breaks down.

Adding instructions for PRINTF and executing its associated instructions also changes the instruction flow and timing. This alone could interact with complex timing algorithms to create errors. Using PRINTF intrudes into the real-time interaction of software and hardware. It is an intrusive debug strategy in real-time applications.

## 6.6.2 Core-Level Debug

For SoC design, nonintrusive monitor capability is a must. Transient faults will often happen only after hours (or days) of running the code. An example would be a digital cable settop box. The possible error conditions may relate to a combination of conditions that rarely occur. Nonintrusive debug can monitor processor status and break on these infrequent conditions.

A simulator should have extensive test vector files. They cannot, during simulation, mimic all possible combinations of hardware and software events. Simulation is a tool, not a total solution. It lacks the capability to detect real-time or transient faults.

Another important factor with real-time design is that events are happening in real time. Some method is required to trap and record them. In a way it is similar to a quantum system. Any attempt at measurement alters its state. A way is needed for the processor to track what is happening and report on it without altering its real-time nature.

A solution is to incorporate core-level debug as a functional block within the SoC design. It is then an integral function of the core and SoC. It functions in concert with the other logic and can monitor processor activities without interrupting the real-time flow. ARM defines an EmbeddedICE Macrocell for this purpose (Figure 6.14).

**FIGURE 6.14**   ARM EmbeddedICE Macrocell (Reprinted by
permission of ARM Limited.)

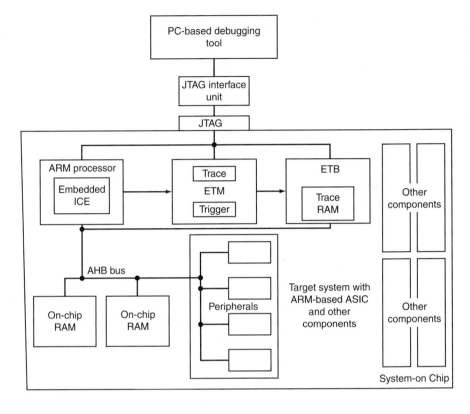

### 6.6.3  JTAG/EJTAG Specification

ARM (JTAG) and MIPS (EJTAG), two major designers of embedded cores, have both developed
functional debug blocks. They rely on the industry JTAG/EJTAG specification ([Extended] Joint
Test Action Group), Test Access Port (TAP) standard (Figure 6.15). It is a test specification for a
method to access integrated circuits mounted on a PCB for testing purposes. It is a de facto industry standard, and both ARM and MIPS incorporate it in their debug blocks.

**FIGURE 6.15**   ARM JTAG TAP
(Reprinted by permission of
ARM Limited.)

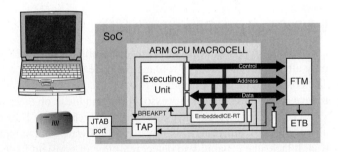

## 6.7        ARM SoC DEBUG

There are two aspects to ARM's implementation of on-chip debug. The EmbeddedICE function
monitors the address, data, and control buses of the processor. This is at the point between the
processor and any other logic. This means that the state of the execution unit can be determined
for any given moment in time.

From the IDE, a hardware address with a specific data value can be loaded to the EmbeddedICE function. When these breakpoint conditions are matched, an interrupt condition will halt the processor. Control will pass to the IDE for further analysis.

In a real-time environment, you want to be able to trace the instructions that led to the break point condition. This will enable you to reconstruct what occurred during program execution. Most important, the instruction, data, and control tracing should be done nonintrusively. Figure 6.16 shows the addition of a trace memory buffer.

**FIGURE 6.16** Trace Buffer Memory (Reprinted by permission of ARM Limited.)

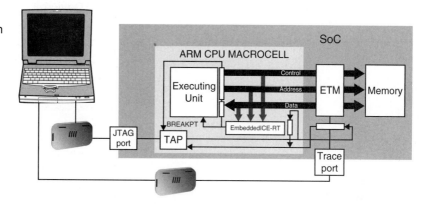

Trace buffer memory can range from 256 k words or more. With a transient real-time event, it could be hours or days for the break point to happen. The more information available on how the processor got to that point the better. The processor is executing hundreds of thousands of instructions per second. Even a 512 k words trace buffer represents only a few milliseconds of real-time instruction execution.

The instructions in the trace buffer will be stored in their hex code compiled format. The IDE tool suite provides a method to reconstruct the instruction sequence to its original form. This allows for source-level debug, a powerful feature of advanced IDEs (Figure 6.17).

**FIGURE 6.17** Green Hills Source-Level Debug for MIPS (Reprinted by permission of Green Hills Software, Incorporated.)

There is a growing demand for SoCs with multiple core processors (Figure 6.18). This is particularly well suited for high-performance Internet switching and router applications. The same basic JTAG TAP concept can be extended to multiple cores. There are constraints, as always, but it brings a level of debug to leading-edge multicore SoC design.

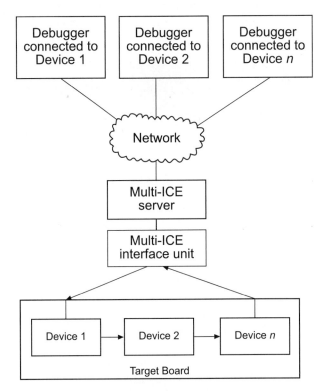

**FIGURE 6.18**   ARM Multi Core Debug (Reprinted by permission of ARM Limited.)

## 6.8    MIPS SoC DEBUG

MIPS, with its partners, developed an "extended" version of JTAG for system debug named EJTAG. It defines three basic components: test access port (TAP), debug control register, and hardware break point unit. They are implemented with a minimal amount of incremental logic to the core processor (Figure 6.19).

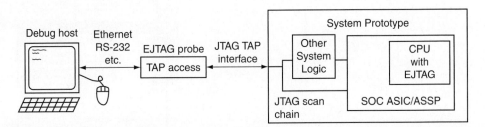

**FIGURE 6.19**   MIPS EJTAG-Based Debug System (Reprinted by permission of MIPS Technologies Incorporated.)

### 6.8.1 EJTAG Functions

Five basic functions are defined by MIPS for EJTAG:

Off-board EJTAG memory
Hardware breakpoints
Single-step execution
System access via the EJTAG TAP
Debug break point instruction

MIPS focuses additional resources on the TRACE capability. They define an extended JTAG physical interface (Figure 6.20). This allows trace information to be brought out in parallel rather than serially. This significantly increases debug performance when dealing with large trace buffers.

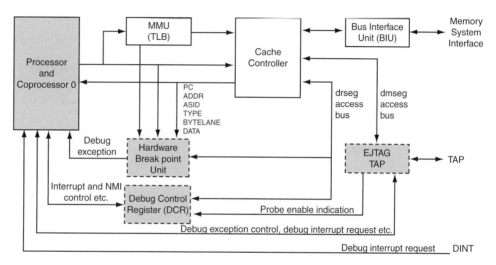

**FIGURE 6.20**   MIPS EJTAG TAP Interface (Reprinted by permission of MIPS Technologies Incorporated.)

There are always trade-offs to be made when using any design tool. Debugging a design is a complex task. The COTS microcontroller design methodology has been developed and "debugged" over the past 30 years. It is a well-worn path with tried and true ways to get from product inception to product out the door.

SoC design is an evolving technology process providing debug capability to a core processor embedded in a sea of logic gates. It is a daunting task. Both MIPS and ARM have developed debug strategies to cope with this challenge. Coupled with robust IDEs, the JTAG TAP approach is used by both companies. Embedded microcontroller debug strategies will continue to evolve with the technological advances in SoC design.

## QUESTIONS AND PROBLEMS

1. What is meant by robust software?
2. Name three reasons why it is important to have first silicon success.
3. What does it mean when a probe is nonintrusive?
4. List the five debug steps.

5. Why use a software editor?
6. What is the primary purpose of a simulator?
7. What is meant by a transient bug?
8. Why use a trace buffer memory?
9. How is a break point used?
10. What is meant by source-level debugging?
11. List the five basic functions of EJTAG.

## SOURCES

*ARM Embedded Trace Macrocell, ETMv3.4 architectural specification.* 2004–2007 ARM Limited.

*ARM Multi-ICE user guide.* Version 2.2. 2004–2007. ARM Limited.

*Development tools data sheet.* 2008. Microchip Technology.

Dienstbeck, R. March 2006. *Tracking the virtual world.* White Paper. RTOS Integrations, Lauterbach, Gmbh.

*EJTAG specification.* Document Number MD00047 Revision 3.10. MIPS Technologies, July 2005.

*Green Hills Probe data sheet.* 2008. Green Hills Software, Inc.

*Green Hills Probe user's guide.* 2008. Green Hills Software, Inc.

*Microprocessor debug interface (MDI) specification.* Document Number MD00412, Revision 02.12. MIPS Technologies, July 2005.

MPLAB Intergrated Development Environment (IDE) brochure, April 2008.

Orme, W. *ARM Embedded Trace Macrocells presentation.* ARM Limited.

*PDtrace interface specification.* Document Number MD00136, Revision 3.01. MIPS Technologies, May 2003.

16-bit embedded control developer's resource. February 2008.

# CHAPTER 7

# Serial Data Communications

**OBJECTIVE: AN INTRODUCTION TO SIGNIFICANT PROTOCOLS USED IN EMBEDDED CONTROLLERS FOR SERIAL COMMUNICATION**

The reader will learn:

1. The industry standard UART controller function.
2. Introduction to the CAN/LIN protocol for automotive networking.
3. Use of $I^2C$ and SPI in multichip system design.
4. Introduction to $I^2S$ for audio applications.
5. Universal peripheral interface (USB).
6. Bluetooth.

## 7.0  SERIAL DATA COMMUNICATION

Serial data communication is vital to embedded controller design. Unlike microprocessors that may have 900 pins, package size and pin count can be a key factor in product design. The ability to communicate with other system devices with minimum pin and package count becomes vital to keeping system costs low.

Figure 7.1a gives a comparison of the basic serial communication busses. It lists the pros and cons using the different serial communication busses in a microcontroller-based design. It is always a good design technique to choose the simplest and lowest cost for high-volume designs. Figure 7.1b shows a comparison of the different transmission distances and data rates.

## 7.1        UART

A universal asynchronous receiver transmitter (UART) is the most basic type of serial interface in use today. Its main function is to simply transmit or receive serial data. It is most commonly used for communication between a modem and a PC. It is also often used as a direct connection to computer-based equipment such as an Internet router. The Freescale microcontrollers implement the UART function via the serial communication interface, or SCI.

| UART | CAN | USB | SPI | I²C |
|---|---|---|---|---|
| • Well Known<br>• Cost effective<br>• Simple | • Secure<br>• Fast | • Fast<br>• Plug&Play HW<br>• Simple<br>• Low cost | • Fast<br>• Universally accepted<br>• Low cost<br>• Large portfolio | • Simple<br>• Well known<br>• Universally accepted<br>• Plug&Play<br>• Large portfolio<br>• Cost effective |
| • Limited functionality<br>• Point to point | • Complex<br>• Automotive oriented<br>• Limited portfolio<br>• Expensive firmware | • Powerful master required<br>• No Plug&Play SW—Specific drivers required | • No Plug&Play HW<br>• No "fixed" standard | • Limited speed |

(a)

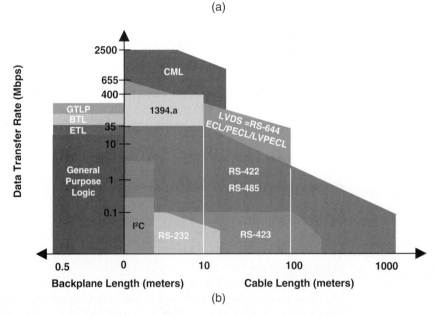

(b)

**FIGURE 7.1** (a) Pros and Cons of Different Serial Busses; (b) Serial Bus Data Rates (Reprinted by permission of NXP Semiconductors.)

A version with a synchronous mode, the USART (universal synchronous asynchronous receiver transmitter), is also available but rarely used in synchronous mode. Common ethernet networks have replaced it for high-speed serial computer-to-computer data transmission.

### 7.1.1 Asynchronous Mode

Because no clocking signal is used in asynchronous mode, the receiver needs a way to establish a method of synchronizing the data transfer. This is accomplished by having a fixed bit rate (bps) and the use of START and STOP bits. The receiving and transmitting UARTs are set for the same asynchronous mode and bit rate. This is shown in Figure 7.2.

**FIGURE 7.2**  UART Frame
Format

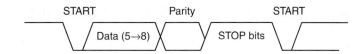

The general-purpose UART designed for common applications have multiple options for the format of the data word. For the industry standard function of the National Semiconductor 16550, the options are

5 to 8 data bits
1 or 2 stop bits
Odd or even parity bit

For the signal shown in Figure 7.2, the signal stays low for the duration of the START bit. The Tx/Rx signal is in the normally high mode for this example. Data is transmitted with the least significant bit (LSB) first. Five to eight bits of data can be transmitted. After the last data bit, the STOP bit, which is always a logic one, is sent. The transmission of the data therefore ends with the Tx pin high. After the STOP bit has been completed, the START bit of the next transmission can occur, as shown in the figure by the dotted lines.

There are several things to note about the waveform in Figure 7.2, which represents the signal on the Tx or Rx pins of the microcontroller. The START bit is a logic "zero," and the STOP bit is a logic "one." Importantly, the data is sent LSB first, so the bit pattern looks backward in comparison to the way it appears when written as a binary number. The data is not inverted, even though RS-232 uses negative voltages to represent a logic one. Generally, when using the UART for RS-232 communications, the signal must be inverted and level-shifted through a transceiver chip.

For example, the Microchip PIC family of microcontrollers implements a 9-bit data mode. The 9-bit mode is useful when parity or an extra STOP bit is needed. To implement parity, the ninth bit is set to make the total number of data bits either even or odd, depending on whether even or odd parity is being used. If two STOP bits are needed, the ninth data bit is set to 1 so that the signal stays high for 2-bit periods after the first eight data bits.

Parity is the addition of a bit to the total number of "logic one" bits being sent to create an even or odd number of ones. Because it is desirable to have at least 1 bit sent if using parity, odd parity is preferable. Examples of parity are shown in Figures 7.3a and 7.3b.

**FIGURE 7.3**  (a) Odd Parity;
(b) Even Parity

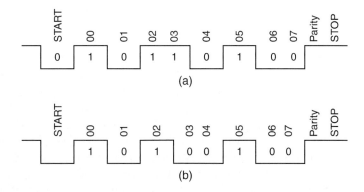

Because signal integrity is so high today, and transmission from a display terminal or modem is very reliable, parity is not typically used. For example, to connect to a CISCO router serial port from a Windows-based PC using Hyperterminal, you would use 8N1:

eight data bits, no parity, and one stop bit. The total transmission per character would then be 10 bits.

## 7.1.2  Transmit/Receive Buffers

To send the data bits, the parallel binary word needs to be loaded to a shift register and then clocked out at the established bit rate. The parallel to serial transmit shift register of the Microchip family is shown in Figure 7.4. It is loaded from a reserved register location called TXREG in the figure.

**FIGURE 7.4**  UART Transmit Buffer

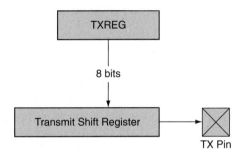

Reception of the serial data stream uses a serial-to-parallel shift register. However, it takes a finite amount of time to read the data from the shift register. During this time, a new data bit could require processing from the RX pin. To avoid the timing problem, a buffer is placed between the serial-to-parallel shift register and the memory register location, as shown in Figure 7.5.

**FIGURE 7.5**  UART Receive Buffer

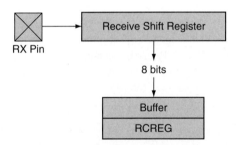

The NSC 16650D UART function uses a 16-byte-deep FIFO (first in first out) buffer for both the transmitter and receive functions. This enhances the reliability and allows for higher bit rates by decreasing the need for interrupts to the microcontroller. Figure 7.6 shows how this is implemented in the National Semiconductor 16550D with FIFOs.

There are the main error conditions reported by the UART: overrun error, framing error, and parity error. Using the FIFO helps eliminate overrun error, a situation in which an incoming data byte is not processed fast enough and is overrun by the next transmitted byte.

A parity error will occur when the number of active bits, 1s, does not agree with the specified parity of odd or even. Odd parity is best, if parity is used, because it will force at least one transition in the transmitted frame.

Finally, a framing error occurs when the start or stop bits are not properly recognized. In such a case, the receiver will not be in the proper data reception mode when the "stop bit" is encountered.

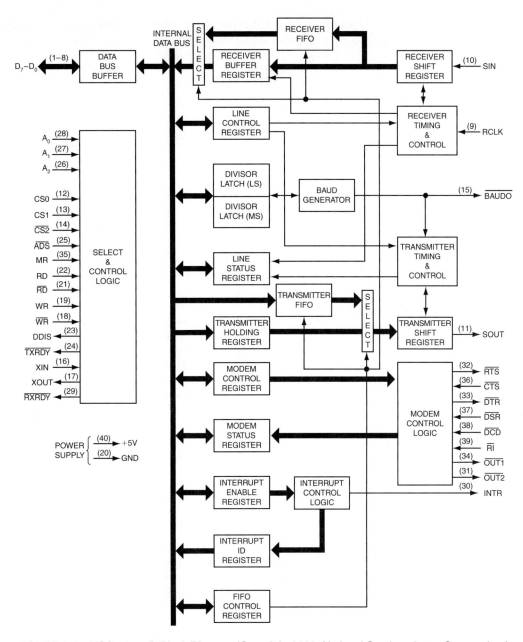

**FIGURE 7.6**  NSC 16550D Block Diagram (Copyright 2003, National Semiconductor Corporation.)

## 7.2    SPI—SERIAL PERIPHERAL INTERFACE

Serial bus systems for interchip communication are becoming increasingly popular. SPI (serial peripheral interface, a name coined by Freescale) is but one of many. Other important serial protocols are $I^2C$ , CAN/LIN, USB, and $I^2S$. These are discussed in more detail later in this chapter.

SPI is a synchronous serial interface that can transfer data in 8-bit byte format a single bit at a time. It can be used to interface to peripheral devices and another microcontroller with an SPI interface. This allows for the possibility of multimicrocontroller system designs.

A key aspect of SPI is the use of the master/slave paradigm, as shown in Figure 7.7. The serial SPI bus requires only four signal lines: two control lines and two data lines, as shown in Figure 7.7. Figure 7.8 lists signal definitions.

**FIGURE 7.7**   Serial Peripheral Interface (SPI)

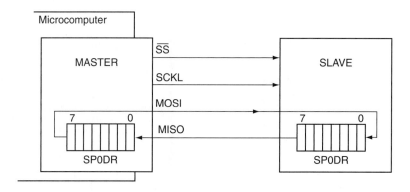

**FIGURE 7.8**   SPI Signal Definitions

| SCKL | Serial clock | Control |
|------|--------------|---------|
| SS | Slave chip select | Control |
| MISO | Serial data out | Data |
| MISI | Serial data in | Data |

Depending on the implementation of the system design, there can be multiple slave chip select control lines, as shown in Figure 7.9.

**FIGURE 7.9**   SPI Master/Slave System (Copyright 2004 Microchip Technology Incorporated. Reprinted with permission.)

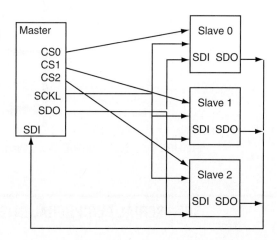

Figure 7.10 shows the Microchip PIC18F45xx family implementation of the SPI interface. No formal public domain specification of the SPI is defined. It does not have an IEEE designation. Instead, the Freescale design specifications have been widely adopted as the standard. Notice the PIC, SDO, and SDI are defined instead of MISO and MISI; however, they carry the same functionality.

**FIGURE 7.10** Microchip PIC18F45xx Family SPI Implementation (Copyright 2004 Microchip Technology Incorporated. Reprinted with permission.)

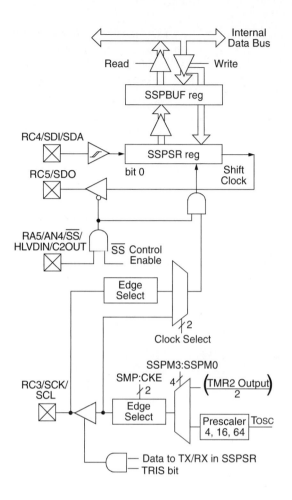

Figure 7.11 shows an example of the timing diagram between SPI signals. Note that in this example, the LSB is being shifted out first. After eight clock cycles, a full byte has been transmitted or received, and the SS (or slave chip select) line can return to the normally high state.

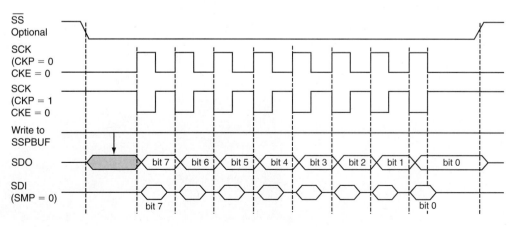

**FIGURE 7.11** SPI Timing Diagram (Reprinted by permission of NXP Semiconductors.)

Data transfer rates can range up to several megabits per second. Basically it is only limited by the specifications of the master and slave SPI interface signals. A significant problem with the interface is the lack of addressing capability. Individual slave select lines are required for each slave device. However, for simple master/slave designs, it is an excellent choice.

## 7.3      I²C—INTER-IC BUS

The Inter-IC bus, commonly known as the I²C ("eye-squared-see") bus, is a control bus that provides the communications link between integrated circuits in a system (Figure 7.12). Developed by Philips in the early 1980s, this simple two-wire bus with a software-defined protocol has evolved to become a worldwide standard for system control. It is used in everything from temperature sensors and voltage level translators to electrically erasable programmable read-only memory (EEPROMs), general-purpose input/output (I/O), analog-to-digital (A/D) and D/A converters, coder-decoders (CODECs), and microcontrollers of all types.

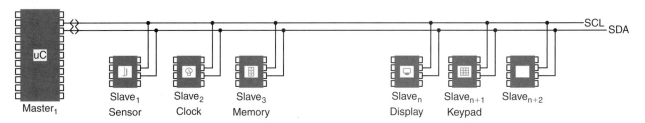

**FIGURE 7.12**   I²C Bus Configuration (Reprinted by permission of NXP Semiconductors.)

There are several reasons why the I²C bus has endured for over two decades. It has kept pace with performance of microcontrollers and now provides three data rates: up to 100 kbps in standard mode, up to 400 kbps in fast mode, and up to 3.4 Mbps in high-speed mode (see Figure 7.13). Hubs, bus repeaters, bidirectional switches, and multiplexers have increased the number of devices the bus can support, extending bus capacitance well beyond its original maximum of 400 pF. Also, software-controlled collision detection and arbitration prevent data corruption and ensure reliable performance, even in complex systems.

**FIGURE 7.13**   I²C Data Rate Comparison (Reprinted by permission of NXP Semiconductors.)

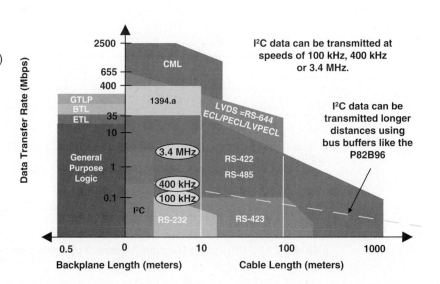

Two simple lines connect all the ICs in a system. Any I$^2$C device can be attached to a common I$^2$C bus, and any master device can exchange information with any slave device. The software-controlled addressing scheme eliminates the need for address-decoding hardware, and there's no need to design and debug external control logic because it's already provided by the I$^2$C protocol.

Design can proceed quickly from block diagram to final hardware, simply by clipping new devices and functions to an existing bus. The I$^2$C bus also saves space and lowers overall cost. The two-line structure means fewer trace lines, so the PCB can be much smaller. Debug and test are easier too because there are fewer trace lines and fewer information sources to verify. As the system evolves over several generations, I$^2$C devices can easily be added or removed without affecting the rest of the system.

## 7.3.1 How the I$^2$C Bus Works

Any I$^2$C device can be attached to an I$^2$C bus, and every device can talk with any master, passing information back and forth. There needs to be at least one master (e.g., microcontroller or DSP) on the bus, but there can be more than one, with all masters having equal priority. Devices can be easily added to and removed from the I$^2$C bus.

Total bus capacitance needs to be less than 400 pF, which works out to be about 20 to 30 standards device or 10 meters of signal trace. This is required to meet the drive requirement of 3 mA for a logic low level and 0.4 mA on an open-drain bus with pull-ups in the range of 2 K to 10 K ohms (see Figure 7.14).

**FIGURE 7.14**  Example of I$^2$C Bus Signals

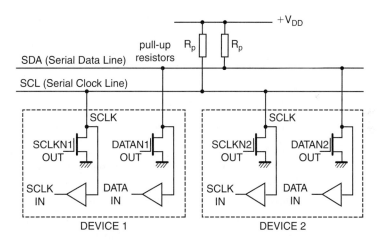

A typical microcontroller (Microchip PIC18F4520) implementation of the I$^2$C bus is shown in Figure 7.15. Here it is treated as an internal peripheral device.

In standard configuration, each device has a unique 7-bit address (Figure 7.16) so that the master knows specifically with whom it is communicating. Typically, the four most significant bits are fixed and assigned to specific categories of devices (e.g., 1010 is assigned to serial EEPROMS). The three less-significant bits (e.g., A2, A1, A0) are programmable through hardware address pins allowing up to eight different I$^2$C address combinations and, therefore, allowing up to eight of that type of device to operate on the same I$^2$C bus. These pins are held high to VCC (logic 1) or held low to GND (logic 0). The 7-bit addressing (Figure 7.16) allows up to 128 devices on the same bus, but some of these addresses are reserved for special commands, so the practical limit is around 120.

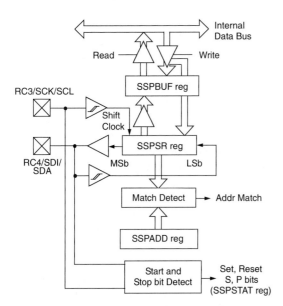

**FIGURE 7.15**   Microchip PIC18F4520 I²C Function Block (Copyright 2004 Microchip Technology Incorporated. Reprinted with permission.)

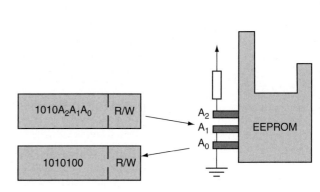

**FIGURE 7.16**   I²C 7-bit Addressing Mode

An extended address mode is also supported with a 10-bit address. This allows a maximum of 1024 devices to be connected. It does not change the format for addresses defined in the I²C bus specification, using addresses reserved in the existing specification. The 10-bit addressing does not affect the existing 7-bit addressing, allowing devices with 7-bit or 10-bit addresses to be connected to the same I²C bus, and both types of devices can be used in standard, fast, or high-speed mode systems.

### 7.3.2  I²C Bus Terminology

The following section defines the basic terminology of the I²C bus, as shown in the block diagram of Figure 7.17.

**Transmitter**—the device that sends to the bus. A transmitter can either be a device that puts data on the bus of its own accord (a "master-transmitter"), or in response to a request from the master (a "slave-transmitter").

**Receiver**—the device that receives data from the bus. A receiver can either be a device that receives data on its own request ("master-receiver"), or in response to a request from the master (a "slave-receiver").

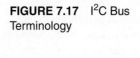

**FIGURE 7.17**   I²C Bus Terminology

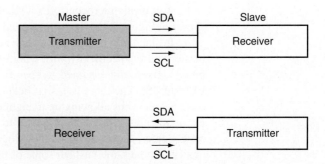

**Master**—the component that initializes a transfer (Start command), generates the clock (SCL) signal, and terminates the transfer (Stop command). A master can be either a transmitter or a receiver.

**Slave**—the device addressed by the master. A slave can be either receiver or transmitter.

**Multimaster**—the ability for more than one master to coexist on the bus at the same time without collision or data loss. Typically "bit-banged" software-implemented masters are not multimaster capable. Parallel to $I^2C$ bus controllers provide an easy way to add a multimaster hardware $I^2C$ port for digital signal processors (DSPs) and application-specific integrated circuits (ASICs).

**Arbitration**—the prearranged procedure that authorizes only one master at a time to take control of the bus.

**Synchronization**—the prearranged procedure that synchronizes the clock signals provided by two or more masters.

**Serial data signal**—data signal line (SDA)

**Serial clock signal**—clock signal line (SCL)

$I^2C$ address of the targeted device is sent in the first byte, and the LSB of this initial byte indicates if the master is going to send (write) or receive (read) data from the receiver, called the slave device. Each transmission sequence must begin with the Start condition and end with the Stop or ReStart condition. If there are two masters on the same $I^2C$ bus, there is an arbitration procedure if both try to take control of the bus at the same time by generating the Start command at the same time. Once a master (e.g., microcontroller) has control of the bus, no other master can take control until the first master sends a Stop condition and places the bus in an idle state.

### 7.3.3  Terminology for Bus Transfer

The following section describes the basic timing and terminology for a bus transfer. Figure 7.18 shows the basic timing for a bus transfer.

**F (FREE)**—the bus is free or idle: the data line SDA and SCL clock are both in the high state.

**S (START) or R (RESTART)**—data transfer begins with a Start condition. The level of the SDA data line changes from high to low, while the SCL clock line remains high. When this occurs, the bus becomes "busy."

**C (CHANGE)**—while the SCL clock line is low, the data bit to be transferred can be applied to the SDA data line by a transmitter. During this time, SDA may change its state, as long as the SCL line remains low.

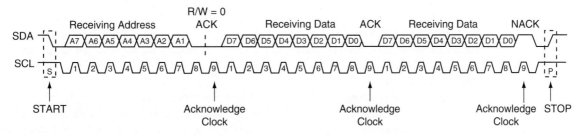

**FIGURE 7.18**   $I^2C$ Bus Transfer

**D (DATA)**—a high or low bit of information on the SDA line is valid during the high level of the SCL clock line. This level must be kept stable during the entire time that the clock remains high to avoid misinterpretation as a Start or Stop condition.

**P (STOP)**—data transfer is terminated by a Stop condition. This occurs when the level on the SDA data line passes from the low state to the high state, while the SCL clock line remains high. When the data transfer has been terminated, the bus is free once again.

## 7.4    CAN—CONTROLLER AREA NETWORK

Today's vehicles contain hundreds of circuits, sensors, and many other electrical components. Communication is needed among the many circuits and functions of the vehicle. For example, when the driver presses the headlight switch on the dashboard, the headlights react. For this to occur, communication is needed between the dashboard switch and the front of the vehicle.

In current vehicle systems, this type of communication is handled via a dedicated wire through point-to-point connections. If all possible combinations of switches, sensors, motors, and electrical devices in full-feature vehicles are calculated, the resulting number of connections and dedicated wiring is enormous. Networking provides a more efficient method for today's complex in-vehicle communications.

In-vehicle networking (IVN), also know as multiplexing, is a method for transferring data among distributed electronic modules via a serial data bus. Without serial networking, intermodule communications require dedicated, point-to-point wiring resulting in bulky, expensive, complex, and difficult-to-install wiring harnesses. Applying a serial data bus reduces the number of wires by combining the signals on a single wire through time division multiplexing. Information is sent to an individual control module that controls each function, such as antilock braking, turn signals, and dashboard displays (see Figure 7.19).

In-vehicle networking provides many system-level benefits, many of which are only beginning to be realized. A decreased number of dedicated wires is required for each function and thus reduces the size of the wiring harness. System cost, weight, reliability, serviceability, and installation are improved. Common sensor data, such as vehicle speed and engine temperature, are available on the network, so data can be shared, thus eliminating the need for redundant sensors.

Networking allows greater vehicle content flexibility because functions can be added through software changes. Existing systems require an additional module or additional I/O pins for each function added. A list of CAN I/O characteristics is shown in Figure 7.20.

For more complex networking to expand into higher-volume economy class vehicles, the overall system benefits need to outweigh the incremental costs. Standardized protocols enable this expansion. Automotive manufacturers and various automotive industry standards organizations have been working for many years to develop standards for in-vehicle networking. CAN, LIN, and SAE J1850 are becoming the predominant standards.

Both SAE J1850 and CAN 2.0 are CSMA/CR (carrier sense, multiple access with collision resolution) arbitration protocols. Through a multimaster architecture, prioritized messages are sent on a serial bus. When two or more try to transmit a message at the same time, the protocols handle message contention by arbitration. The lowest priority modes lose, but the highest priority message successfully reaches its destination without being destroyed by the collision, hence "collision resolution."

One of the key benefits of networking is the ability to add functions without adding new hardware or decreasing reliability. As the networking capability becomes common on mid- and low-priced automobiles, the car manufacturers will be able to easily offer functionality found today only on high-end vehicles.

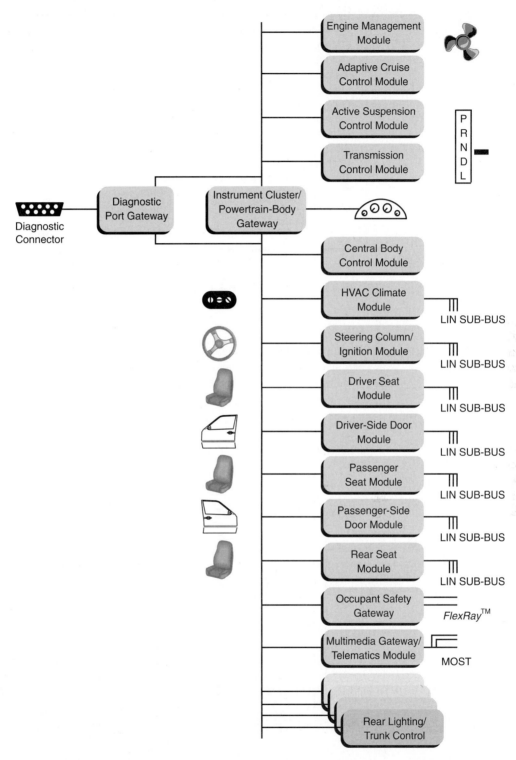

**FIGURE 7.19** CAN Functions (Copyright of Freescale Semiconductor, Inc. 2007, used by permission.)

| CANbus Signal Type | Digital Interface |
|---|---|
| Output voltage (high) | $V_{oh}$ +4 volts min, +5.5 volts max |
| Output voltage (low) | $V_{ol}$ +0 volts min, +1.5 volts max |
| Output voltage | +16 volts (Absolute Max) |
| Output current | 100 mA |
| Impedance | 124 ohm termination between +/− terminals |
| Circuit type | Differential |
| Bit times | 1 uS@ 1 Mb/s; 2 uS@ .5 Mb/s; 4 uS@ .2.5 Mb/s |
| Encoding format | Non-return-to-zero (NRZ) |
| Transmit/receive frequency | 1 Mb/s @ 40 meters |
| Topology | Point-to-point |
| Medium | Shielded twisted pair (STP) @ 9 pin D-sub |
| Access control | CSMA/CD |
| | Nondestructive bitwise arbitration |

**FIGURE 7.20**    CAN/LIN I/O Bus Characteristics

Not all networks are created equal. The need for speed and low latency are critical for power train control and vehicle dynamics. However, the driver compartment functions, such as power windows and instrumentation, only require a speed sufficient to exceed human perception. To optimize the costs and control data access, multiple networks in a single vehicle are becoming common. For example, a CAN network (Figure 7.21) running at 500 Kb/s may link the engine,

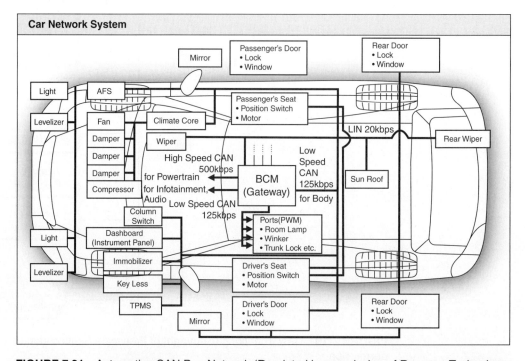

**FIGURE 7.21**    Automotive CAN Bus Network (Reprinted by permission of Renesas Technology Corporation. These specifications may be changed without notice. When using these materials for the design of other products, the most recent version of the associated specifications should be consulted.)

transmission, and ABS, whereas a slower CAN network or J1850 network would link the doors, instrumentation, and other body electronics. A gateway would transfer required diagnostic information between the multiple networks.

Not only is there a benefit to a standardized protocol at the data link and physical layers, but also system designers are seeing the benefits of standardized application layer protocols. These standards allow system designers to avoid low-level protocol details and focus on the application itself. However, the impact of this type of standardization is increased demand on the microcontrollers and protocol devices, and thus the need for efficient message handling and standardized protocol devices will become even more important.

## 7.5   LIN—LOCAL INTERCONNECT NETWORK

A key feature of LIN is its focus on low cost. It is a single-wire network implementation with speeds limited to 20 Kbits/s maximum. This is done for limitations on EMI (electro-magnetic interference). LIN uses a single dedicated master and multiple slave concept (Figure 7.22). This eliminates the need for arbitration protocols and reduces the controller complexity.

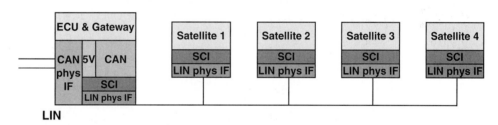

**FIGURE 7.22**   LIN Network Topology (Copyright of Freescale Semiconductor, Inc. 2007, used by permission.)

LIN utilizes the common UART/SCI interface hardware that is present in almost all general-purpose microcontrollers (Figure 7.23). This again eliminates incremental costs. Also, LIN does not require a crystal or ceramic resonator in the slave node, which is another significant cost saver.

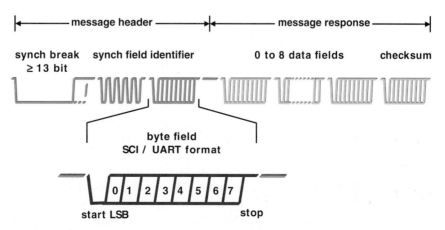

**FIGURE 7.23**   LIN Message Format (Copyright of Freescale Semiconductor, Inc. 2007, used by permission.)

## 7.6    I²S—INTER-IC SOUND

Inter-IC Sound (I²S) is a serial bus design for digital audio devices and technologies such as compact disc players, digital sound processors, and digital TV sound. The digital audio signals in these systems are being processed by a number of VLSI integrated circuits, such as:

A/D and D/A converters
Digital signal processors
Error correction for compact disc and digital recording
Digital filters
Digital I/O interfaces

Standardized communication structures are vital for both the equipment and the IC manufacturer because they increase system flexibility. To this end, Philips has developed the Inter-IC Sound (I²S) bus—a serial link especially for digital audio.

The I²S design handles audio data separately from clock signals. By separating the data and clock signals, time-related errors that cause jitter do not occur, thereby eliminating the need for antijitter devices.

An I²S bus design consists of three serial bus lines: a line with two time-division multiplexing (TDM) data channels, a word select line, and a clock line. Figure 7.24 shows the interconnection of a transmitter and receiver with the clock (SCK), data SD.

**FIGURE 7.24**   I²S Transmitter as Master

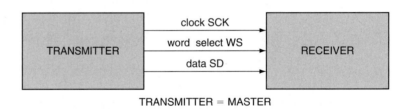

The bus only has to handle audio data, whereas the other signals, such as subcoding and control, are transferred separately. To minimize the number of pins required and to keep wiring simple, a three-line serial bus is used consisting of a line for two time-multiplexed data channels, a word select line, and a clock line (Figure 7.25).

**FIGURE 7.25**   I²S Receiver as Master

Because the transmitter and receiver have the same clock signal for data transmission, the transmitter, as the master, has to generate the bit clock, word-select signal, and data. In complex systems, however, there may be several transmitter and receivers, which makes it difficult to define the master. In such systems, there is usually a system master controlling digital audio data flow between the various ICs. The transmitters then have to generate data under the control of external clock, and so act as a slave. Figure 7.26 illustrates a simple system configuration where the controller acts as the bus master. Note that the system master can be combined with a transmitter or receiver, and it may be enabled or disabled under software control or by pin programming.

**FIGURE 7.26**  I²S Controller as
Master

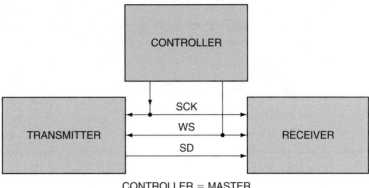

CONTROLLER = MASTER

### 7.6.1 I²S Serial Data

Serial data is transmitted in 2 complement form with the most-significant bit (MSB) first. The MSB is transmitted first because the transmitter and receiver may have different word lengths. It isn't necessary for the transmitter to know how many bits the receiver can handle, nor does the receiver need to know how many bits are being transmitted.

When the system word length is greater that the transmitter word length, the word is truncated (least-significant data bits are set to "0") for data transmission. If the receiver is sent more bits than its word length, the bits after the LSB are ignored. On the other hand, if the receiver is sent fewer bits than its word length, the missing bits are set to zero internally. And so the MSB has a fixed position, whereas the position of the LSB depends on the word length. The transmitter always sends the MSB of the next word one clock period after the word select (WS) changes.

Serial data sent by the transmitter may be synchronized with either the trailing (HIGH-to-LOW) or the leading (LOW-to-HIGH) edge of the clock signal. However, the serial data must be latched into the receiver on the leading edge of the serial clock signal, and so there are some restrictions when transmitting data that is synchronized with the leading edge.

### 7.6.2 I²S Word Select

The WS line indicates the channel being transmitted:

$$WS = 0, \text{ channel 1(left);}$$
$$WS = 1, \text{ channel 2 (right).}$$

WS may change either on a trailing or leading edge of the serial clock, but it doesn't need to be symmetrical. In the slave, this signal is latched on the leading edge of the clock signal. The WS line changes one clock period before the MSB is transmitted. This allows the slave transmitter to derive synchronous timing of the serial data that will be set up for transmission. Furthermore, it enables the receiver to store the previous word and clear the input for the next word.

### 7.6.3 I²S Bus Timing

In the I²S format, any device can act as the system master by producing the necessary clock signals. A slave will usually derive its internal clock signal from an external clock input. This means that by taking into account the propagation delays between master clock and the data and/or WS signals, the total delay is simply the sum of

The delay between the external (master) clock and the slave and
The delay between the internal clock and the data and/or WS signals.

For data and WS inputs, the external-to-internal clock delay is of no consequence because it only lengthens the effective setup time. The major part of the time margin is to accommodate the difference between the propagation delay of the transmitter and the time required to set up the receiver.

All timing requirements are specified relative to the clock period or to the minimum allowed clock period of a device. This means that higher data rates can be used in the future.

## 7.7        IrDA—INFRARED DATA ASSOCIATION

Infrared Data Association (IrDA) has long been used as a transmission medium for TV/VCR controllers, calculators, printers, and PDAs. In late 1993 an industrial group spearheaded by HP, IBM, and Sharp was founded to promote an industrial standard for Infrared communications. A short two and half years later, this group, the Infrared Data Association, grew to 130 members. The membership is international and includes component manufacturers, OEMs, hardware, and software companies.

By 1995, many IrDA-compliant products were in end users' hands. This included IR-equipped notebook PCs, PDAs, printers, as well as IR adapters for PCs, printers, etc. Unlike earlier IR predecessors, which used proprietary protocols, the newer compliant IrDA-based equipment was interoperative across applications, manufacturers, and platforms (Figure 7.27). The key features of IrDA standard are

Simple and low-cost implementation,
Low power requirement,
Directed, point-point connectivity, and
Efficient and reliable data transfer.

**FIGURE 7.27**  IrDA Appliances (Reprinted by permission of ZiLOG Corporation.)

Communications protocols deal with many issues and so are generally broken into layers, each of which deals with a manageable set of responsibilities and supplies needed for capabilities of the layers above and below. When you place the layers on top of each other, you get what is called a protocol stack, rather like a stack of pancakes or a stack of plates. An IrDA protocol stack is the layered set of protocols particularly aimed at point-to-point infrared communications and the applications needed in that environment.

### 7.7.1 IrDA Stack

The IrDA protocol stack is a five-layer model composed of eight basic functions, some of which some are optional. These are shown in their relative position in Figure 7.28. The definitions for each layer are given in the table in Figure 7.29.

**FIGURE 7.28**  IrDA Stack

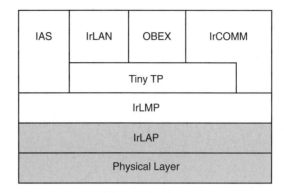

**FIGURE 7.29**  IrDA Stack Definitions

| | |
|---|---|
| Physical Layer | Framer/Physical Layer |
| IrLAP | Link access protocol |
| IrLMP | Link management protocol |
| IAS | Information access service |
| Tiny TP | Tiny transport protocol |
| IrLAN | LAN access |
| OBEX | Object exchange protocol |
| IrCOMM | Serial and parallel port emulation |

Features like automatic selection of compatible communication parameters and service "yellow pages" make the IrDA protocols well suited to embedded devices. This is true even in the consumer market, where the device must communicate simply (like a TV remote control) to gain widespread acceptance.

The IrDA standards documents and additional information about the Infrared Data Association are available at http://www.irda.org. The IrDA Web site also includes links to suppliers of hardware and software.

## 7.8        USB—UNIVERSAL PERIPHERAL BUS

The universal bus interface (USB) was initially defined by seven companies in 1994 as a simple interface between phones and computers. Once the definition of the specification was in progress, the original participants formed the USB Implementers Forum (USB-IF). It now has well over 1,000 members. The board of directors is formed from the original group of members: Agere Sytems, Hewlitt-Packard, Intel, Microsoft, NEC, and Philips.

The USB 2.0 specification calls for support of data transfer rates to 480 Mbps. This is adequate for picture quality scanners and large backup hard disk drives where 60 Mbps is a sufficiently fast transfer rate. It is backward compatible to USB 1.1 with the low-speed 1.5 Mbps and medium-speed 12 Mbps transfer rates.

USB covers a wide range of applications. Example products that have adopted USB for data interchange are listed in Figure 7.30. For small form factor products, like digital cameras, an enhanced miniature USB connector has been developed under the USB-OTG (On The Go) specification.

**FIGURE 7.30**   USB Products
(Copyright of Freescale
Semiconductor, Inc. 2007, used
by permission.)

| USB2.0 – High Speed 480 mbit/sec | | |
|---|---|---|
| **External Mass Storage** – CDRW, DVD, HD, Storage Media – CF, SD/MMC, Memory Stick, PCMCIA | **Audio/Video** – WebCam, MP3 player, Digital Wallet, DSC & others | **Peripherals** – Scanner, Printer, USB to comm bridge & others |

## 7.8.1 USB Topology

USB uses a tiered star, or rooted tree, topology, as shown in Figure 7.31. The highest tier contains the host, which controls the rest of the tree. The host is typically connected to a hub, which divides a bus segment so several nodes can be connected to one segment. Every hub can connect to several nodes and hubs.

**FIGURE 7.31**   USB Rooted
Tree Topology

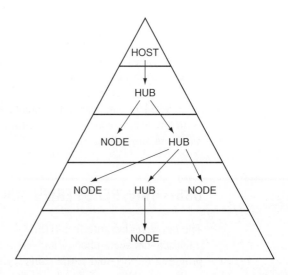

One or more nodes can be packaged with a hub such as a keyboard with two USB ports. These devices are called compound devices. Regardless of how many devices are connected, the host is always bus master. A hop between a hub and another hub or node is limited to 5 meters, with a maximum of five hubs between the host and a given node. The bus is limited to a total of 127 devices, plus the host.

## 7.8.2 USB Architecture

USB communicates with devices on the bus by sending the bits as an encoded sequence, at the rate determined by the destination node. USB 2.0 high-speed devices communicate at 480 Mbps. The host, which is root of the topology, is the bus master. That means that the host schedules and initiates all transactions.

USB supports asynchronous and isochronous data transfer modes. Asynchronous transport does not guarantee that a transmission will be completed within any given time, but it does guarantee that the data will be transferred accurately. Isochronous transport guarantees that a transmission will be completed within a given amount of time, but it does not guarantee that the transmission will be received error free.

The host and device view the transmission medium in different ways. The host sees the device through pipes, which are bundles of unidirectional transport links. An application will communicate with a device function through different pipes. Meanwhile, the device sees the host through endpoints, which are unidirectional data sources and sinks. Each endpoint has certain properties that determine which endpoint communicates with the host.

USB also has the advantage of providing a power source to devices through the bus. A device is allowed to sink a certain amount of power, based on the configuration of the network. If a device requires more than the maximum available power (500 mA at 5 V), it must provide its own power source. A self-powered device is one that has its own power source, whereas a bus-powered device uses the USB bus for its power supply.

USB provides for self-identification and enumeration of devices. When a device is added to the network, it is queried for certain characteristics by the host and is assigned an address. The host then starts the driver for the device and communicates the appropriate information to that driver. To make the USB devices hot-pluggable, USB device drivers must be able to load and unload dynamically. The device-specific driver still communicates through the operating system (OS). USB drivers transfer data through the appropriate pipe.

The USB transmission begins at the bit level, which is a bit-stuffed NRZI encoding scheme. This means that bits are encoded such that a transition must occur on the USB data lines once every seven bit-times. This method of encoding generates enough transitions to allow clock recovery by the device. Figure 7.32 shows the overall data communications scheme.

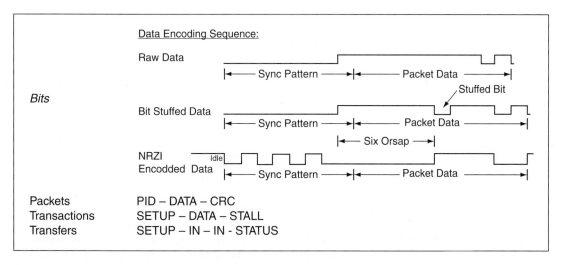

**FIGURE 7.32**   USB Data Communications Scheme

Bits form packets, which are the atomic transfers on the USB bus. A packet is composed of a synchronization sequence, followed by a packet identifier (a 4-bit code followed by its inverse) called the PID. The data payload and cycle redundancy check (CRC) follow the PID. The type of CRC varies based on the type of transmission.

Packets form transactions, which relate a command or data. All transactions have a token packet. If a data payload is being transferred, a data packet follows. If the pipe is asynchronous, a handshake follows.

Based on the type of pipe transferring the data, the transfer can be from one transaction long to many transactions long. Control transfers are at least three transactions long, whereas an isochronous transfer is only one transaction long.

### 7.8.3  USB Physical Connection

A standard USB 2.0 cable is terminated by type "A" (female) and type "B" (male) connectors. For small form factor devices, Mini-A and Mini-B type connectors are defined by the OTG Supplement 1.0a to the USB 2.0 specification.

The USB cable is composed of four wires encased in a shielded sleeve. There are two wires for power, +5 volts (red) and ground (brown). A twisted pair of wires (yellow and blue) carry the data signals. The power wires can support up to 500 milliamps of power at 5 volts, which is sufficient to run, for example, external 2.5-inch hard disk drives. The data wires have a maximum data rate of 480 Mbps with the USB 2.0 specification.

### 7.8.4  USB Interface

USB devices are hot-swappable, meaning you can plug and unplug them from a host or hub at any time. This is a key feature that makes USB very convenient to use. An example would be an external hard disk drive used for backing up several computers.

Another feature of USB is the ability to daisy chain devices. A keyboard may include USB ports that support input peripherals like a mouse. Further, this keyboard may be attached to a USB port on the monitor. This is an attempt with USB to reduce the number of wired connections directly to the computer.

Many embedded microcontrollers include a USB 2.0 host interface function. This greatly simplifies the system design with a USB device for embedded applications. An example of a device with incorporated USB functionality is the Microchip PIC18F4550 shown in Figure 7.33.

### 7.8.5  USB 2.0 Specification

The USB 2.0 specification runs more than 600 pages. It is too long and complex to define in detail in this book. For more information on the specification, refer to the USB Implementers Forum Web site at http://www.usb.org.

## 7.9          BLUETOOTH

Bluetooth is essentially a term used to describe the protocol of a short-range ($<$ 10 meter) frequency-hopping radio link between two electronic devices. The devices are then termed to be Bluetooth enabled. Bluetooth is intended to get around the problems that come with both infrared and cable synchronizing systems. This is accomplished by incorporating a very small radio module into the Bluetooth-enabled device.

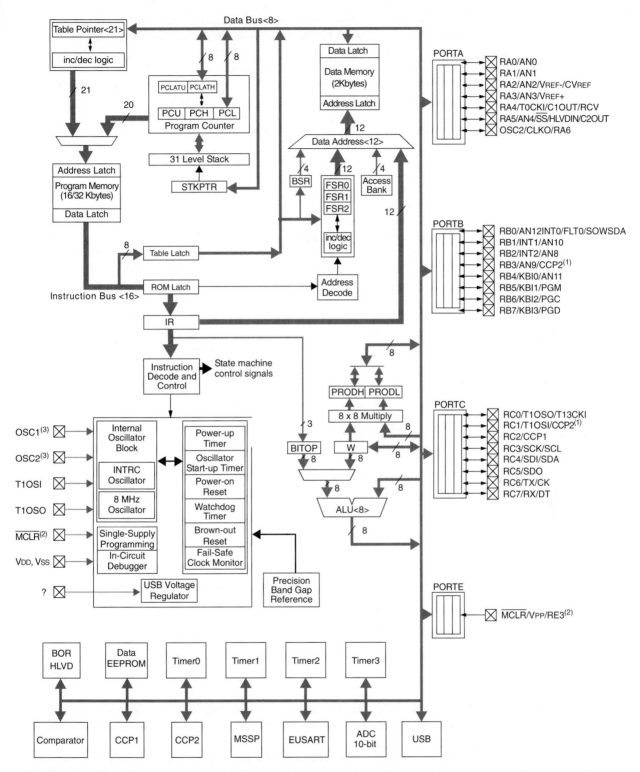

**FIGURE 7.33** PIC18F2550 with USB 2.0 (Copyright 2004 Microchip Technology Incorporated. Reprinted with permission.)

Bluetooth is defined in two sections: the Bluetooth specification and the Bluetooth profiles. The Bluetooth specification describes the basic protocol architecture. The Bluetooth profiles describe how the technology is used. That is to say, how different parts of the specification can be used to implement a desired function for a Bluetooth device.

When computers, entertainment systems, or telephones are used, they make up a network of electronic devices. These devices communicate with each other using a variety of wires, cables, radio signals, and infrared light with an even greater variety of connectors, plugs, and protocols. Bluetooth is wireless and automatic, which can simplify the network implementation.

## 7.9.1  Bluetooth Architecture

There are already a couple ways to get around using wires. One is to carry information between the components in the infrared spectrum. Infrared is commonly used in television and DVD with the IrDA standard. For most of these devices, infrared is used in a simple digital mode. The signal is pulsed on and off very quickly to send data from one point to another.

The second alternative to wires, cable synchronizing, is more troublesome than infrared. In synchronizing, you attach a portable device, for example, to a computer via a cable, press a button, and the data in both devices will be synchronized. It is a technique that works, but is burdensome to make sure you have the correct cables or docking gear for each device.

Bluetooth is intended to get around these problems with infrared and cable synchronizing. Hardware vendors including Intel, Siemens, Freescale, and Ericsson, among others, developed the Bluetooth specification. From the user's point of view there are three key advantages:

It's wireless.
It's inexpensive.
You don't have to think about it.

## 7.9.2  Bluetooth Frequency

Bluetooth communicates on a frequency of 2.45 GHz, which has been set aside by international agreement for the use of industrial, scientific, and medical devices (see Figure 7.34). A number of devices use the radio-frequency ISM (industrial, scientific, medical) band. Baby monitors, garage door openers, and cordless phones all use this ISM band. Making sure that Bluetooth and these other devices do not interfere with one another is a crucial part of the architecture.

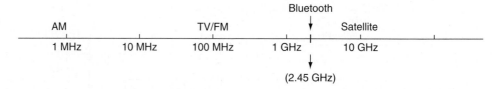

**FIGURE 7.34**   Radio Frequency Allocation

One of the ways Bluetooth devices avoid interfering with other systems is by sending out very weak signals of 1 milliwatt. By comparison, the most powerful cell phones can transmit a signal of 3 watts. The low power limits the range of a Bluetooth device to about 10 meters, reducing the chance of interference. The walls of a house will not stop a Bluetooth signal, making it useful for controlling several devices in different rooms.

Bluetooth uses the 2.4 GHz spectrum to communicate a 1 megabit connection between two devices for both a voice channel and a 768 k data channel. The entire channel has a total capacity of 1 megabit per second. Headers and handshaking information consume about 20 percent of this capacity. In the United States and Europe, the frequency range is 2,400 to 2,483.5 MHz, with 79 1-MHz radio frequency (RF) channels. In practice, the range is 2,402 MHz to 2,480 MHz.

A data channel hops randomly 1,600 times per second between the 79 (or 23) RF channels. Each channel is divided into time slots 625 microseconds long. A piconet has a master and up to seven slaves. The master transmits in even time slots and slaves in odd time slots. Packets can be up to five time slots wide. Data in a packet can be up to 2,745 bits in length.

Bluetooth uses a technique of spread-spectrum frequency hopping to limit interference. In this technique, a device will use 79 individual, randomly chosen frequencies within the designated range, changing from one to another on a regular basis. In the case of Bluetooth, the transmitters change frequencies 1,600 times every second. Each channel is allowed time slots of 625 microseconds long, meaning that more devices can make full use of the limited slice of the radio spectrum available. Because every Bluetooth transmitter uses spread-spectrum transmitting automatically, it is unlikely that two transmitters will be on the same frequency at the same time. This same technique minimizes the risk that portable phones or baby monitors will disrupt Bluetooth devices because any interference on a particular frequency will only last a tiny fraction of a second.

When Bluetooth devices come within range of each other, an electronic conversation takes place to determine whether they have data to share or whether one needs to control the other. The user does not have to press a button or give a command; the electronic conversation happens automatically. Once the conversation has occurred, the devices form a network. Bluetooth systems create a personal-area network (PAN), or piconet, that may fill a room or may encompass no more distance than that between a cell phone and headset. Once a piconet is established, the members randomly hop frequencies in unison so they stay in touch with one another and avoid other piconets that may be operating in the same room.

### 7.9.3 Bluetooth Network

The manufacturer programs each Bluetooth-enabled device with an address that falls into a range of addresses it has established for a particular type of device. For example, the cordless phone has a Bluetooth transmitter in the base and one in the handset. When the base is turned on, it sends radio signals asking for a response from any units with an address in a particular range. Because the handset has an address in the proper range, it responds, and a piconet (PAN) is formed. A piconet has a master and up to seven slaves. Now, if even one of the devices should receive a signal from another system, it will ignore it because it is not from within the network.

A living room could have three separate PANs established, each one made up of devices that know the address of transmitters it should listen to and the address of receivers it should talk to. Because each network is changing the frequency of its operation thousands of times a second, it is unlikely that any two networks will be on the same frequency at the same time. If it turns out that they are, then the resulting confusion will only cover a tiny fraction of a second. The software is designed to correct for such errors and sorts out the confusing information.

## QUESTIONS AND PROBLEMS

1. Define a USART.
2. What is the maximum number of bits in a USART data word?
3. Why use odd parity instead of even parity?
4. Why use a buffer in the USART receive logic?
5. What is the Freescale implementation of the USART called?
6. Identify the signal wires required for an SPI bus.
7. Describe the primary application for the CAN network.

8. What is the primary reason to use LIN?
9. List three key applications for $I^2S$.
10. What are the bus signals in $I^2C$?
11. Why would you use IrDA in a product design?
12. How many layers are in the IrDA stack?
13. What has made USB so popular?
14. What types of data transfer modes does USB support?
15. What does NRZI mean for USB?
16. What problem is Bluetooth designed to solve?
17. What is meant by frequency hopping?

## SOURCES

*Automotive controller area network (CAN) applications.* www.freescale.com. *Freescale Semiconductor.* September 2006.

*CAN specification, Version 2.0.* September 1991. Robert Bosch GmbH.

Core Specification V2.1 + EDR. July 2007. http://www.bluetooth.com

*Impact of new bus technologies.* National Instruments Presentation, NIWEEK. August 2000.

*Intel Automotive introduction to in-vehicle networking.* 2008. www.intel.com. Intel Corporation.

Irazabal, J. M., and Blozis, S. *$I^2C$ bus overview, Philips Presentation, DesignCon 2003.* January 2003.

Layton, Jalia, and Curt Franklin. June 2000. http://electronics.howstuffworks.com/bluetooth1.htm

*Network CAN.* 2008. Renesas Technology.

*PC16550D universal asynchronous receiver/transmitter with FIFOs data sheet. National Semiconductor Corporation.* June 1995.

*PIC18F2420/2520/4420/4520 data sheet 28/40/44-pin enhanced flash microcontrollers with 10-bit A/D and nanoWatt technology. Microchip Technology.* July 2007.

*Simple infrared remote control reference design, application note AN024001-1105.* 2008. ZiLOG Corporation.

# CHAPTER 8

## Analog to Digital Conversion

**OBJECTIVE: A BASIC UNDERSTANDING OF HOW ANALOG TO DIGITAL
CONVERSION IS ACCOMPLISHED**

The reader will learn:

1. The basic fundamentals of analog to digital conversion.
2. Different conversion algorithms and implementations.
3. How to condition input signals.
4. System design interface considerations.

## 8.0 ANALOG-TO-DIGITAL CONVERSION

The real world is analog, from the daily temperature to a person's weight and the sound we hear
from home theater speakers. Microcontrollers, like mainframe computers, operate in the digital
domain. This chapter covers the basics of converting the analog signals of the real world to digital signals the microcontroller can use for its intended purpose.

## 8.1     ANALOG-TO-DIGITAL CONVERSION OVERVIEW

Signals can basically be put into two distinct categories: analog and digital. Analog signals,
x(t), can be defined in a continuous time domain. Digital signals, x(n), can be represented as
a sequence of numbers in a discrete-time domain. The time index n of a discrete-time signal
x(n) is an integer number defined by sampling interval T (usually in samples per second).
Thus, a discrete-time signal, x*(t), can be represented by a sample continuous-time signal
x(t) as:

$$X^x(t) = \sum_{-\infty}^{\infty} x(t)\delta(t - nT)$$

$$\delta(t) = 1, t = 0$$

$$0, \text{ elsewhere}$$

Equation 8.1

A practical analog-to-digital (A/D) converter transforms x(t) into a discrete-time digital signal x*(t), where each sample is expressed with finite precision. Each sample is approximated by a digital code (i.e., x(t) is transformed into a sequence of finite precision or quantized samples x(n)). Refer to Figure 8.1.

**FIGURE 8.1**  Generalized Analog-to-Digital Conversion Process

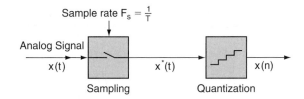

Most A/D converters can be classified into two groups according to the sampling rate criteria: "Nyquist rate" converters, such as SAR or FLASH and oversampling converters like the sigma-delta. Nyquist rate converters have maximum frequencies slightly less than the Nyquist frequency, $f_N = f_s/2$, where $f_s$ is the sampling frequency. Oversampling converters have sampling frequencies at significantly higher rates where $f_N \ll F_s$.

Figure 8.2 illustrates the conventional A/D conversion process that transforms an analog input signal x(t) into a sequence of digital codes x(n) at a sampling rate of fs = 1/T, where T denotes the sampling interval. Because &(t − nT) in Eqn. 8.1 is a periodic function with period T, it can be represented by a Fourier series.

**FIGURE 8.2**  Conventional Analog-to-Digital Process

Equation 8.1 can then be interpreted that the act of sampling (i.e., the sampling function) is equivalent to modulating the input signal by carrier signals having frequencies at 0, fs, 2fs, ... . The sampled signal can be expressed in the frequency domain as the summation of the original signal component and signals frequency modulated by integer multiples of the sampling frequency, as shown in Equation 8.2.

$$X^x(t) = \sum_{-\infty}^{\infty} x(t)\delta(t - nT)$$

Equation 8.2

Thus, input signals above the Nyquist frequency (twice the sampling frequency) cannot be properly converted, and they create new signals in the base-band, that were not present in the original signal. This is frequently referred to as aliasing.

To prevent aliasing, the input signal needs to be passed through a low-pass filter up to the Nyquist frequency. This filter is also referred to as an anti-aliasing filter. It must have a flat response over the frequency band of interest (base-band) and attenuate the frequencies above the Nyquist frequency enough to put them under the noise floor. Because the analog anti-aliasing filter is the limiting factor in controlling the bandwidth and phase distortion of the input signal, a high-performance anti-aliasing filter is required to obtain high resolution and minimum distortion. The higher-performance requirements of the anti-aliasing filter add unwanted cost to the design.

In addition to an anti-aliasing filter, a sample-and-hold circuit is required. Although the analog signal is continuously changing, the output of the sample-and-hold circuitry must be constant between samples so the signal can be properly quantized. This allows the converter enough time to compare the sampled analog signal to a set of reference levels. If the output of the sample-and-hold circuit varies during T, it can limit the performance of the A/D converter.

Each of these reference levels is assigned a digital code. Based on the results of the comparison, a digital encoder generates the code of the level the input signal is closest to. The resolution of such a converter is determined by the number and spacing of the reference levels that are predefined. For high-resolution Nyquist samplers, establishing the reference voltages is a serious challenge.

For example, a 16-bit A/D converter, which is the standard for high-accuracy ADC, requires $2^{16} - 1 = 65535$ different reference levels. If the converter has a 2 V input dynamic range, the spacing of these levels is only 30 mV apart. This is beyond the limit of component matching tolerances of very-large-scale integration (VLSI) technologies. New techniques, such as laser trimming or self-calibration, can be used to boost the resolution of a Nyquist rate converter beyond normal component tolerances. However, these approaches result in additional fabrication complexity, increased circuit area, and higher cost.

## 8.2 TRANSDUCERS

We live in an analog world (ignoring quantum mechanical effects for the moment!). Today's temperature, wind speed, direction, barometric pressure, and other parameters must be converted to digital numbers for the computer to generate a weather forecast. At some point in the circuitry, they are converted from an analog voltage of the transducer to a digital string of 1s and 0s for the computer.

**FIGURE 8.3** Transducer Interface

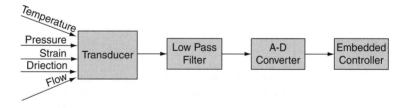

There are many different types of transducers. Some common examples are shown in Figure 8.4.

**FIGURE 8.4** Transducer Types

| Electromechanical | Electroacoustic | Photoelectric | Thermoelectric |
|---|---|---|---|
| Acutator | microphone | photodiode | thermocouple |
| Galvanometer | earphone | LED | thermister |
| Servomechanism | loudspeaker | solar cell | |

Transducers convert one type of energy to another. Sensors are a form of transducers. For electronic applications with microcontrollers, they convert the energy being sensed to a corresponding voltage (or current). A typical example of this is a Hall effect keyboard switch.

It is an important consideration when using sensors that they be sensitive to the property being measured. They should also be insensitive to any other property, and the sensor itself should not affect the property being measured. The closer to these goals, the more ideal is the sensor. Figure 8.5 shows a temperature sensor with an ideal linear output voltage versus temperature.

**FIGURE 8.5**   Linear Sensor Output

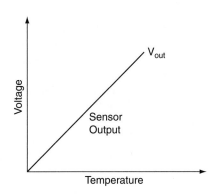

---

## 8.3      LOW-PASS FILTER

Low-pass filters are used to limit the input analog signal frequency to the quantizer. Typical analog-to-digital converters (ADCs) (except the sigma-delta) use low-pass analog filters and sampling at twice the Nyquist rate.

**FIGURE 8.6**   Passive Low-Pass Filter

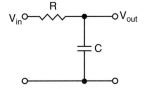

An idealized low-pass filter response is shown in Figure 8.7(a). The critical cutoff frequency, $f_c$, is defined to be –3 db below the passband frequency. Note that the passband frequency rolls off at a –20 dB slope. This can be improved with a multipole design. Multipole implies multiple RC networks. Figure 8.7(b) shows what the frequency response looks like in a practical circuit. Notice the ripple in the passband frequency response.

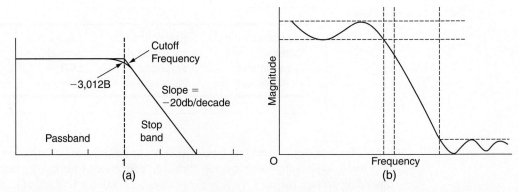

**FIGURE 8.7**   (a) Ideal Low-Pass Filter Frequency Plot; (b) Practical Low-Pass Filter Response

### 8.3.1 Active Filter

An active low-pass filter is implemented with an op-amp. A capacitor, C, is wired in parallel with the feedback transistor R2, as shown in Figure 8.8.

**FIGURE 8.8** Active Low-Pass Filter

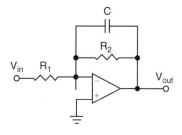

This gives a critical frequency of:

$$f_c = \frac{1}{2\pi R_2 C}$$

Equation 8.3

## 8.4 SAMPLING

A key part of the analog-to-digital conversion process is the sampling time period. The frequency at which the converted analog values are sampled can have a significant impact on the accuracy of the resulting digital output. The samples of the analog signal for seasonal rainfall are shown in Figure 8.9.

**FIGURE 8.9** Seasonal Rainfall Amounts

| Season | Average Rainfall |
|--------|------------------|
| Spring | 2.0 |
| Summer | 3.0 |
| Fall | 2.5 |
| Winter | 1.0 |

Four samples are collected, one per season, to determine average rainfall. They are plotted in Figure 8.10. A line can be connected to them to show the average rainfall over a full year. From this, it can extrapolate the monthly totals, as shown in Figure 8.11.

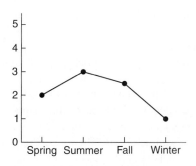

**FIGURE 8.10** Seasonal Rainfall Plot

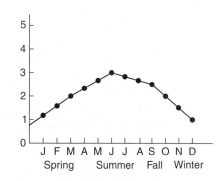

**FIGURE 8.11** Monthly Rainfall Plot

It appears from Figure 8.11 that the monthly rainfall average changes smoothly over the year. This is a result of sampling the rainfall once per season. However, is this correct? From the chart for monthly rainfall (see Figure 8.13), it is clear that the monthly rainfall numbers from Figure 8.12 and those in the chart do not agree. In fact, a significant change in rainfall for the month of July is incorrectly forecast.

These problems are a result of the sampling frequency. The more accurate the results, the higher the sampling frequency needs to be. This gives a greater accuracy to the digital output, as shown in Figure 8.13. Here, the dip in July rainfall is clearly seen.

| Month | Rainfall |
|-------|----------|
| Jan | 2.0 |
| Feb | 2.5 |
| Mar | 3.0 |
| Apr | 4.0 |
| May | 4.0 |
| June | 3.5 |
| July | 3.0 |
| Aug | 4.5 |
| Sept | 3.0 |
| Oct | 2.5 |
| Nov | 2.0 |
| Dec | 1.5 |

**FIGURE 8.12**   Monthly Rainfall Table

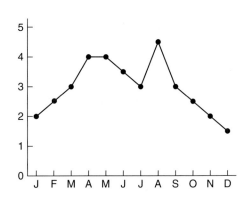

**FIGURE 8.13**   Rainfall Plot by Month

## 8.5          SHANNON'S SAMPLING THEOREM

Shannon's original statement of the Shannon sampling theorem:

*If a function contains no frequencies higher than w CPS, then it is completely determined by giving its ordinates at a series of points spaced 1/2 w apart.*

In other words, when a signal is to be sampled (digitized), the minimum sampling frequency must be twice the signal frequency.

As per Shannon's theorem, it is necessary to sample at a minimum twice the frequency of the analog signal. This is known as the Nyquist rate. At this rate, the signal can be exactly reproduced. If it is sampled at less than the Nyquist rate, it is said to be undersampled. When an undersampled signal is converted back into an analog signal, frequencies not in the original sample will appear. The higher frequency sine waves will now have an "alias," lower-frequency sine waves, not present in the original signal.

Figure 8.14(a) shows an analog sine wave with samples taken at an interval of $T_s$. This corresponds to a sampling period of $1/F_s$. In Figure 8.14(b), the sine wave frequency is defined to be 100 Hz with a corresponding time interval for measurement of 0.005 seconds. For the Nyquist rule, the sampling frequency needs to exceed 200 Hz (double the analog signal frequency).

Figure 8.15 shows a basic analog signal. The sampling rate is not sufficient, and a higher rate signal can also be derived from the sample points. The higher frequency signal is called an alias because it is not present in the original signal.

**FIGURE 8.14** (a) Analog Signal Sampling Period; (b) Analog Signal Sample

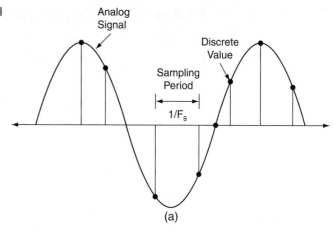

(a)

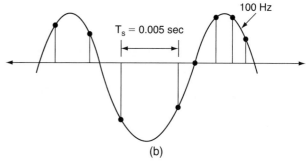

(b)

**FIGURE 8.15** Aliased Signal

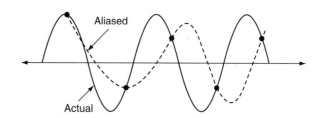

## 8.6      WHAT IS AN ADC?

Because an analog-to-digital converter (ADC) has both analog and digital functions, it is considered to be a mixed-signal device. It can be considered to be a very simple device that provides an output that digitally represents the input voltage or current level. This discussion will focus on ADCs that convert input voltages to binary-coded digital outputs.

Basically the ADC is a divider. It has an analog reference voltage against which the analog input is compared. The digital output represents what fraction of the reference voltage is the input voltage. The input/output transfer function is given by the formula:

$$\text{Output} = 2^n \times G \times A_{IN}/V_{REF}$$

The "G" term represents gain, which is normally unity, so the formula is often written without this term. However, some manufacturers, such as National Semiconductor, have introduced ADCs with other gain factors.

### 8.6.1 ADC Converter Resolution

For a binary-coded output of an ADC there are $2^n$ codes, where n is the number of bits of conversion. The 3-bit ADC shown in Figure 8.16 has an 8-volt reference. There are $2^3$ or 8 possible codes, which are

$$2^0 = 000$$
$$2^1 = 001$$
$$2^2 = 010$$
$$2^3 = 011$$
$$2^4 = 100$$
$$2^5 = 101$$
$$2^6 = 110$$
$$2^7 = 111$$

**FIGURE 8.16** Eight-Bit A/D Converter

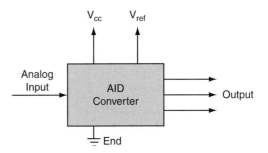

Assuming that the output response has no errors, each time you increase voltage at the input by 1 volt, the output code will increase by 1 bit. This means in the example that the least significant bit (LSB) represents 1 volt, which is the smallest increment this converter can resolve.

Decreasing the reference voltage allows you to measure a smaller range of voltages with greater accuracy with the same number (n) of codes. With a reference voltage of 0.8 V, the LSB would represent just 100mV. This is a method to increase precision without using higher precision (i.e., more expensive) ADC converters.

### 8.6.2 LSB and MSB Defined

The least and most significant bits (LSB and MSB) are simply the bits that have the least weight (LSB) and most weight (MSB) in the digital binary-coded world. The LSB has a weight of 1 ($2^0$). For an n-bit word, the MSB has a weight of $2^{(n-1)} = 2^n/2$. Figure 8.17 shows the LSB to MSB from right to left, as is the normal engineering convention for an 8-bit converter.

| MSB $2^{(n-1)}$ | $2^{(n-2)}$ | $2^{(n-3)}$ | $2^{(n-4)}$ | $2^{(n-5)}$ | $2^{(n-6)}$ | $2^{(n-7)}$ | LSB $2^{(n-8)}$ |
|---|---|---|---|---|---|---|---|
| 0 | 1 | 1 | 0 | 0 | 1 | 1 | 0 |

**FIGURE 8.17** MSB-LSB

Because one LSB is equal to VREF/2 n, it stands to reason that better accuracy (lower error) can be realized if either (or both) of two things happen: (1) use a higher resolution converter and/or (2) use a smaller reference voltage.

### 8.6.3 Quantization

An analog signal cannot be converted to a digital value instantaneously. It takes a finite amount of time to convert an analog value to a digital code. This is accomplished with the sample and hold circuitry (see Figure 8.18). The analog signal is sampled and then latched and held until the analog-to-digital conversion can be accomplished.

A typical analog signal is shown in Figure 8.19. The time increments represent the hold time required for the analog-to-digital conversion. This is referred to in samples per second, which represents the conversion capability of the ADC.

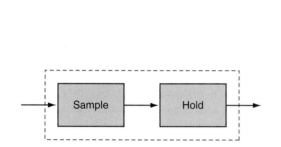

**FIGURE 8.18**   Sample and Hold

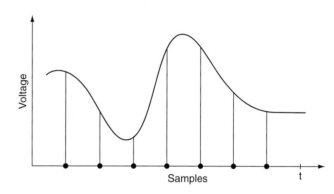

**FIGURE 8.19**   Time Varying Analog Signal

Quantization of the signal is shown in Figure 8.20, where the hold time is shown per sample period. It is clear from this diagram that many points on the analog signal cannot be converted to a digital code. This limit to the conversion time is reflected in the ability of the ADC to accurately represent the input analog signal.

**FIGURE 8.20**   Quantized Analog Signal

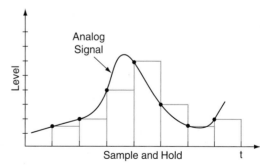

The digital portion of the ADC produces a quantization value that is reflective of the amplitude of the analog signal. Figure 8.21 shows the quantization of an analog signal with the corresponding 3-bit ADC. A 3-bit ADC can quantize $2^3$ or 8 values, as represented in Figure 8.22.

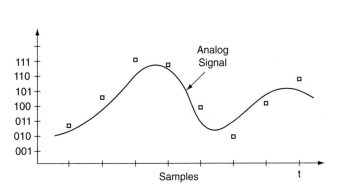

**FIGURE 8.21**    Quantized Signal

| Sample Interval | Quantization Value |
|:---:|:---:|
| 1 | 000 |
| 2 | 010 |
| 3 | 100 |
| 4 | 101 |
| 5 | 100 |
| 6 | 011 |
| 7 | 100 |
| 8 | 101 |
| 9 | 101 |
| 10 | 100 |

**FIGURE 8.22**    Example of Quantized Values

Figure 8.23(a) shows the result of reconstructing the input signal with a 3-bit ADC. It is clear that the reconstructed signal is not representative of the input signal. This is the result of the quantization being of such low resolution.

Increasing the number of steps in the quantization will improve the accuracy of the signal. With a 4-bit ADC operating at twice the frequency, more points of the analog signal are converted with a corresponding reduction in sample and hold time. A 4-bit ADC gives 16 quantization steps, as shown in Figure 8.23(b).

**FIGURE 8.23**    (a) Quantization with 3-bit ADC; (b) Quantization with 4-bit ADC

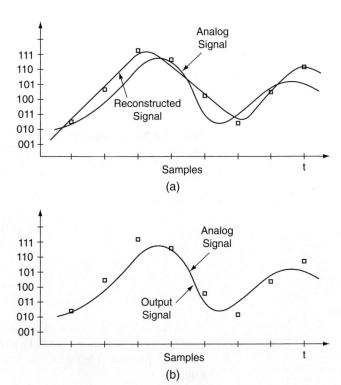

## 8.6.4  Quantization Error

For the simple case of a 3-bit ADC, with an ADC input of zero, the output code is zero (000). As the input voltage increases toward $V_{REF}/8$, the error also increases because the input is no longer zero, but the output code remains at zero because a range of input voltages is represented by a single output code. When the input reaches $V_{REF}/8$, the output code changes from 000 to 001, where the output exactly represents the input voltage and the error reduces to zero. As the input voltage increases past VREF/8, the error again drops to zero. This process continues through the entire input range, and the error plot is a sawtooth, as shown in Figure 8.24.

**FIGURE 8.24**  Quantization Error (Copyright 2003, National Semiconductor Corporation.)*

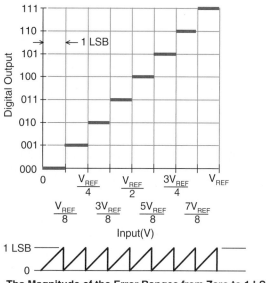

The Magnitude of the Error Ranges from Zero to 1 LSB

The maximum error here is 1 LSB. This 0 to 1 LSB range is known as the quantization uncertainty because a range of analog input values could have caused any given code, and it is uncertain exactly what that input voltage was. The maximum quantization uncertainty is also known as the quantization error. This error results from the finite resolution of the ADC. That is, the ADC can only resolve the input into $2^n$ discrete values. The converter resolution is then $2^n$.

So, for an 8-volt reference (with unity gain factor), a 3-bit converter resolves the input into VREF/8 = 8V/8 = 1 volt steps. Quantization error then is a round off error.

An error of 0–1 LSB is not as desirable as is an error of +1/2 LSB, so an offset is introduced into the A/D converter to force an error range of +1/2 LSB (see Figure 8.25). If 1/2 LSB offset to the ADC input is added, the output code will change 1/2 LSB before it otherwise would. The output changes from 000 to 001 with an input value of 1/2 LSB, rather than 1 LSB, and all subsequent codes change at a point 1/2 LSB below where they would have changed without the offset.

With an input voltage of zero, the output code is zero (000), as before. As the input voltage increases toward the 1/2 LSB level, the error increases because the input is no longer zero, but the output code remains at zero. When the input reaches 1/2 LSB, the output code changes from 000 to 001. The input is not yet at the 1 LSB level, but only at 1.2 LSB, so the error is now −1/2 LSB.

**FIGURE 8.25** ½ LSB Offset Error (Copyright 2003, National Semiconductor Corporation.)

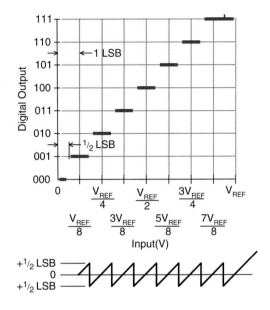

As the input increases past 1/2 LSB, the error becomes less negative, until the input reaches 1 LSB, where the error is zero. As the input increases beyond 1 LSB, the error increases until the input reaches 1½ LSB, where output code is increased by one and the sign of the error again becomes negative. This process continues through the entire input range.

Note that each code transition point decreased by 1/2 LSB compared with the no offset previous page, so that the first code transition (000 to 001) is at +1/2 LSB, and the code transition (from 110 to 111) is at 1½ LSB below VREF. The output of the ADC should not rotate with an over range input as would a digital counter that is given more than enough clock cycles to cause a full count.

### 8.6.5 Offset Error

In an ideal ADC, an input voltage of q/2 will just barely cause an output code transition from zero to a count of one, as shown in Figure 8.26. Any deviation from this is called "zero scale

**FIGURE 8.26** Offset Error (Copyright 2003, National Semiconductor Corporation.)

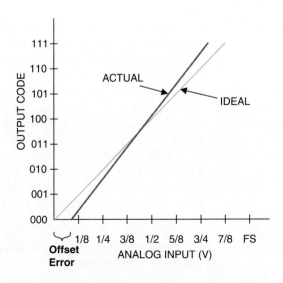

offset error," or "offset error." This error is positive or negative when the first transition point is higher or lower than ideal, respectively. Offset error is a constant and can easily be factored or calibrated out. Offset error may be expressed in percent of full-scale voltage, volts, or in LSB.

### 8.6.6 Differential Nonlinearity

In an ideal converter, the code-to-code transition points are exactly 1 LSB apart. In an 8-bit ADC, for example, these changes are separated from each other by 1 LSB, or 1/256 of full scale. The difference between the ideal 1 LSB and the worst-case actual input voltage change between output code transitions is called differential nonlinearity (DNL), as shown in Figure 8.27.

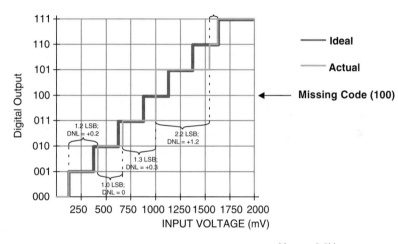

**FIGURE 8.27** Differential Nonlinearity (Copyright 2003, National Semiconductor Corporation.)

$V_{REF} = 2.0V$

DNL can be illustrated using the transfer function of a 3-bit digital-to-analog converter (DAC) shown in Figure 8.27. Each input step should be precisely 1/8 of full scale. In the preceding example, the first code transition (from 000 to 001) is caused by an input change of FS/8 (250 mV for the 2-volt reference example shown here), where FS is the full-scale input. This is exactly as it should be. The second transition, from 001 to 010, has an input change that is 1.2 LSB, so is too large by 0.2 LSB.

The input change for the third transition is exactly the right size. The digital output remains constant when the input voltage changes from 1000 mV to beyond 1500 mV, and the code 101 can never appear at the output. It is missing. To avoid missing codes in the transfer function, DNL should be greater (more positive) than –1.0 LSB.

DNL indicates the deviation from the ideal 1 LSB step size of the analog input signal corresponding to a code-to-code increment. DNL, a static specification, relates to signal-to-noise (SNR), a dynamic specification. However, noise performance cannot be predicted from NNL performance, except to say that SNR tends to become worse as DNL departs from zero.

### 8.6.7 Missing Codes

When no value of input voltage will produce a given output code, such that the code in question never appears in the output, that code is missing from the transfer function and is know as a missing code (Figure 8.28).

**FIGURE 8.28**   Missing Codes
(Copyright 2003, National
Semiconductor Corporation.)

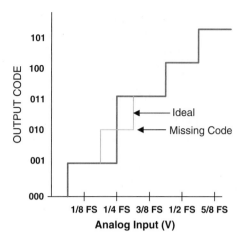

The transfer function is for a 3-bit A/D converter. The first code transition, from 000 to 001, takes place when the input voltage is 1/2 LSB, which is correct for an ADC. The second transition takes place when the input voltage reaches 1/4 FS, so the differential linearity error at that point is +1/2 LSB. The second transition has a differential linearity error of 1 LSB, causing the output code to jump from 001 to 011, and 010 is a missing code.

Any time DNL is –1.0, there is a possibility of one or more missing codes. A DNL that is less than –1.0 means that part of the transfer function takes on a negative slope. Many ADC data sheets specify "no missing codes," as this specification can be critical in some applications, such as in servo systems.

## 8.6.8  SNR—Signal-to-Noise Ratio

Signal-to-noise ratio (SNR) is the ratio of the output signal amplitude to the output noise level, not including harmonics or DC. A signal level of 1 VRMS and a noise level of 100 uVRMS yields an SNR of 104 or 80 dB. The noise level is integrated over half the clock frequency (refer to Figure 8.29).

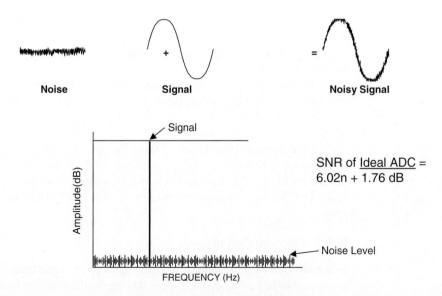

SNR of $\underline{\text{Ideal ADC}}$ = 6.02n + 1.76 dB

**FIGURE 8.29**   Signal-to-Noise Ratio (Copyright 2003, National Semiconductor Corporation.)

SNR usually degrades as frequency increases because the accuracy of the comparator(s) within the ADC degrades at higher input slew rates. This loss of accuracy shows up as noise at the ADC output.

In an ADC, noise comes from four sources:

Quantization noise
Noise generated by the converter itself
Application circuit noise
Jitter

Quantization noise results from the quantization process of assigning an output code to a range of input values. The amplitude of the quantization noise decreases as resolution increases because the size of an LSB is smaller at higher resolutions, which reduces the maximum quantization error. The theoretical maximum signal-to-noise ratio for an ADC with a full-scale sine-wave input derives from the quantization noise and is defined as

$$20 * \log(2(N - 1) * \mathrm{sqrt}(6))$$

or about

$$6.02\,n + 1.76\,\mathrm{dB}$$

SNR increases with increasing input amplitude until the input gets close to full scale. The SNR increases at the same rate as the input signal until the input signal approaches full scale. That is, increasing the input signal amplitude by 1 dB will cause a 1 dB increase in SNR. This is because the step size becomes a smaller part of the total signal amplitude as the signal amplitude increases. When the input amplitude starts approaching full scale, however, the rate of increase of SNR versus input signal decreases.

## 8.7 ANALOG-TO-DIGITAL CONVERSION ALGORITHMS

There are many different algorithms for an analog-to-digital conversion. The most popular five for microcontroller applications are

Successive-Approximation-Register (SAR)
Parallel (Flash)
Dual-slope (Integrating)
Pipeline
Sigma-Delta

Except for the sigma-delta, these ADCs are often referred to as "Nyquist samplers". This is from the fact that most of them sample at a frequency twice the maximum input frequency. They are used in conjunction with analog low-pass filters to limit the maximum frequencies of the input signal to the A/D, as well as the sample and hold circuits.

The sigma-delta ADC uses a low-resolution A/D converter (1-bit quantizer) with noise shaping and a correspondingly high oversampling rate. The low-resolution A/D converter does not require high-precision components as in the analog filter. It thus is attractive for mixed-signal devices with analog and digital functions.

The chart in Figure 8.30 shows the number of bits of the ADC and resultant resolution where full scale is 1 volt. One can see from the chart that past 16 bits, the quantized voltage is less than 0.4 uV per step. This can be very difficult to achieve with standard analog filters requiring expensive high-tolerance pars. Here is where the sigma-delta converter has an advantage.

**FIGURE 8.30**  ADC Resolution
Chart

| # of Bits | $2^n$ | LSB (FS = 1V) |
|---|---|---|
| 8 | 256 | 3.91 mv |
| 10 | 1024 | 977 uV |
| 12 | 4096 | 244 uv |
| 14 | 16384 | 61 uv |
| 16 | 65536 | 15.3 uV |
| 18 | 262144 | 3.81 uV |
| 20 | 201048576 | 954 nv |

Figure 8.31 shows the resolution of the ADC as it relates to speed of conversion. It is clearly shown that as the sampling rate (in samples per second) increases, the corresponding resolution decreases. This is an effect of the speed capability of the quantizer and sample and hold circuitry.

**FIGURE 8.31**  ADC Speed
Comparison

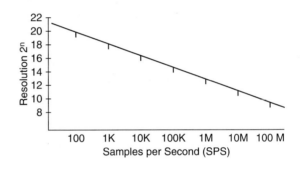

Figure 8.32 shows a comparison of seven key aspects of functional differences among different types of ADCs. Depending on cost and sampling speed requirements, different ADCs offer better advantages. For lower cost and high speed needed for consumer products (DVD), the single-bit sigma-delta converter is the type most often chosen.

### 8.7.1  Successive Approximation

Successive-approximation-register (SAR) analog-to-digital converters (ADCs) are frequently the chosen architecture for medium-to-high resolution applications with sample rates under 5 mega samples per second (Msps). SAR ADCs most commonly range in resolution from 8 to 16 bits and provide low-power consumption as well as a small form factor. This combination makes them ideal for a wide variety of applications, such as portable/better-powered instruments, pen digitizers, industrial controls, and data/signal conversion.

As the name implies, the SAR ADC basically implements a binary search algorithm. Therefore, although the internal circuitry may be running at several MHz, the ADC sample rate is a fraction of that number due to the successive-approximation algorithm.

### 8.7.2  SAR ADC Architecture

Although there are many variations in the implementation of a SAR ADC, the basic architecture is quite simple (Figure 8.33). The analog input voltage (VIN) is held on a track/hold. To implement

| | FLASH (Parallel) | SAR | DUAL SLOPE (Integrating ADC) | PIPELINE | SIGMA DELTA |
|---|---|---|---|---|---|
| **Pick This Architecture if you want:** | Ultra-High Speed when power consumption not primary concern? | Medium to high resolution (8 to 16bit), 5Msps and under, low power, small size. | Monitoring DC signals, high resolution, low power consumption, good noise performance ICL7106. | High speeds, few Msps to 100+ Msps, 8 bits to 16 bits, lower power consumption than flash. | High resolution, low to medium speed, no precision external components, simultaneous 50/60Hz rejection, digital filter reduces anti-aliasing requirements. |
| **Conversion Method** | N bits - 2^N-1 Comparators Caps increase by a factor of 2 for each bit. | Binary search algorithm, internal circuitry runs higher speed. | Unknown input voltage is integrated and value compared against known reference value. | Small parallel structure, each structure, each stage works on one to a few bits. | Oversampling ADC, 5-Hz - 60Hz rejection programmable data output. |
| **Encoding Method** | Thermometer Code Encoding | Successive Approximation | Analog Integration | Digital Correction Logic | Over-Sampling Modulator, Digital Decimation Filter |
| **Disadvantages** | Sparkle codes/metastability, high power consumption, larger size, expensive. | Speed limited to ~5Msps. May require anti-aliasing filter. | Slow Conversion rate. High precision external components required to achieve accuracy. | Parallelism increases throughput at the expense of power and latency. | Higher order (4th order or higher) - multibit ADC and multibit feedback DAC. |
| **Conversion Time** | Conversion Time does not change with increased resolution. | Increases linearly with increased resolution. | Conversion time doubles with every bit increase in resolution. | Increases linearly with increased resolution. | Tradeoff between data output rate and noise free resolution. |
| **Resolution** | Component matching typically limits resolution to 8 bits. | Component matching requirements double with every bit increase in resolution. | Component matching does not increase with increase in resolution. | Component matching requirements double with every bit increase in resolution. | Component matching requirements double with every bit increase in resolution. |
| **Size** | 2^N-1 comparators, Die size and power increases exponentially with resolution. | Die increases linearly with increase in resolution. | Core die size will not materially change with increase in resolution. | Die increases linearly with increase in resolution. | Core die size will not materially change with increase in resolution. |

**FIGURE 8.32** Comparison of Analog-to-Digital Converter

the binary search algorithm, the N-bit register is first set to midscale (that is, 1000 . . . . 00, where the MSB is set to "1"). This forces the DAC output (VDAC) to be VREF/2, where VREF is the reference voltage provided to the ADC.

A comparison is performed to determine if VIN is less than or greater than VDAC. If VIN is greater than VDAC, the comparator output is set to a logic "1" and the MSB of the N-bit register remains at "1". Conversely, if VIN is less than VDAC, the comparator output is a logic low, and the MSB of the register is cleared to logic "0". The SAR control logic then moves to the next bit down, forces that bit high, and does another comparison. The sequence continues all the way down to the LSB. Once this is done, the conversion is complete, and the N-bit digital word is available in the register.

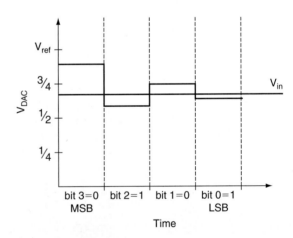

**FIGURE 8.33**   SAR ADC Block Diagram

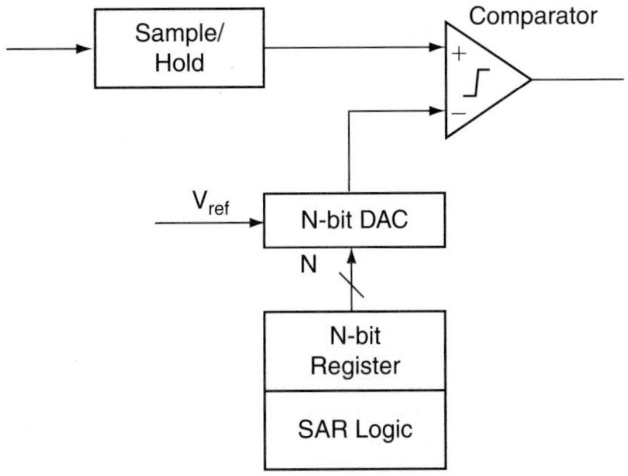

Figure 8.34 shows an example of a 4-bit conversion. The $y$-axis (and the bold line in the figure) represents the DAC output voltage. In the example, the first comparison shows that VIN < VCAD. Thus, bit 3 is set to "0". The DAC is then set to "0100," and the second comparison is performed. As VIN > VDAC, bit 2 remains at "1". The DAC is then set to "0110," and finally the third comparison is performed. Bit 1 is set to "0," and the DAC is then set to "0101" for the final comparison. Finally, bit 0 remains at "1" because VIN > VDAC.

**FIGURE 8.34**   SAR Sample and Hold

Note that four comparison periods are required for the 4-bit ADC. Generally speaking, an N-bit SAR ADC will require N comparison periods and will not be ready for the next conversion until the current one is complete. This explains why these types of ADCs are power- and space-efficient, yet are rarely seen in speed-and-resolution combinations beyond a few Msps at 14 to 16 bits. Some of the smallest ADCs available commercially are based on the SAR architecture.

Two critical components of the SAR ADC are the comparator and the DAC itself. The track/hold function is often embedded in the DAC and may not be an explicit circuit. A SAR ADC's speed is limited by:

The settling time of the DAC, which must settle to within the resolution of the overall converter, for example, 1/2 LSB
The comparator, which must resolve small differences in Vin and VDAC within the specified time
The logic overhead

The SAR converters are typically available in resolutions up to 16 bits. Flash ADCs, by contrast, are typically limited to around 8 bits. The slower speed also allows the SAR ADC to be much lower in power. For example, a typical 8-bit SAR converter uses 100 uA at 3.3 V with a conversion rate of 25 ksps. A high-performance Flash ADC can dissipate 5.25 watts. This is about 16,000 times higher power consumption, but is also 40,000 times faster in terms of its maximum sampling rate.

The SAR architecture is also inexpensive compared to other ADCs. An 8-bit SAR in 1 k volume typically sells for around $1. By contrast, the 8-bit Flash ADC sells for around $375. In addition, SAR ADCs come in low pin count packages, further saving on costs.

### 8.7.3 Flash ADC

Flash analog-to-digital converters, also known as parallel ADCs, are the fastest way to convert an analog signal to a digital signal. They are suitable for applications requiring very large bandwidths. However, flash converters consume a lot of power, have relatively low resolution, and can be quite expensive. This limits them to high-frequency applications that typically cannot be addressed any other way. Examples include data acquisition, satellite communication, radar processing, sampling oscilloscopes, and high-density disk drives.

Figure 8.35 shows the typical flash ADC architecture. For an "N" bit converter, the circuit employs $2^{N-1}$ comparators. A resistive divider and $2^N$ resistors provide the reference voltage. The reference voltage for each comparator is one LSB greater than the reference voltage for the comparator immediately below it. Each comparator produces a "1" when the analog input voltage is

**FIGURE 8.35**   Flash ADC
Architecture

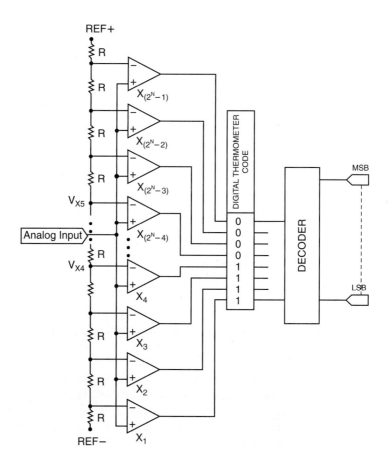

higher than the reference voltage applied to it. Otherwise, the comparator output is "0". Thus, if the analog input is between V × 4 and V × 5, comparators ×1 through ×4 produce "1s," and the remaining comparators produce "0s". The point where the code changes from ones to zeros is the point where the input signal becomes smaller than the respective comparator reference voltage levels.

This is known as thermometer code encoding, so named because it is similar to a mercury thermometer, where the mercury column always rises to the appropriate temperature and mercury is present above the temperature. The thermometer code is then decoded to the appropriate digital output code.

The comparators are typically a cascade of wideband low-gain stages. They are low gain because at high frequencies, it's difficult to obtain both wide bandwidth and high gain. They are designed for low-voltage offset, such that the input offset of each comparator is smaller than an LSB of the ADC. Otherwise, the comparator's offset could falsely trip the comparator, resulting in a digital output code not representative of the thermometer code. A regenerative latch at each comparator output stores the result. The latch has positive feedback, so that the end state is forced to either a "1" or "0".

### 8.7.4 Integrating ADCs

Integrating analog-to-digital converters provides high resolution and can provide good line frequency and noise rejection. These converters have been around for many years but are still applicable to today's applications. The integrating architecture provides a novel yet straightforward approach to converting a low-bandwidth analog signal into its digital representation. This type of converter often includes built-in drivers for LCD or LED displays and is found in many portable instrument applications, including digital panel meters and digital multimeters.

**8.7.4.1 Single-Slope Architecture**   The simplest form of an integrating ADC uses a single-slope architecture (Figure 8.36a and b). Here, an unknown input voltage is integrated and the value compared against a known reference value. The time it takes for the integrator to trip the comparator is proportional to the unknown voltage ($T_{INT}/V_{IN}$). In this case, the known reference voltage must be stable and accurate to guarantee the accuracy of the measurement.

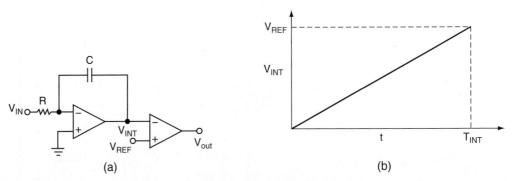

**FIGURE 8.36**   (a) Single-Slope ADC; (b) Single-Slope Output

One drawback to the single-slope approach is that the accuracy is also dependent on the tolerances of the integrator's R and C values. Thus, in a production environment, slight differences in each component's value change the conversion result and make measurement repeatability quite difficult to attain. To overcome this sensitivity to the component values, the dual-slope integrating architecture is used.

**8.7.4.2 Dual-Slope Architecture**   A dual-slope ADC integrates an unknown input voltage ($V_{IN}$) for a fixed amount of time ($T_{INT}$), then deintegrates ($T_{DE-INT}$), using a known reference voltage ($V_{REF}$) for a variable amount of time (see Figure 8.36).

Where:

$$\frac{V_{IN}}{T_{INT}} = \frac{V_{REF}}{T_{DE-INT}}$$

Equation 8.4a

$$T_{INT} = Fixed$$

Equation 8.4b

$$T_{DE-INT} \propto \frac{V_{IN}}{V_{REF}}$$

Equation 8.4c

**FIGURE 8.36**   Dual-Slope Timing Diagram

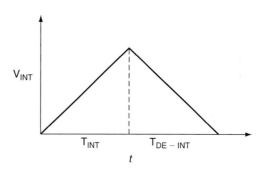

The key advantage of this architecture over the single-slope is that the final conversion result is insensitive to the errors in the component values. That is, any error introduced by a component value during the integrate cycle will be cancelled out during the deintegrate phase.

In form:

$$\frac{V_{IN}}{V_{REF}} = \frac{T_{CHARGE}}{T_{DISCHARGE}}$$

Equation 8.5

**FIGURE 8.37**   Dual-Slope Timing Formula

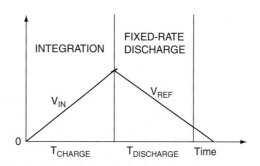

From this equation, we see that the deintegrate time is proportional to the ration of $V_{IN}/V_{REF}$. A block diagram of a dual-slope converter is shown in Figure 8.38.

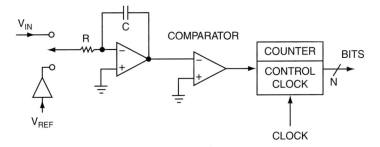

**FIGURE 8.38**   Dual-Slope ADC Block Diagram

As an example, to obtain 10-bit resolution, you would integrate for 1024 ($2^{10}$) clock cycles, then deintegrate for up to 1024 clock cycles (giving a maximum conversion of $2 \times 2^{10}$ cycles). For more resolution, increase the number of clock cycles. This trade-off between conversion time and resolution is inherent in this implementation. It is possible to speed up the conversion time for a given resolution with moderate circuit changes. However, there is a trade-off between accuracy and conversion speed.

### 8.7.5  Pipeline ADC

Because pipeline ADCs provide an optimum balance of size, speed, resolution, power dissipation, and analog design effort, they have become increasingly attractive to engineers. Also known as subranging quantizers, pipeline ADCs consist of numerous consecutive stages, each containing a track/hold (T/H), a low-resolution ADC and DAC, and a summing circuit that includes an interstage amplifier to provide gain.

Target applications for pipeline ADCs include communication systems, in which total harmonic distortion (THD), spurious-free dynamic range (SFDR), and other frequency-domain specifications are significant; (change-coupled-device CCD-) based imaging systems, in which favorable time-domain specifications for noise, bandwidth, and fast transient response guarantee quick settling; and data-acquisition in systems, in which time- and frequency-domain characteristics are both important (i.e., low spurs and high-input bandwidth).

Fast and accurate N-bit conversions can be accomplished using at least two or more steps of subranging (also called pipelining). A coarse, M-bit A/D conversion is executed first (Figure 8.39). Then using a DAC with at least N-bit accuracy, the result is converted back to one of 2 M analog levels and compared with the input. Finally, the difference is converted with a "fine" K-bit flash converter, and two (or more) output stages are combined.

The following inequality should be met to correct for overlapping errors:

$$(L*M) + K > N$$

where L is the number of stages, M is the coarse resolution of the subsequent stages in the ADC/MCAD circuit, K is the fine resolution of the final ADC stage, and N is the pipeline ADC's overall resolution. Most pipeline ADCs include digital error-correction circuitry that operates between the stages.

Some pipeline quantizers feature a calibration unit that compensates for unwanted side effects, such as temperature drift or capacitor mismatch in the multiplying DAC. This digital calibration is usually performed for several (not all) of the pipeline's consecutive stages, using two adjacent codes that cause a transition equal to VREF at the MDAC output. Any deviation from this ideal step is an error that can be measured. When all errors have been acquired and accumulated

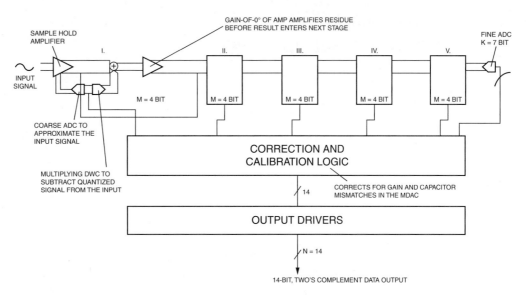

**FIGURE 8.39** Functional diagram of a pipelined ADC

by the subsequent converter stages, they are stored in an onboard memory. Then the results are fetched from RAM during normal operation to redeem gain and capacitor mismatches in the MDAC stages of the pipeline.

The new pipeline architectures simplify ADC design and provide other advantages as well:

> Extra bits per stage optimize correction for overlapping errors.
> Separate track-and-hold (T/H) amplifiers for each stage release each previous T/H to process the next incoming sample, enabling conversion of multiple samples simultaneously in different stages of the pipeline.
> Lower power consumption.
> Higher-speed ADCs (fCONV < 100 ns, typical) entail less cost and less design time effort.

But pipeline ADCs also impose difficulties:

> Complex reference circuitry and biasing schemes.
> Pipeline latency, caused by the number of stages through which the input signal must pass.
> Critical latch timing, needed for synchronization of all outputs.
> Greater sensitivity to board layout, compared with other architectures.
> Sensitivity to process imperfections that cause nonlinearities in gain, offset, and other parameters.

## 8.7.6 Sigma-Delta

A block diagram of a first-order sigma-delta converter is shown in Figure 8.40. It includes a difference amplifier, an integrator, and a comparator with feedback loop that contains a 1-bit DAC. (This DAC is simply a switch that connects the negative input of the difference amplifier to a positive or a negative reference voltage.) The purpose of the feedback DAC is to maintain the average output of the integrator near the comparator's reference level.

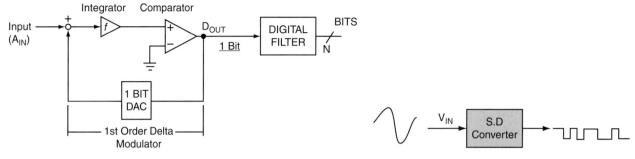

**FIGURE 8.40**   Sigma-Delta Converter Diagram          **FIGURE 8.41**   Sigma-Delta Output Signal

The density of "ones" at the modulator output is proportional to the input signal. For an increasing input the comparator generates a greater number of "ones," and vice versa for a decreasing input. By summing the error voltage, the integrator acts as a low-pass filter to the input signal and a high-pass filter to the quantization noise.

Sigma-delta converters offer high resolution, high integration, and low cost, making them a good choice for applications such as process control and weighing scales. The analog side of a sigma-delta converter (a 1-bit ADC) is very simple. The digital side, which is what makes the sigma-delta ADC inexpensive to produce, is more complex. It performs digital filtering and decimation.

## 8.8        OVERSAMPLING

First, consider the frequency-domain transfer function of a traditional multibit ADC with a sine-wave input signal. This input is sampled at a frequency Fs. According to Nyquist theory, Fs must be a least twice the bandwidth of the input signal.

When observing the result of an FFT analysis on the digital output, a single tone and significant random noise can be seen extending from DC to Fs/2 (Figure 8.42). Known as quantization noise, this effect results from the following consideration: the ADC input is a continuous signal with an infinite number of possible states, but the digital output is a discrete function whose number of different states is determined by the convert's resolution. So, the conversion from analog to digital loses some information and introduces some distortion into the signal. The magnitude of this error is random, with values up to $+/-$ LSB.

**FIGURE 8.42**   Frequency Domain                    Frequency Domain

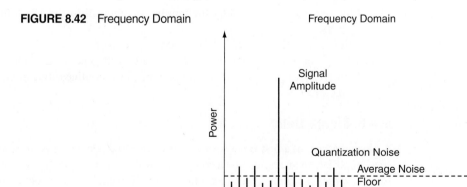

If we divide the fundamental amplitude by the root mean square (RMS) sum of all the frequencies representing noise, we obtain the signal-to-noise ratio (SNR). For an N-bit ADC, SNR = 6.02 N + 1.76 dB. To improve the SNR in a conventional ADC (and consequently the accuracy of signal reproduction), you must increase the number of bits.

Consider again the preceding example, but with a sampling frequency increased by the oversampling ratio k, to kFs (Figure 8.43). An fast fourier transform (FFT) analysis shows that the noise floor has dropped. SNR is the same as before, but the noise energy has been spread over a wider frequency range. Sigma-delta converters exploit this effect by following the 1-bit ADC with a digital filter (Figure 8.44). The RMS noise is less, because most of the noise passes through the digital filter. This action enables sigma-delta converters to achieve wide dynamic range from a low-resolution (1-bit) ADC.

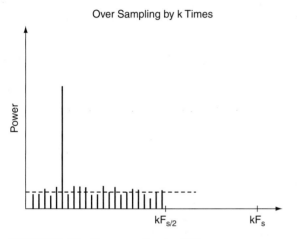

**FIGURE 8.43**   Oversampling by K Times

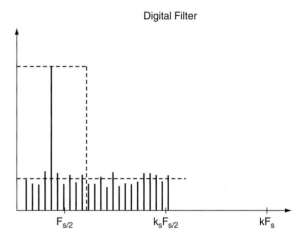

**FIGURE 8.44**   The Digital Filter

## QUESTIONS AND PROBLEMS

1. What are the two types of signals?
2. How do you represent a discrete time signal?
3. What are the two major groups of A/D converters?
4. Describe aliasing.
5. What type of filter is used to prevent aliasing?
6. Describe how a sample and hold circuit functions.
7. What does a transducer do?
8. What key consideration should be taken when using sensors?
9. What is the roll off slope of an ideal low-pass filter?
10. What is the critical frequency of an active bypass filter?
11. How does the sampling frequency affect the A/D conversion accuracy?
12. What does Shannon's theorem say about sampling frequency?
13. What happens if the input signal is not sampled at a high enough rate?
14. Define an A/D converter.
15. List the possible binary codes for a 4-bit A/D converter.
16. What results when you decrease the reference voltage?
17. What is the MSB weight for a 4-bit A/D converter?
18. Describe the process of quantization.

19. Describe what happens if the quantization is too low.
20. Describe offset error.
21. Name the four sources of A/D converter noise.
22. What does the signal-to-noise ratio mean?
23. List the most popular A/D converter types.
24. Why use different types of A/D converters?
25. When would you choose not to use a flash A/D converter?
26. What is the key advantage of a dual-slope A/D converter?
27. Compare the advantages and disadvantages of pipelined A/D converters.
28. How do you determine the converted values of a sigma-delta converter?

## SOURCES

Carr, Joseph J. 1978. *Digital interfacing with an analog world*. Blue Ridge Summit, PA: TAB Books.

*Demystifying sigma-delta ADCs*. Application Note 1870. January 31, 2003, Dallas Semiconductor.

eFunda. *Sampling theorem and Nyquist rate*. Retrieved from http://www.efunda.com.

Engineering Staff. 1986. *Analog-digital conversion handbook, analog devices*. Englewood Cliffs, NJ: Prentice Hall.

Gray, Nicholas. 2003. *ABCs of ADCs: Analog-to-digital converter basics, presentation*. National Semiconductor Corporation.

*Nyquist-Shannon sampling theorem*. Retrieved from http://www.worldhistory.com.

Park. Sangil.1998. *Principles of sigma-delta modulation for analog-to-digital converters*, presentation. Motorola Semiconductors.

*Pipeline ADCs come of AGE*. March 21, 2000, Dallas Semiconductor.

U.C. Berkeley. *The Nyquist-Shannon sampling theorem*. http://ptolemy.eecs.berkeley.edu.

*Understanding flash ADCs*. Application Note 810. October 2, 2001, Dallas Semiconductor.

*Understanding integrating ADCs*. Application Note 1041. May 2, 2002, Dallas Semiconductor.

*Understanding SAR ADCs*. Application Note 1080. March 1, 2001, Dallas Semiconductor.

Wikipedia. *Analog-to-digital converter*. Retrieved from http://en.widipedia.org.

# CHAPTER 9

# Digital Signal Processing

**OBJECTIVE: AN INTRODUCTION TO THE CONCEPTS AND USE OF DIGITAL SIGNAL PROCESSING IN EMBEDDED MICROCONTROLLER SYSTEMS**

The reader will learn:

1. The definition of DSP.
2. Basic analog signaling techniques.
2. Types of analog to digital converters.
3. Transformation of analog signals to digital representation.
4. Basic types of digital filters.
5. System design constraints.

## 9.0 DIGITAL SIGNAL PROCESSING

Signal processing (SP) is a broad topic usually considered a branch of applied mathematics. Once a problem has been reduced to algorithmic form, SP equations can be applied to it. SP encompasses many overlapping areas, as shown in Figure 9.1.

Many applications use SP, including

Mechanical problems (ABS)
Acoustic engineering (voice recognition)

**FIGURE 9.1** DSP Overlapping Application Areas

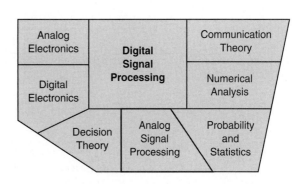

Biomedical engineering (CAT scanner)
Military (radar/sonar)
Telephony (cell phones)
Industrial (oil exploration)

A common consumer electronic item that uses digital signal processing (DSP) is the MP3 player, as shown in Figure 9.2.

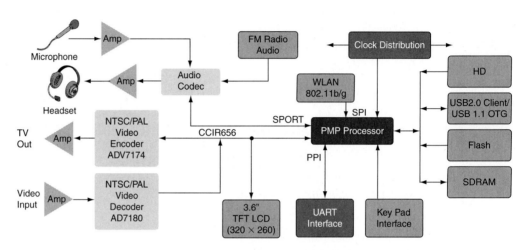

**FIGURE 9.2**    Portable Media Player (Reprinted by permission of Analog Devices Incorporated.)

## 9.1    WHAT IS A DSP?

The world of DSP is entered when an analog signal (a continuous voltage, for example) is converted to a numeric representation. This is accomplished via analog-to-digital converters, (ADC) as described in Chapter 8. There are constraints and limitations defined by signal processing equations that define how much and what can be done to an analog signal in deriving information from it.

Figure 9.3 shows a typical representation of a DSP system. The input signal, X[t], is initially passed through an anti-aliasing filter. This filter is designed to band limit the signal to one-half the sampling frequency (Nyquist rate).

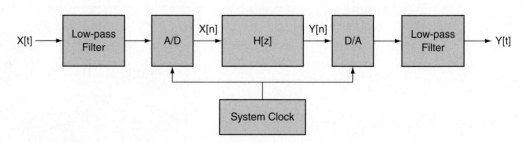

**FIGURE 9.3**    Signal Processing Flow

The filtered signal is then digitized by the ADC at the sample rate determined by the clocking frequency to produce the discrete time signal, X[n]. A system transfer function, H[z], is applied, normally in the time domain, to produce the output signal, Y[n]. The output signal Y[n] is then passed through a digital-to-analog converter (DAC) and low-pass filter to produce the continuous time signal, Y[t].

With DSP, almost everything can be accomplished that can be done by traditional means of analog signal processing (ASP). However, the key point is that with DSP, more can be done than by pure analog means alone. This is the power of DSP. For example, equations can be constructed of physical circuit behavior, like inductors, capacitors, and resistors. However, the opposite is not necessarily true. Some signal processing equations cannot be represented with physical circuitry. This is not a constraint for processing a signal digitally. Thus applications (see table in Figure 9.4) can be created that would otherwise not normally be possible. This is the real power of DSP.

| DSP creates new products | DSP enhances existing products |
|---|---|
| High-Speed Modems | Automotive Engine Control |
| Speech and Image Recognition | Automotive Active Suspension |
| Medical Imaging Equipment | Answering Machines |
| Active Noise Cancellation (ANC) | Portable Phones |
| CD (Music), CD-ROM | Cellular Phones |
| DVD | Radio |
| MP3 Player | Hard Disk Electronics |
| | Electronic Music |
| | Speech Synthesis |
| | Television |

**FIGURE 9.4** DSP Applications

### 9.1.1 Filtering and Synthesis

It takes multiple differing functions to create a working DSP system. Filtering a signal is often necessary to isolate the frequencies that are of interest and reject others. Once the signal is acquired, analysis needs to be performed to determine what action to take. Often the digitized signal needs to be transformed, for example, in speech applications such as a digital answering machine.

Synthesis of a signal can also be done digitally to reproduce sounds. A common and well-known application of this is the music synthesizer. Another aspect of processing is correlation, where parts of signals are compared to look for patterns. And finally, in many applications, there is the control aspect as interface to a device such as a hard disk drive.

### 9.1.2 DSP Performance

There are also practical limitations to what can be done to process a signal digitally. Perhaps the overriding point is performance. A process needs to convert the signal fast enough to meet the demands of the application. Clearly, a cell phone cannot process signals at the same speed as a large, custom mainframe computer. This also correlates to market pricing. Cell phones have to be affordable, no matter what potential signal processing enabled features may be able to be accomplished through engineering.

### 9.1.3 Analog Signal Conversion

It is an analog world. Temperature changes, water flow, and wind direction are analog processes. They can be measured via transducers that provide continuous analog signals, most often as a voltage. The signal can be used to activate motors and other devices in response to the voltage level against a reference voltage.

For example, a microphone is shown in Figure 9.5. Voltage references determine the minimum and maximum voltage of the transducer. They are labeled as Vref+ and Vref−. Often, Vref− is set to ground.

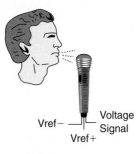

**FIGURE 9.5**   Microphone

**FIGURE 9.6**   Analog Voltage versus Frequency

Different frequency bands can be digitally displayed on an LCD panel with numeric display capability. As was seen in the example of Figure 9.2, an audio codec can also be used to convert the voltage signal to a digital number applying DSP techniques. DSP allows this time-varying signal to be converted into the frequency domain and displays the results on a spectrum analyzer.

**FIGURE 9.7**   Audio MP3 Spectrum Analyzer

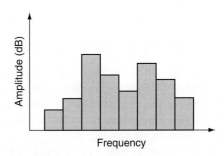

## 9.2     DSP CONTROLLER ARCHITECTURES

Figure 9.8 shows a functional block diagram of the traditional Harvard architecture used in reduced instruction set computing (RISC) processors. In Figure 9.9, a more detailed internal functional diagram of a Harvard RISC processor is shown. Note the separate arithmetic functional units. They can be extended to handle the basic functions of the DSP algorithms and accelerate the processing speed. An additional multiplier unit is provided to support the MAC (multiply-accumulate) requirements of DSP algorithms.

A specific example of the DSP implementation by Analog Devices' Blackfin processor is shown in Figure 9.10. It contains parallel MAC units, as shown in Figure 9.11. These speed up

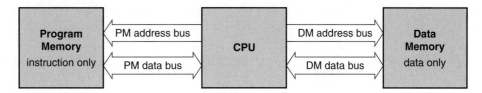

**FIGURE 9.8**   Harvard Architecture

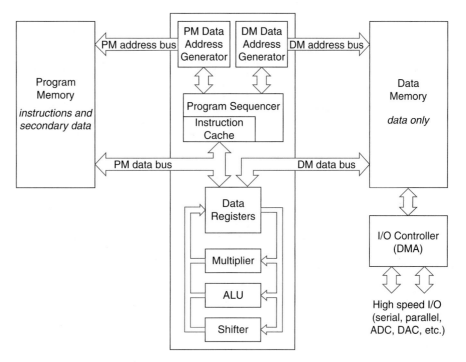

**FIGURE 9.9**   Harvard Architecture Block Diagram

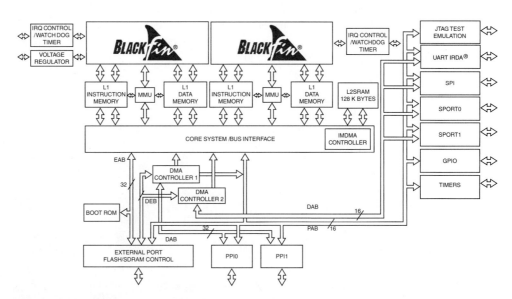

**FIGURE 9.10**   Blackfin ADSP-BF561 Dual-Core Block Diagram (Reprinted by permission of Analog Devices Incorporated.)

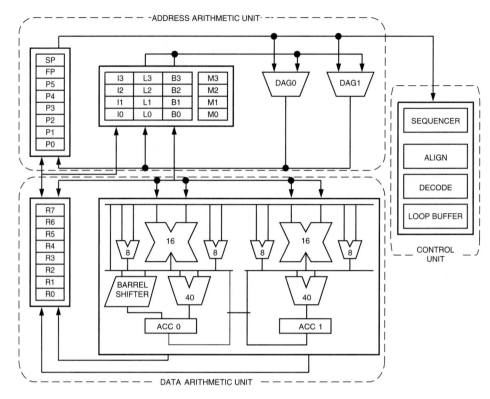

**FIGURE 9.11**    Blackfin ADSP-BF533 Dual MAC Units (Reprinted by permission of Analog Devices Incorporated.)

the discrete Fourier transform calculation, which is composed of repetivie multiply and add operations, as we shall see later in the chapter. It is the use of these MAC units that allows single-chip embedded controllers to perform complex DSP functions at an economical price.

MIPS corporation has extended its 24KE series to include MAC capability through the (MDU) functional block, as shown in Figure 9.12. The MIPS 24KE core is an extension of the 24K series of IP. This brings the ability to incorporate powerful DSP functionality on a custom-designed system on a chip (SoC). Previously this would have required a separate chip at incremental cost and space requirements.

**FIGURE 9.12**    MIPS 24KE with ASE Extension (MDU) (Reprinted by permission of Analog Devices Incorporated.)

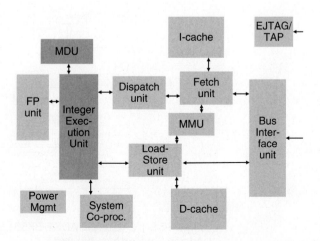

## 9.3         ANALOG FILTERS

Before the basics of digital signal processing are covered, analog low-pass filters need to be addressed. These are the classic passive resistor-capacitor (RC) circuit and the active type based on a tuned op-amp. The Chebyshev and Butterworth filters will be compared for their frequency characteristics. By digitizing the filter with mathematical algorithms, the signals can be manipulated in ways not possible with passive or op-amps-based circuits only.

### 9.3.1  Filter Performance Measurements

Generally, the gain response of a filter is used for comparison among filter types. Figure 9.13 shows the typical response and performance measurements of a low-pass filter.

**FIGURE 9.13**  Low-Pass Filter Measurements

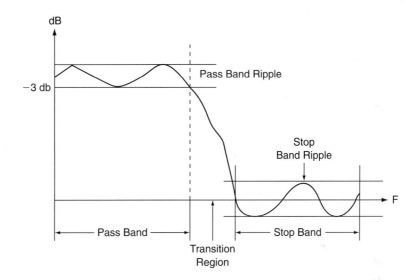

The pass band is defined as the range of frequencies over which signals pass virtually unattenuated through the circuit. The pass band extends to the point where the response drops off by 3 dB, which is known as the cutoff frequency ($f_{3db}$). It is possible to design filters that have no ripple over the pass band, but a certain level of ripple is accepted in this region in exchange for a faster roll-off of gain with frequency in the transition region.

The transition region is the area between the pass band and stop band. The roll-off in this band has been previously discussed. This rate of change of gain with frequency is another important filter performance criterion.

The stop band is chosen by the designer depending on requirements. For example, it may be defined as where the gain response falls below $-4$ dB. In another application, the stop band may be where the filter's response falls below $-80$dB. The important feature is that the response throughout the stop band should always be below the design specification. Several types of filters have ripple in the stop band, but as long as it is below the required specification, it is of little importance.

It is unrealistic to describe a filter solely by the variation of its gain with frequency. The other important characteristic is its phase response, as shown in Figure 9.14. Phase is important because it is directly related to the time delay of the various frequencies passing through the filter.

**FIGURE 9.14**    Filter Phase
Response

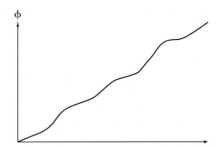

A filter with a linear phase response (see Figure 9.15) delays all frequencies by the same amount of time (see Figure 9.16). Conversely, a filter with a nonlinear phase response causes all frequencies to be delayed by different periods of time, meaning that the signal emerges distorted at the filter output.

**FIGURE 9.15**    Linear Phase
Response

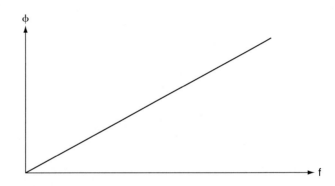

**FIGURE 9.16**    Linear Phase
Delay

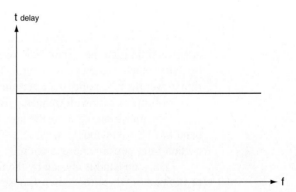

Realistically, when designing filters, linear phase is actually important only over the pass band and transition band of the filter because all frequencies beyond that are attenuated. Also, as most designs are a compromise, a small variation in the phase response to obtain better performance in some other characteristic is tolerable.

### 9.3.2 Time Domain Response

A filter can also be described using its time domain response. Figure 9.17 shows a typical response for a low-pass filter when a sudden step change in voltage is applied to its input.

**FIGURE 9.17**   Filter Step
Response

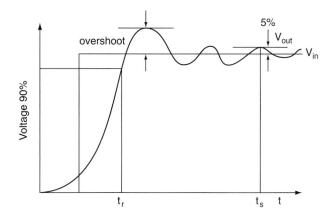

The common measures of performance for the response of the filter in the time domain are

| | |
|---|---|
| Rise Time | The time taken for the output to reach 90% of full scale. |
| Settling Time | The time taken to settle within 5% of the final value. |
| Overshoot | The maximum amount by which the output momentarily exceeds the desired value after level transition. |
| Ringing | Oscillation about the final (mean) value. |

### 9.3.3 Analog Low-Pass Filter

As a reminder, a simple analog low-pass filter is shown in Figure 9.18. It can be used to compare to the FIR and IIR filters that will be examined later in the chapter. How the passive low-pass filter works on a step function signal will also be reviewed.

**FIGURE 9.18**   Analog Low-Pass
RC Filter

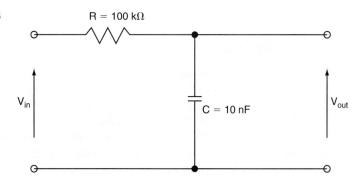

The time constant is given by the formula:

$$T = CR$$
$$T = 10 \times 10^{-9} \times 100 \times 10^{3}$$
$$T = 10^{-3}$$
$$T = 1ms$$

The time constant represents the time it takes for the capacitor to charge to about 63% of its final value. This is shown in the output filtering of the graph in Figure 9.19(a). The low-pass filter acts to smooth the square function input signal voltage level, as shown in Figure 9.19(b).

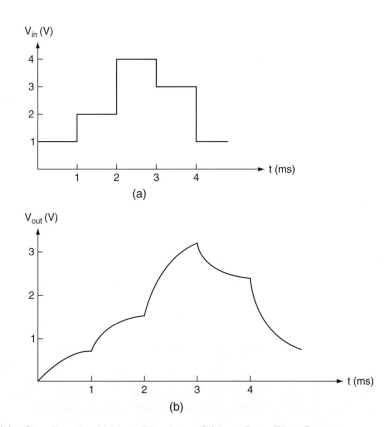

**FIGURE 9.19(a)**    Step Function Voltage Filter Input; **(b)** Low-Pass Filter Output

## 9.3.4 Active Analog Filters

An active filter is an op-amp circuit that can be tuned for specific frequencies. There are four basic types of active filters: low-pass, high-pass, band-pass, and band-stop (notch). The frequency response curves for the filters are shown in Figure 9.20(a, b, c, d).

Several terms are used to describe active filters. A pole is simply an RC circuit. A one-pole filter contains one RC circuit, and a two-pole filter contains two RC circuits, for example. The term *order* is used to identify the number of poles. For example, a first-order filter contains one pole, and a second-order filter contains two poles.

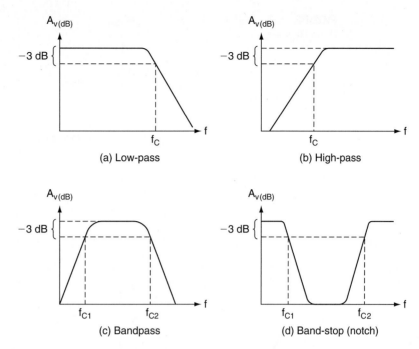

FIGURE 9.20(a, b, c, d)   Active Filter Types

The more poles an active filter has, the higher the gain roll-off rate outside the pass band. One type of active filter, the Butterworth filter, has a roll-off rate of 20 cB/decade per pole. The table in Figure 9.21 shows the relationship among order, poles, and gain roll-off for Butterworth filters.

| Filter Type | Number of Poles | Total Gain Roll-Off |
|---|---|---|
| First Order | 1 | 20 dB/decade |
| Second Order | 2 | 40 dB/decade |
| Third Order | 3 | 60 dB/decade |

FIGURE 9.21   Butterworth Filter Table

## 9.3.5  Active Filter Comparison

The Butterworth filter has a relatively flat response across its pass band, as shown in Figure 9.22.

The Chebyshev filter has a higher initial roll-off rate (per pole) than a Butterworth filter at frequencies just outside the pass band. As the operating frequency moves further outside the pass band, the two filters have equal roll-off rates. The gain of a Chebyshev filter is not constant across its pass band. Figure 9.23(a) shows the response curve of a first-order Chebyshev filter. The ripple width is the difference between the minimum and maximum values of the midband gain. A comparison of the Butterworth and Chebyshev filters is shown in Figure 9.23(b).

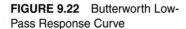

**FIGURE 9.22** Butterworth Low-Pass Response Curve

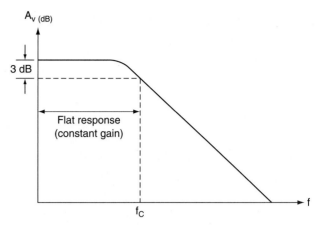

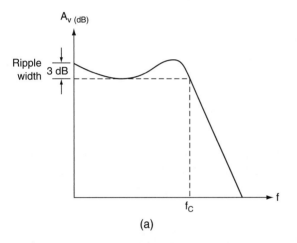

(a)

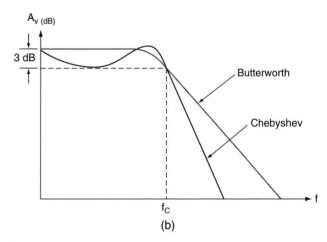

(b)

**FIGURE 9.23(a)** Chebyshev Filter Frequency Response; **(b)** Response Curve Comparison

## 9.4 DIGITAL FILTERS

Digital filters are one of the most common applications for digital signal processors. Digitizing a design ensures that it can be reproduced repeatedly with exactly the same characteristics. There are two other significant advantages with respect to filters. First, it is possible to reprogram the DSP and drastically alter the filter's gain or phase response. For example, a system can be reprogrammed from low-pass to high-pass filter without discarding existing hardware. Second, it can be updated for the filter coefficients while the program is running (i.e., an "adaptive" filter).

There are two basic forms of digital filter: the finite impulse response (FIR) filter and the infinite impulse response (IIR) filter. The initial descriptions are based on low-pass filters but can be converted to other types such as band-pass or high-pass.

### 9.4.1 Finite Input Response Filter

The block diagram of an FIR is shown in Figure 9.24. The input signal, $x(n)$, is a series of discrete values obtained by sampling an analog waveform. In this series, $x(0)$ corresponds to the input value at $t = 0$, $x(1)$ is the value at $t = t_s$, $x(2)$ is the value at $t = 2 t_s$, and so on. The value $t_s$ is the sampling period, where $t_s = 1/fs$.

**FIGURE 9.24** FIR Filter Diagram

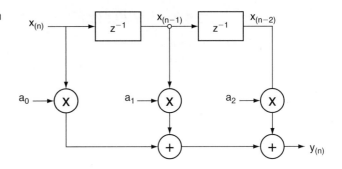

The "z" in Figure 9.24 is the Z-transform. It can be thought of as the block shown as z-1 as being the time delay of one sampling period ($t_s$), also known as the unit delay. It can then be seen that the value shown as x(n-1) is actually the value of x(n) one time period before the previous input. Similarly, x(n-2) is the value of the input two sampling periods beforehand. This simple fact makes the FIR filter circuit easier to understand.

In the example shown in Figure 9.24, the output signal y(n) is therefore always a combination of the last three input samples. In the diagram, each of the samples is multiplied by a coefficient, $a_R$, to give

$$y(n) = a_0 \times (n) + a_1 \times (n - 1) + a_2 \times (n - 2)$$

As an example, the following table lists the midday temperatures over a period of a week. These values shall be used as the input samples (i.e., x(n)). The value for n = 0 is taken to be the daytime temperature on Sunday:

| Day | Period | x(n) | Temperature (°F) |
|---|---|---|---|
| Sunday | 0 | x(0) | 70 |
| Monday | 1 | x(1) | 70 |
| Tuesday | 2 | x(2) | 75 |
| Wednesday | 3 | x(3) | 78 |
| Thursday | 4 | x(4) | 65 |
| Friday | 5 | x(5) | 85 |
| Saturday | 6 | x(6) | 75 |

A histogram of the temperatures versus the days is plotted in Figure 9.25.

**FIGURE 9.25** Histogram of Monthly Temperatures

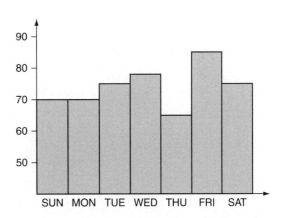

## 9.4.2 FIR Filter Implementation

For the purpose of this example, it is assumed that the set of values for the FIR filter are as follows:

| $a_n$ | Value |
|-------|-------|
| $a_0$ | 0.25  |
| $a_1$ | 0.50  |
| $a_2$ | 0.25  |

The FIR filter implementation then becomes (Figure 9.26).

**FIGURE 9.26**    FIR Filter

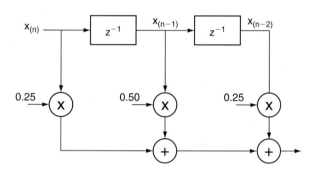

Using the formula, $x(n)$, $x(n-1)$, and $x(n-2)$ at $n = 0$ gives:

$$x(0) = 70$$
$$x(-1) = 0$$
$$x(-2) = 0$$

Completing the formula by performing the multiplications and the additions, the following equation is obtained:

$$y(0) = 0.25 \times 70 + 0.50 \times 0 + 0.25 \times 0$$
$$= 17.5$$

For the next sample period, this is repeated on the same operations, except it is now offset in the calculations by 1 day. Now, $x(0)$ becomes Monday; Sunday becomes $x(n-1)$ or the old value, one sampling period behind.

$$y(0) = 0.25 \times 0 + 0.50 \times 70 + 0.25 \times 70$$
$$= 52.5$$

After 3 days, the results are now three samples for the equation, which gives the full result of:

$$y(0) = 0.25 \times 75 + 0.50 \times 70 + 0.25 \times 70$$
$$= 77.25$$

With the next sample period, it is seen that Sunday's temperature value has dropped out of the equation. This reflects the fact that the filter is limited to three taps, or points, of calculation.

$$y(0) = 0.25 \times 78 + 0.50 \times 75 + 0.25 \times 70$$
$$= 74.5$$

Shown in Figure 9.27 is the output of the filter assuming it has passed through a DAC. The temperatures are connected at the conversion points.

**FIGURE 9.27**  DAC Filter Output

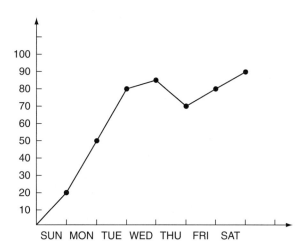

Now it can be seen that the daily temperature changes are averaged out similar to what we accomplished with the low-pass analog filter. However, here it has been done digitally.

It can also be seen from this example why this is called a finite impulse response filter. The calculation used three taps. The calculation is based on y(0) only on the present (n) and previous two (n-1, n-2) samples. This limits the total number of samples to three. There were a finite number of samples for the purposes of calculation.

## 9.4.3 Convolution

This is the original equation for the FIR filter. As has been shown previously, it is basically a low-pass filter with three taps described by the following formula:

$$y(n) = a_0x(n) + a_1x(n - 1) + a_2x(n - 2)$$

It can be seen that this is visually weighted as the input response of the filter. This is shown when the impulse function is applied to the filter with the input value at t = 0. Figure 9.28 shows the resultant pulse train if this is repeated with each new input value.

Once the weighted impulse functions have been defined, the calculation can be performed for the output by multiplying the input by the filter impulse response. Figure 9.29(a) shows the convolved temperatures by day, and Figure 9.29(b) shows the weighted impulse functions.

Finally, the description of the impulse equation can be completed. The initial equation was

$$y(n) = a_0x(n) + a_1x(n - 1) + a_2x(n - 2)$$

**FIGURE 9.28**   Filter Output for
Unity Impulse Input

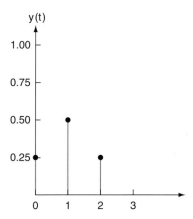

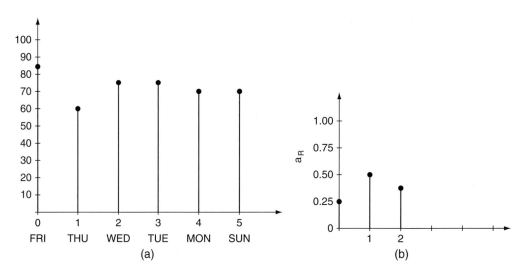

**FIGURE 9.29(a)**   Convolved Input Sequence; **(b)** Filter Coefficient Values

The unit delay is defined as $Z^{-1}$. From this it is defined as the delayed input samples as:

$$x(n-1) = x(n)(z^{-1})$$
$$x(n-2) = x(n)(z^{-1})(z^{-1})$$
$$= x(n)(z^{-2})$$

This equation can be written for the filter as:

$$y(n) = (a_0 + a_1 z^{-1})(x(n))$$

Alternatively it can be written:

$$H(n) = \frac{y(n)}{x(n)} = a_0 + a_1 z^{-1} + a_2 z^{-2}$$

The transfer function H(n) is the mathematical representation of the impulse response of the digital filter. An important aspect of the equation is that if $z^{-n}$ appears (where n is any number), then the filter is inherently stable. This means that, unlike a typical analog filter design, the FIR filter will always act like a filter, never an oscillator.

## 9.4.4 Infinite Impulse Response Filter

There are two basic forms of digital filters. The second is the infinite impulse response (IIR) filter. A simple example is shown in Figure 9.30.

**FIGURE 9.30** Infinite Impulse Response Filter

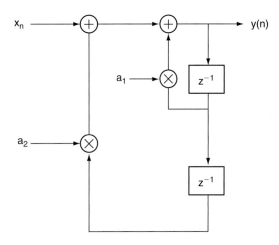

Using similar notation to that for the FIR filter, we can show that

$$y(n) = x(n) + a_1 y(n - 1) + a_2 y(n - z)$$
$$= x(n) + [a_1 z^{-1} + a_2 z^{-2}] \cdot y(n)$$
$$= x(n)\frac{1}{1 - a_1 z^{-1} - a_2 z^{-2}}$$

One thing immediately apparent in the first equation is that the second and third terms include terms based on $y(n-_t)$. In other words, they are recursive, making the equation infinite.

The calculation to create the transfer function can be shown as:

$$H(n) = \frac{y(n)}{x(n)} = \frac{1}{1 - a_1 z^{-1} - a_2 z^{-2}}$$

From the previous equation, it can be shown that each output, $y(n)$ is dependent on the input value $x(n)$ and the two previous outputs, $y(n-1)$ and $y(n-2)$. Starting where there are no previous input samples and $n = 0$, then simply:

$$y(0) = x(0)$$

At the next sample period:

$$y(1) = x(1) + a_1 y(0)$$
$$= x(1) + a_1 x(0)$$

Then when $n = 2$:

$$y(2) = x(2) + a_1 y(1) + a_2 y(0)$$
$$= x(2) + a_1[x(1) + a_1 x(0)] + a_2 x(0)$$

The equation is complete when n = 3:

$$y(3) = x(3) + a_1 y(2) + a_2 y(1)$$
$$= x(3) + a_1[x(2) + a_1[x(1) + a_1 x(0)] + a_2 x(0)] + a_2 x(0)]$$
$$+ a_2[x(1) + a_1 x(0)]$$

What it shown here is that the output is dependent on all the previous inputs. The equation just keeps getting longer. It is an infinite equation and hence the name of infinite impulse response filter.

One issue with IIR filters is possible instability. The filter is actually a series of feedback loops. Under certain conditions, feedback loops can become unstable. Because of this, careful design is required to ensure a stable filter.

## 9.5    SIGNAL TRANSFORMATION

Two aspects of DSP, the FIR and IIR filters, have been covered. Another method is transformation of signals from the time domain to the frequency domain. The most obvious example is the spectrum analyzer. Other examples are telecommunication channel bandwidth analysis and digital speech recognition. All three applications require transformation of a time-varying signal to the frequency domain.

### 9.5.1 Phasor Model

A basic method to describe a signal is the phasor model. A phasor is a rotating vector in the complex plane, with a magnitude of A and a rotational speed of w radian/sec, as shown in Figure 9.31.

**FIGURE 9.31**    Phasor Model

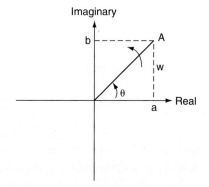

The phasor model can be extended to discrete time, where signals occur only at specific intervals in time defined by the sampling interval $T_s$. Instead of the continuous variable time t, a discrete variable n is used so that the phasor advances in steps of $T_s$.

The cosine signal can be represented by two phasors that form a conjugate pair. This means that they have the same real value (a) and equal and opposite values of b. From this it is possible to conclude an interesting fact that all real signals must be made up of conjugate pairs of phasors so that the sum of the vectors always lie on the real axis.

## 9.5.2 Fourier Series

A single frequency can be represented with sine and cosine waves. However, more intricate waveforms can also be represented this way. For example, a rectangular pulse train consists of an infinite number of sine waves of varying amplitude. In the same way, any complex periodic signal can be described as the sum of many phasors.

One method of describing signals this way is called the Fourier series. It assumes that the set of phasors have frequencies that are multiples of some fundamental frequency, $f_0$. Any periodic signal can be represented by a Fourier series provided that N is big enough. The individual frequency components are known as harmonics. Normally, in real-world applications, the complex signal is not periodic. The Fourier series can be generalized by using phasors whose frequencies are not harmonically related, which is generally the case when the complex signal is not periodic (most cases in real applications):

$$x(t) = \sum_{k=-N}^{N} c_\kappa e^{j(w_k t)}$$

Any arbitrary waveform may be represented by a Fourier series of this general type.

## 9.5.3 Discrete Fourier Series

Digital signal processors run in the digital domain. The continuous equations need to be transformed into the discrete or digital domain so that some useful formulas can be derived. What is done is to replace the continuous function, t, with one that progresses in steps of $w_0 T_s$.

It can be shown that the frequency response of a discrete signal is periodic with a period of $1/T_s$. This is an important feature of digital signal processing. The phasor model has now been used to describe a general discrete signal.

## 9.5.4 Fourier Transform

In real-world applications, signals are rarely periodic in nature. The Fourier series needs to be modified and developed to comprehend this fact. This can be done through integration. Given the general Fourier series, where all the frequencies are harmonically related, that is:

$$w_K = K w_0$$

The fact that the final signal is not periodic may be represented by:

$$w_0 \rightarrow 0$$

This equation simply states that there is no "least common denominator" in the frequencies of the separate phasors. The number of phasors tends toward infinity, and our summation becomes an integral.

$$X(t) = \frac{1}{2\pi} \int_{-\infty}^{+\infty} X(w) \cdot e^{j(wt)} \cdot dw$$

In the preceding equation, it is assumed that the signal amplitude can be defined as a function of frequency (w) (i.e., X(w)). The "reverse" equation defining X(w) is given by:

$$X(w) = \int_{-\infty}^{+\infty} X(t) \cdot e^{j(-wt)} \cdot dt$$

This is an equation that allows the calculation of the amplitude response of a continuous signal in the frequency domain. The two equations taken together are called the Fourier transform pair. Although they are great for mathematicians, they clearly can't be implemented directly on a DSP. For that a discrete form is required.

### 9.5.5 Discrete Fourier Transform (DFT)

To determine the discrete equivalent of the Fourier transform, the continuous time variable t must be replaced by the discrete variable $nT_s$. Outside $\pm\pi/T_s$ the spectrum repeats itself, making the variable of integration $wT_s$ an integral.

The reverse transform still uses a discrete summation. This is because of x(n) being only valid at $nT_s$ instants of time. A pair of equations how allow the transformation of digital signals between the time and frequency domains.

At this point, a method has been shown to describe a signal that varies with time, called the phasor model. This model has then been developed to show the Fourier series. This, in turn, has been shown how to produce the DFT pair. Now there exist a pair of equations that can provide a method to transform between the time and frequency domains for any discrete signal.

There is a problem with the discrete DFT equation with X(w). It cannot be done without an infinite summation in the real world. Also, in practical terms, there is a limit to the number of calculations that can be performed for each frequency. It is simple enough to eliminate the infinities by limiting the samples to x(n). This is referred to as "windowing." The resulting spectrum is given by the equation:

$$X_N(w) = X(w) * W(w)$$

where $X_N(w)$ represents the windowed spectrum with N as the number of samples used, W(w) denotes the spectrum of the window, and the * denotes the convolution of W(w) and X(w). This is the same concept of convolution that was covered earlier in the chapter. Thus the equation can be rewritten:

$$X_N(w) = \sum_{r=0}^{N-1} X_{(r)} \cdot w(N - r)$$

Now suppose a windowing function has been chosen. What must be known is the effect of using only a limited number of frequencies. How many frequencies are required to maintain a certain degree of accuracy must also be known. Unfortunately there is no simple answer. In general, the optimum number of phasors is equal to the number of initial points in x(n) (i.e., N). The simplest way to imagine this is to assume that the windowed portion of x(n) consists of one period of a long sequence (Figure 9.32), with period NTs and frequency ws/N. If this can be done, then the sequence can be treated as a Fourier series, and it can be seen that the spectrum will consist of N phasors (Figure 9.33).

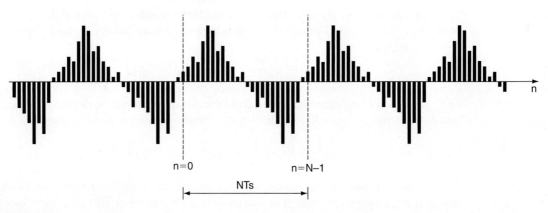

**FIGURE 9.32**   Window for DFT

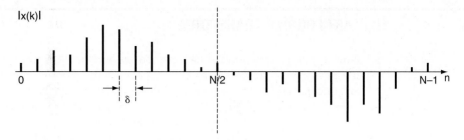

**FIGURE 9.33** Phasor Spacing

For practical application of the DFT formula, another "variable" called the twiddle factor is generally substituted. This can be derived by noting that the spectrum of a DFT is repetitive with period $w_s$, so that:

$$N\delta = w_s$$

The frequency scale can then be digitized so that the spectrum can be rewritten in terms of k instead of w:

$$X(k\delta) = K(\delta) = \sum_{n=0}^{N-1} x(n) \cdot e^{j(-k\delta T_{sn})}$$

Noting that:

$$w_s = \frac{2\pi}{T_s}$$

and:

$$\delta = \frac{w_s}{N}$$

the equation can be rewritten as:

$$X_N(k) = \sum_{n=0}^{N-1} x(n) \cdot e^{-j\left(\frac{2\pi kn}{N}\right)}$$

This then is the form of the practical DFT formula that is generally used.
The "twiddle factor", called $W_N$, is defined as:

$$W_N = e^{-j\left(\frac{2\pi}{N}\right)}$$

The DFT pair of formula can now be rewritten in their most common form:

$$X_N(k) = \sum_{n=0}^{N-1} X(n) \cdot w_N^{kn}$$

and:

$$X(n) = \frac{1}{N} \sum_{k=0}^{N-1} X_N(K) \cdot W_N^{-kn}$$

## 9.6    FAST FOURIER TRANSFORM

Now that the discrete form of the equation has been derived, the practical issue of compute time must be faced. From the equations it can be seen that there is a repetitive sequence of additions and multiplications. This is because both indices, k and n, must progress through N values to produce the full range of output phasors and, therefore, $N^2$ computations must be performed.

A 1000-point DFT (where N = 1000) will require 1,000,000 instruction cycles to calculate. This is because each cycle will be a multiply and add function. This is where the MAC (MULTIPLY-ADD-ACCUMULATE) function of the DSP comes from. If each cycle can be done in 1 ns, this would still take 0.001 seconds, and the maximum sample rate would be 1kHz.

To solve this problem, the fast Fourier transform (FFT) was developed. It is a method to speed up the calculation by taking advantage of inherent redundancy in the DFT to reduce the number of calculations required. This is because the twiddle factor $W_n$ is periodic with a limited number of distinct values, and the same values of $W_n$ are calculated many times.

### 9.6.1  FFT Implementation

There are many different forms of the FFT. It is not a single formula. It is often optimized specifically to a certain DSP hardware and/or software combination. There are high-level language versions that differ from assembly-level language versions.

The basic concept to the FFT is to reduce the input signal x(n) into several shorter interleaved sequences. This is usually referred to as decimation-in-time. We can take our original signal with N values and split it into odd and even numbered sequences:

$$X_N(k) = \sum_{n=0}^{N-1} X(n) \cdot W_N^{kn}$$

which, after appropriate substitution, becomes:

$$X_N(k) = \sum_{r=0}^{\frac{N}{2}-1} x(Zr) \cdot (W_N^Z)^{rk} + W_N^k \sum_{r=0}^{\frac{N}{2}-1} x(Zr+1) \cdot (W_N^Z)^{rk}$$

and using the original definition of $W_N$:

$$X_N(k) = \sum_{r=0}^{\frac{N}{2}-1} x(Zr) \cdot WN_{N/2}^{rk} + W_N^K \sum_{r=0}^{\frac{N}{2}-1} x(Zr+1) \cdot W_{N/2}^{rk}$$

which can be rewritten:

$$X_N(k) = G(K) + W_N^K 1 + (K)$$

G(k) becomes the even-numbered points of the DFT and H(k) the odd-numbered points. Notice that the original DFT is in the form of two smaller DFTs of length N/2. The calculation with N = 1000 now is a calculation that requires $500^2 + 500^2 + 500 = 500{,}500$ instruction cycles, rather than the original 1,000,000.

If the number of samples, N, is a power of 2, the decimation can be taken a step further and broken into halves. In fact, this process can be repeated until only 2-point DFTs are needed. This is one of the most commonly used types of FFTs, the radix-2. For example, for N = 8, x(n) is defined for:

$$n = \{0,1,2,3,4,5,6,7\}$$

then decimation gives two sequences:

$$n = \{0,2,4,6,\} \text{ and } \{1,3,5,7\}$$

and decimation again gives four sequences:

$$n = \{0,4\} \{2,6\} \{1,5\} \text{ and } \{3,7\}$$

The FFT can now be calculated by performing four 2-point DFTs on the preceding pairs of sample values. This simplifies the mathematics significantly; however, the "twiddle factor" arises each time the sequence is decimated. The twiddle factors are also different at each stage of the calculation. The FFT algorithm must be able to incorporate these factors without making the calculation too complex for the DSP.

Then the FFT becomes:

$$X_4(k) = [x(0) + x(2) \cdot W_4{}^{2k}] + W_4^k[x(1) + x(3) \cdot W_4{}^{2k}]$$

Writing out all the values for k longhand generates the following equations:

$$X_4(0) = [x(0) + x(2) \cdot w_4^0] + W_4^0[x(1) + x(3) \cdot W_4^0]$$
$$X_4(1) = [x(0) + x(2) \cdot W_4^2] + W_4^1[x(1) + x(3) \cdot W_4^2]$$
$$X_4(2) = [x(0) + x(2) \cdot W_4^0] + W_4^2[x(1) + x(3) \cdot W_4^0]$$
$$X_4(3) = [x(0) + x(2) \cdot W_4^2] + W_4^3[x(1) + x(3) \cdot W_4^2]$$

### 9.6.2 DFT "Butterfly"

Figure 9.34 is the signal flow graph for the 4-point DFT just calculated. It shows the classic "butterfly." The numbers in the circle represent the power of $W_4$ that is required to be calculated at that stage.

**FIGURE 9-34**  Four-Point Butterfly

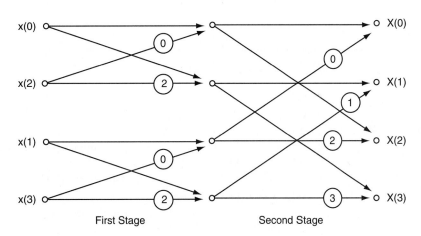

Each section of the graph is made up of a number of butterflies, as shown in Figure 9.35. This is the generally accepted form of the butterfly when the multiplication factors are $x = 0$ and $y = -1$.

**FIGURE 9.35**  FFT Butterfly

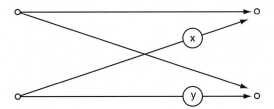

## 9.7          TABLE ADDRESSING

For the FFT implementation, two tables can be used, one for input values and one for the corresponding output values. This is not a problem for a 4-point DFT; however, for a 1024-point or larger DFT, it is not so simple. Most general-purpose DSPs include the capability for bit-reversed addressing. This is an addressing mode where the input values are stored in ascending order and the processor selects them in the correct sequence. For example, for an 8-point DFT:

$$\text{start at } 000 = x(0)$$
$$000 + 100 = 100 = x(4)$$
$$000 + 100 = 100 = x(4)$$
$$100 + 100 = 010 = x(2)$$
$$010 + 100 = 110 = x(6)$$
$$110 + 100 = 001 = x(1)$$
$$001 + 100 = 101 = x(5)$$
$$101 + 100 = 011 = x(3)$$
$$011 + 100 = 111 = x(7)$$

Many new high-performance DSPs have separate address generation units that eliminate any time penalty for accessing the input table data values. This increases overall performance of the DSP dramatically.

## QUESTIONS AND PROBLEMS

1. Define the term DSP.
2. Diagram the five steps of the DSP process flow.
3. Why use transducers in a DSP system?
4. Which processor architecture is used to implement DSPs and why?
5. What does a MAC functional block do?
6. What is the difference between the pass and stop bands?
7. Why is signal phase important?
8. Diagram the typical output waveform of a low-pass filter.
9. What are the four active filter types?
10. How many RC circuits does a fourth-order filter have?
11. Draw a histogram for monthly rain totals for a year period.
12. What is meant by taps in a FIR filter?
13. Write the transfer function for a three-tap filter.
14. Why is an IIR filter infinite?
15. What is an example of a transformation from the time domain to the frequency domain?
16. Define a discrete Fourier series.
17. Why was the FFT developed?
18. What is the advantage of bit-reversed addressing?

## SOURCES

ADSP-BF533 Blackfin® processor hardware reference revision 3.1. May 2005. Part Number 82-002005-01. Analog Devices.

ARM1022E technical reference manual, revision: r0p2. 2005. ARM Limited.

Francis, Hedley. 2001, May. ARM DSP-enhanced extensions, ARM White Paper. ARM Ltd.

Grover, Dale & John R. Deller. 1999. *Digital signal processing and the microcontroller*, Upper Saddle River, NJ: Prentice Hall.

Marven, Craig & Gillian Ewers. 1996. A *simple approach to digital signal processing*. New York: John Wiley & Sons, Inc.

MC9S12DT128 device user guide 9S12DT128DGV2/D, V02.15 05. October 2005. Motorola, Inc.

MIPS32 4KE processor core family software user's manual, Document Number: MD00103, Revision 01.08. January 30, 2002. MIPS Technologies.

Smith, Steven W. 1999. *The scientist and engineer's guide to digital signal processing*, San Diego, CA: California Technical Publishing.

TMS320C55x DSP, programmer's guide, preliminary draft, TI Literature Number: SPRU376A. August 2001.

Xtensa architecture and performance. October 2005. White Paper. Tensilica, Inc.

# CHAPTER 10

## Fuzzy Logic

**OBJECTIVE: AN INTRODUCTION TO THE CONCEPT OF FUZZY LOGIC AS APPLIED TO EMBEDDED CONTROLLER-BASED SYSTEMS**

The reader will learn:

1. What is fuzzy logic.
2. Basic mathematics behind fuzzy logic.
3. Application for applying fuzzy logic.

## 10.0 FUZZY LOGIC

Most often, measurements can be thought of in terms relative to each other. Outside temperature can be hot, cold, or mild; relative terms are based on current measurements. Mild can change, for example, depending on which part of the season you are experiencing. What is mild in Phoenix may be hot in Stockholm. Fuzzy logic is a methodology to deal with relative terms.

Dr. Lotfi Zadeh at Berkeley first developed fuzzy logic theory in 1962. He built a mathematical foundation for dealing with the relative nature of the real world. Applied to control situations, it is a way to implement control algorithms that can converge to the desired point faster than if typical "on-off" binary algorithms were used. An example of the control logic for a microwave oven controller is shown in Figure 10.1.

Fuzzy logic was conceived as a better method for sorting and handling data, but has proven to be an excellent choice for many control system applications because it mimics human control logic. It can be built into anything from small, handheld products to large computerized process control systems. It uses an imprecise but very descriptive language to deal with input data more

**FIGURE 10.1** The Fuzzy Logic Control-Analysis Method (Copyright of Freescale Semiconductor, Inc. 2007, used by permission.)

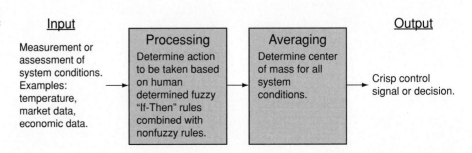

**Input**
Measurement or assessment of system conditions. Examples: temperature, market data, economic data.

**Processing**
Determine action to be taken based on human determined fuzzy "If-Then" rules combined with nonfuzzy rules.

**Averaging**
Determine center of mass for all system conditions.

**Output**
Crisp control signal or decision.

like a human operator. It is very robust and forgiving of operator and data input and often works when first implemented with little or no tuning.

Rather than try to create warm water by using a binary method of turning a cold valve and hot valve full on or full off (binary), fuzzy logic attempts to produce warm water by partially turning the valves on or off. In the end, this gives a much finer control of the water temperature and converges to the desired temperature much faster. In fact, with the binary method it may not be possible to get to a tolerable level of accuracy for the combined water temperature.

Fuzzy logic provides a methodology to converge on a final value without oscillating around the point with a binary algorithm. In fact, the binary algorithm may not converge to the desired point and oscillate with an error level outside acceptable levels (see Figure 10.2a). Fuzzy logic provides a faster and more accurate convergence (see Figure 10.2b). With advances in microcontroller performance, fuzzy logic circuitry can be part of the CPU logic function as shown in the Freescale HCS12X microcontrollers (Figure 10.3).

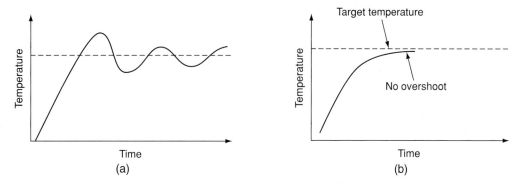

**FIGURE 10.2 (a)** Graph of Temperature versus Time; **(b)** Graph of Temperature with Fuzzy Rule

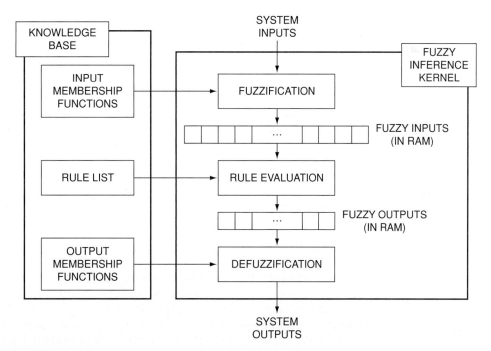

**FIGURE 10.3** Freescale S12XCPU Fuzzy Logic Function (Copyright of Freescale Semiconductor, Inc. 2007, Used by Permission.)

## 10.1    FUZZY LOGIC METHOD

Much of the information you take in is not very precisely defined, such as the speed of a vehicle coming up from behind. This is called a fuzzy input. However, some "input" is reasonably precise and nonfuzzy, such as the speedometer reading. Processing of all this information is not very precisely definable. This is called fuzzy processing. Fuzzy logic theorists would call it using fuzzy algorithms.

Fuzzy logic is similar to the way human brains function, and this can be mimicked so that machines will perform somewhat like humans. This is not to be confused with artificial intelligence, which is an attempt to perform exactly like humans (or better). Fuzzy logic control and analysis systems may be electromechanical in nature or concerned only with data, for example, economic data, in all cases guided by "If-Then rules" stated in human language.

The fuzzy logic method then is a combination of analysis and control, as shown in Figure 10.4. It consists of three basic components:

Receiving of one, or a large number, of measurements or other assessment of conditions existing in some system that is to be analyzed.

Processing all this input according to human-based, fuzzy "If-Then" rules, which can be expressed in plain language words, in combination with traditional nonfuzzy processing.

Averaging and weighing the resulting outputs from all the individual rules into one single output decision or signal that decides what to do or tells a controlled system what to do. The output signal eventually arrived at is a precise appearing, defuzzified, "crisp" value.

| Input | Processing | Averaging | Output |
|---|---|---|---|
| Measurement or assessment of system conditions. examples: temperature, market data, economic data. | Determine action to be taken based on human determined fuzzy "If-Then" rules combined with nonfuzzy rules. | Determine center of mass for all system conditions. | Crisp control signal or decision. |

**FIGURE 10.4**  Fuzzy Logic Control-Analysis Method

## 10.2    FUZZY PERCEPTION

A fuzzy perception is an assessment of physical condition that is not measured with precision, but is assigned an intuitive value. In fact, the fuzzy logic basically assumes that everything is a little fuzzy, no matter how good your measuring equipment is. It will be seen that fuzzy perceptions can serve as a basis for processing and analysis in a fuzzy logic control system.

Measured, nonfuzzy data is the primary input for the fuzzy logic method. Examples of nonfuzzy data include temperature measure by a temperature transducer, motor speed, economic data, and financial markets data. It would not be usual in an electromechanical control system or a financial or economic analysis system, but humans with their fuzzy perceptions could also provide input. There could be a human in the loop.

In the fuzzy logic literature, the term "fuzzy set" is used. A fuzzy set is a group of anything that cannot be precisely defined. Consider the fuzzy set of "old houses." How old is an old house? Where is the dividing line between new houses and old houses? Is a fifteen-year-old house an old house? How about 40 years? What about 39.9 years? The assessment is in the eyes of the beholder. Other examples of fuzzy sets are short men, warm days, high pressure gas, small crowds, and medium viscosity.

When humans are the basis for an analysis, there needs to be a way to assign some rational value to intuitive assessments of individual elements of a fuzzy set. There must be a translation from human fuzziness to numbers that can be used by a computer. This is done by assigning an assessment of conditions a value from zero to 1.0. For "how hot the room is," the human might rate it a 0.2, if the temperature were below freezing, and the other human might rate the room at 0.9, or even 1.0, if it is a hot day in summer with the air conditioner off. These perceptions are fuzzy, just intuitive assessments, not precisely measured facts.

By making fuzzy evaluations, with zero at the bottom of the scale and 1.0 at the top, a basis for analysis rules for the fuzzy logic method can be accomplished. The results seem to turn out well for complex systems or systems where human experience is the only base from which to proceed. This is certainly better than doing nothing at all, which would be the case if not for fuzzy rules.

Fuzzy logic makes use of human common sense. This common sense is either applied from what seems reasonable, for a new system, or from experience, for a system that has previously had a human operator. Objects of fuzzy logic analysis and control may include, among others: physical control, such as machine speed, or operating a cement plant, financial and economic decisions, psychological conditions, physiological conditions, safety conditions, and security conditions.

---

**10.3**          **FUZZY LOGIC TERMINOLOGY**

The following are explanations of some terms that are used in fuzzy logic. This terminology was initially established by Dr. Zadeh, when he originated the fuzzy logic concept.

**Fuzzy**—The degree of fuzziness of a system analysis rule can vary between being very precise, in which case it would not be called "fuzzy," to being based on an opinion held by a human, which would be "fuzzy." Being fuzzy or not fuzzy, therefore, has to do with the degree of precision of a system analysis rule.

A system analysis rule need not be based on human fuzzy perception. For example, a rule could be, "If the boiler pressure rises to a danger point of 600 Psi as measured by a pressure transducer, then turn everything off." That rule is not "fuzzy."

**Principle of Incompatibility**—As the complexity of a system increases, it becomes more difficult and eventually impossible to make a precise statement about its behavior, eventually arriving at a point of complexity where the fuzzy logic method born in humans is the only way to get at the problem.

**Fuzzy Sets**—A fuzzy set is almost any condition that has the words: short men, tall women, hot, cold, new buildings, etc., where the condition can be given a value between 0 and 1. An example is a woman who is 6 feet, 3 inches tall. In general, by most human standards, she would be described as tall. A height rating of 0.98 could be assigned. This line of reasoning can go on indefinitely, rating a great number of things between 0 and 1.

**Degree of Membership**—The degree of membership is the placement in the transition from 0 to 1 of conditions within a fuzzy set. If a particular building's placement on the scale is a rating of 0.7 in its position in newness among new buildings, then its degree of membership in new building can be defined as 0.7.

In fuzzy logic method control systems, degree of membership is used in the following way. A measurement of speed, for example, might be found to have a degree of membership in "too fast of" 0.6 and a degree of membership in "no change needed" of 0.2. The system program would then calculate the center of mass between "too fast" and "no change needed" to determine feedback action to send to the input of the control system.

**Summarizing Information**—Human processing of information is not based on two-valued, off-on, either-or logic. It is based on fuzzy perceptions, fuzzy truths, fuzzy inferences, etc., all resulting in an averaged, summarized, normalized output, which is given by the human as a precise number of decision values, which he or she verbalizes, writes down, or acts on. It is the goal of fuzzy logic control systems to also do this.

The input may be large masses of data, but humans can handle the ability to manipulate fuzzy sets and the subsequent summarizing capability to arrive at an output.

**Fuzzy Variable**—Words like red, blue, etc., are fuzzy and can have many shades and tints. They are just human opinions, not based on precise measurement in angstroms. These words are fuzzy variables. If, for example, speed of a system is the attribute being evaluated by fuzzy, "fuzzy" rules, the "speed" is a fuzzy variable.

**Linguistic Variable**—Linguistic means relating to language, in our case plain language words. Speed is a fuzzy variable. Accelerator setting is a fuzzy variable. Examples of linguistic variables are: somewhat fast speed, very high speed, really slow speed, excessively high accelerator setting, accelerator setting about right, etc.

A fuzzy variable becomes a linguistic variable when it is modified with descriptive words, such as somewhat fast, very high, really slow, etc. The main function of linguistic variables is to provide a means of working with the complex systems mentioned earlier as being too complex to handle by conventional mathematics and engineering formulas. Linguistic variables appear in control systems with feedback loop control and can be related to each other with conditional, "if-then" statements. Example: If the speed is too fast, then back off on the high accelerator setting.

**Universe of Discourse**—Let women be the object of consideration. All the women everywhere would be the universe of women. If women are selected to discourse (talk about), then all the women everywhere would be our Universe of Discourse.

The Universe of Discourse, then, is a way to say all the objects in the universe of a particular kind, usually designated by a single word, that happens to be about or working with in a fuzzy logic solution. A Universe of Discourse is made up of fuzzy sets. The Universe of Discourse for women is made up of professional women, tall women, short women, beautiful women, and so on.

**Fuzzy Algorithm**—An algorithm is a procedure, such as the instructions in a computer program. A fuzzy algorithm is a procedure, usually a computer program, made up of statements relating linguistic variables. An example of a statement would be:

If "green X" is very large, then make "tall Y" much smaller. If the rate of change of temperature of the steam engine boiler is much too high then turn the heater down a lot.

## 10.4    FUZZY EXPERT SYSTEM

A fuzzy expert system is an expert system that uses fuzzy logic instead of Boolean logic. In other words, a fuzzy expert system is a collection of membership functions and rules that are used to reason about data. Unlike conventional expert systems, which are mainly symbolic reasoning engines, fuzzy expert systems are oriented toward numerical processing.

The rules in a fuzzy logic system are usually of a form similar to the following:

If x is low and y is high then z = medium

Where x and y are input variables, z is an output variable, low is a membership function (fuzzy subset) defined on x, high is a membership function defined on y, and medium is a membership function defined on z.

The part of the rule between the "if" and "then" is the rule's premise or antecedent. This is a fuzzy logic expression that describes to what degree the rule is applicable. The part of the rule following the "then" is the rule's conclusion or consequence. This part of the rule assigns a membership function to each of one or more output variables. Most tools for working with fuzzy expert systems allow more than one conclusion per rule. A typical fuzzy expert system has more than one rule. The entire group of rules is collectively known as a rule base or knowledge base.

The standard definitions for fuzzy logic rules are

1. Negate (negation criterion): truth (not x) = 1.0 − truth (x).
2. Intersection (minimum criterion): truth (x and y) = minimum (truth(x), truth(y)).
3. Union (maximum criterion): truth (x or y) = maximum (truth(x), truth(y)).

## 10.4.1 The Inference Process

With the definition of rules and membership functions determined, they can now be applied to specific values of the input variables to compute the values of the output variables. This process is referred to as inferencing. In a fuzzy expert system, the inference process is a combination of four subprocesses:

Fuzzification
Inference
Composition
Defuzzification

Note that the defuzzification process is optional.

Assume the variables x, y, and z all take on values in the interval [0,10], and that the following membership functions and rules are defined. Let t be the variable in the interval [0,10].

$$Low(t) = 1 - t/10$$
$$High(t) = t/10$$

Rule 1: if x is low and y is low then z is high
Rule 2: if x is low and y is high then z is low
Rule 3: if x is high and y is low then z is low
Rule 4: if x is high and y is high then z is high

Notice that instead of assigning a single value to the output variable z, each rule assigns an entire fuzzy subset of low or high. For this example note that:

1. low(t) + high(t) = 1.0 for all t. This is not required, but is common.
2. The value of t at which low(t) is maximum is the same as the value of t at which high(t) is minimum, and vice versa. This is also not required, but is common.
3. The same membership functions are used for all variables. This isn't required.

## 10.4.2 Fuzzification

In the fuzzification subprocess, the membership functions defined on the input variables are applied to their actual values to determine the degree of truth for each rule premise. The degree of truth for a rule's premise is sometimes referred to as its alpha. If a rule's premise has a nonzero degree of truth, then the rule is said to fire.

For example, for four value combinations of x and y the table becomes (alpha one corresponds to the consequence for rule1, etc.):

### 10.4.3 Inference

In the inference subprocess, the truth value for the premise of each rule is computed and applied to the conclusion part of each rule. This results in one fuzzy subset to be assigned to each output variable for each rule.

The are two basic inference methods or inference rules: MIN and PRODUCT. In MIN inferencing, the output membership function is clipped off at a height corresponding to the rule premise's computed degree of truth. This corresponds to the traditional interpretation of the fuzzy logic AND operation. In PRODUCT inferencing, the output membership function is scaled by the rule premise's computed degree of truth.

For example, rule1 will have the values x = 0.0 and y = 5.8. As shown in Figure 10.5, the premise degree of truth works out to be 0.58. For this rule, MIN inference will assign $z$ the fuzzy subset defined by the membership function:

$$\text{rule1}(z) = \{z/10, \text{ if } z <= 5.8$$
$$0.58, \text{ if } z >= 5.8$$

| X | Y | Low(x) | High(x) | Low(y) | High(y) | alpha1 | alpha2 | alpha3 | alpha4 |
|---|---|--------|---------|--------|---------|--------|--------|--------|--------|
| 0.0 | 0.0 | 1.0 | 0.0 | 1.0 | 0.0 | 1.0 | 0.0 | 0.0 | 0.0 |
| 0.0 | 5.8 | 1.0 | 0.0 | 0.42 | 0.58 | 0.42 | 0.58 | 0.0 | 0.0 |
| 3.8 | 6.4 | 0.62 | 0.38 | 0.36 | 0.64 | 0.36 | 0.62 | 0.36 | 0.38 |
| 7.2 | 7.1 | 0.28 | 0.72 | 0.29 | 0.71 | 0.28 | 0.28 | 0.29 | 0.71 |

**FIGURE 10.5**   Fuzzification Rule Table

For the same conditions, PRODUCT inference will assign z the fuzzy subset defined by the membership function:

$$\text{rule1}(z) = 0.58 * \text{high}(z)$$
$$= 0.58 * z/10$$
$$= 0.058 * z$$

### 10.4.4 Composition

In the composition subprocess, all the fuzzy subsets assigned to each output variable are combined together to form a single fuzzy subset for each output variable.

The two major composition rules are MAX composition and SUM composition. In MAX composition, the combined output fuzzy subset is constructed by taking the pointwise maximum over all the fuzzy subsets assigned to the output variable by the inference rule. In SUM composition the combined output fuzzy subset is constructed by taking the pointwise sum over all the fuzzy subsets assigned to the output variable by the inference rule. Note that this can result in truth values greater than one. For this reason, SUM composition is only used when it will be followed by a defuzzification method, such as the CENTROID method, that doesn't have a problem with the odd case.

For example, assume x = 0.0 and y = 5.3. MIN inferencing would assign the following four fuzzy subsets to z:

$$rule1(z) = \{z/10, \qquad \text{if } z <= 5.3$$
$$0.53, \qquad \text{if } z >= 5.3\}$$
$$rule2(z) = \{0.37, \qquad \text{if } z <= 5.3$$
$$1 = z/10, \qquad \text{if } z >= 5.3\}$$
$$rule3(z) = 0$$
$$rule4(z) = 0$$

MAX composition would result in the fuzzy subset:

$$fuzzy(z) = \{0.37, \qquad \text{if } z <= 3.7$$
$$z/10, \qquad \text{if } 3.7 <= z <= 5.3$$
$$0.53, \qquad \text{if } z >= 5.3\}$$

PRODUCT inferencing would assign the following four fuzzy subsets to z:

$$rule1(z) = 0.053*z$$
$$rule2(z) = 0.37 - 0.037*z$$
$$rule3(z) = 0.0$$
$$rule4(z) = 0.0$$

### 10.4.5 Defuzzification

Sometimes it is useful to just examine the fuzzy subsets that are the result of the composition process, but more often, this fuzzy value needs to be converted to a single number, a crisp value. This is what the defuzzification subprocess does.

There are many different methods for defuzzification, including centroid, maximizer, singleton, and weighted average. Two of the most used techniques are the CENTROID and MAXIMUM, with CENTROID being the most popular. In the CENTROID method, the crisp value of the output variable is computed by finding the variable value of the center of gravity of the membership function for the fuzzy value.

Centroid defuzzification computes the "center of mass" for the composite fuzzy sets from the previous stage. This method is numerically intensive and requires computing the integral of the set. It also requires that the output sets overlap to avoid a situation that produces an invalid output.

In the MAXIMUM method, one of the variable values at which the fuzzy subset has its maximum truth value chosen as the crisp value for the output variable. There are several variations of the MAXIMUM method that differ only in what they do when there is more than one variable value at which this maximum truth value occurs. One of these, the AVERAGE-OF-MAXIMA method, returns the average of the variable values at which the maximum truth value occurs.

## 10.5    LINGUISTIC VARIABLES

Linguistic variables are used to express what is important as well as the context. "This room is hot" is specific: it represents an opinion independent of a measuring system, and it has information that most listeners will understand. Linguistic variables are used in ordinary daily activities, including preparation instructions for instant soup mixtures. These instructions are filled with linguistic references. Bring to a boil, stirring constantly, reduce heat, partially cover, simmer, and stirring occasionally are all linguistic variables within the context of soup preparation. The manufacturers of instant soup believe these simple vague instructions clearly tell the consumer how to make their product successfully.

The following example shows some of the ways linguistic variables (shown in italics) can be formally defined and used in application software.

1. *Mix thoroughly* 1 cup shortening and 2/3 cup white sugar.
2. Add 1 teaspoon vanilla *stirring constantly*.
3. Beat in 2 eggs and whip at *medium speed*.
4. *Slowly incorporate* about 2 cups of white flour.

An established customer enters a shop asking to buy a few widgets and wants the best price for them from the salesman. It is up to the salesman to come up with a price given many parameters. Taking this hypothetical case, the following parameters need to be accounted for:

> Cost of the widgets
> Normal markup
> Shelf time of the product
> Shelf life of the product
> Length of the relationship
> Customer payment history
> Quantity of the sale
> Repeat business potential

Computerized record keeping will provide hard (crisp) numbers for many of the parameters. This still leaves the need to combine the numbers in some way to compute a discount from the normal list price to quote to the customer.

All the parameters in the sale except "repeat business potential" are crisp numbers that can be precisely defined with information from a well-organized database. How should each of these parameters affect the final quotation discount? The tables in Figures 10.6(a) and 10.6(b) summarize the parameters based on what and why they are important.

Each of the important items is context dependent. It can be said, "If the quantity is high then profit is higher." Huge is meaningful in the context of quantity, and higher is a consequence of profit. This is an example of a rule: an entire set of rules, as follows, will define the logic establishing a discounted price for the potential customer.

| IF shelf time | IS long | THEN discount IS large |
|---|---|---|
| IF shelf time | IS short | THEN discount IS low |
| IF shelf time | IS short | THEN discount IS high |
| IF shelf time | IS long | THEN discount IS normal |
| IF quantity | IS small | THEN discount IS none |
| IF quantity | IS large | THEN discount IS large |
| IF quantity | IS huge | THEN discount IS high |
| IF customer | IS new | THEN discount IS special |
| IF customer | IS recent | THEN discount IS normal |
| IF customer | IS long term | THEN discount IS large |
| IF shelf life | IS short AND | |
| Shelf life | IS long | THEN discount IS deep |

This group of eleven rules (and perhaps a half-dozen more) can be used to establish computable pricing for many different products. This collection of rules individually describes the relationship between the sale parameters and the discount offered. These rules, when they are all evaluated, will provide a weighted value for the discount.

| Sale Parameter | What is important | Why it's important |
|---|---|---|
| Shelf time | Long, short | Cost of keeping a product in inventory |
| Shelf life | Short, long, forever | When shelf life is reached the product loses value |
| Payment | Prompt, normal, eventually, nagging | Payment history is part of cost of doing business with him |
| Quantity | Small, normal, large, huge | There are shop savings and increased profit to selling larger quantities |
| Customer | New, recent, long term | Looks at potential that a new customer might get special treatment just as long-term customers are required for loyalty |

(a)

| Sale Parameter | What is important | Why it's important |
|---|---|---|
| Repeat business potential | Yes, Maybe, No | Both product and customer dependent |

(b)

**FIGURE 10.6(a)** Crisp Measurement Linguistic Variables; **(b)** Vague, Subjective Measurement, Linguistic Variables

## 10.5.1 Using Linguistic Variables

```
IF room IS cold THEN heat IS on;
IF room IS hot THEN heat IS off;
```

Simple thermostats have been doing this for a hundred years or more. Why are linguistic variables and fuzzy logic needed to operate a simple switch? How is crisp temperature evaluated under such vague terms as hot and cold? What are hot and cold anyway?

The heating control problem sounds quite simple: a temperature is measured for the room, and using two fuzzy logic rules, controls a furnace that heats the room. When the meanings of cold and hot are not precise opposites, the outcome becomes more complex and useful.

Linguistic variables associate a linguistic condition with a crisp variable. A crisp variable is the kind of variable that is used in most computer programs: an absolute value. A linguistic variable, on the other hand, has a proportional nature: in all of the software implementations of linguistic variables, they are represented by fractional values in the range of 0 to 1.

In the preceding example, room and heat are crisp variables, and hot, cold, on, and off are linguistic variables. The linguistic variable on and off in the example are represented in the crisp variable heat as a 1 and a 0 respectively. The hot and cold linguistic variables represent a range of values corresponding to the crisp variable room. This relationship can be represented as shown in Figure 10.7.

**FIGURE 10.7**    Linguistic Variable
HOT

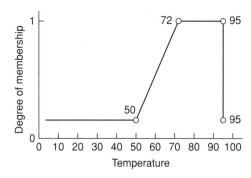

Most linguistic variables can be represented in software with coordinates of 4 points. A crisp variable room is associated with a linguistic variable hot, defined using four break points from the graph.

```
LINGUISTIC room TYPE unsigned int MIN 0 MAX 100
{
    MEMBER HOT { 50, 72, 100, 100 }
}
```

A lot of literature has been written on representation of linguistic variables, but implementations for most applications utilize four points. There are arguments for smooth curves to represent linguistic variables for accuracy, and against smooth curves because of computational intensity. The worst-case error in 4-point presentations is in the corners.

## 10.5.2  Anatomy of a Fuzzy Rule

```
IF room IS cold THEN heat IS on;
```

Each fuzzy rule consists of two parts: predicate and consequence. The predicate determines the rule weight or truth. The result of the room IS cold is a degree of membership (DOM) value between the values of fuzzy zero and fuzzy one.

The DOM of the predicate weighs on the consequence part of a fuzzy rule. In plain language, the urgency to turn the heat on with the preceding rule is determined by how cold the room is. A single fuzzy rule offers nothing over a crisp comparison and action; however, multiple competing rules do.

## 10.5.3  Logically Combining Linguistic Variables

Just as Boolean expressions can be combined to yield a Boolean result that represents the combined result of the expressions, so can linguistic variables. Linguistic variables can be combined with OR, AND, and NOT operators as has been shown. The following C defines can be used in application code:

```
#define F_OR (a,b)    ((a) > (b) ? (a) : (b))
#define F_AND (a,b)   ((a) < (b) ? (a) : (b))
#define F_NOT(a)      (F_ONE + F_ZERO - a)
```

Fuzzy OR is the largest DOM of its arguments, as shown in Figure 10.8. Fuzzy AND is the smallest of its arguments. Fuzzy NOT is the space between the argument and fuzzy 1. If the resolution of linguistic variables is reduced to have only the values of 0 and 1, the

**FIGURE 10.8**  F_OR Operator
(Fuzzy OR)

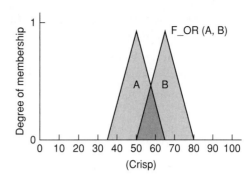

logical definitions for manipulating linguistic variables are the same as conventional Boolean logic.

---

## 10.6          PID CONTROLLER

The classical PID (proportional integrating derivative) controller creates a manipulated variable signal as the sum of three terms: the first is the absolute error multiplied by a constant; the second is the rate of change of error multiplied by a second constant; and the third is the accumulated error multiplied by a constant.

$$mv = (pe \times K1) + \frac{\partial pe}{\partial t} \times K2 + (\Sigma\, pe \times K3)$$

General Equation for a PID Control System

Each of the three parts of the PID control system is (loosely) intended to correct for errors, anticipate potential error conditions, and overcome small accumulated errors (stick bits). This description comes close to what is needed to implement a PID controller using linguistic variables.

This approach in implementing a linguistic variable-based PID controller uses three different control strategies keyed to the size of system error. If the error is large, then the manipulated variable is driven primarily by the error value alone. Smaller errors are dominated by rate-of-error change rules. Finally, with very small errors, control is dominated by the integration of the error.

Linguistic variables can create a control system that is more tolerant of changes in system constants. Most real-word control applications are nonlinear. Some examples include airplane control systems, motor controllers, food and chemical processing; all these have system parameters that vary widely in normal use.

### 10.6.1  Linguistic Time of Day

Many applications can refer to the time of day in linguistic terms. Implementation details vary considerable. This is an example from a home environment application that divides the day into 0.1 hour segments, conveniently storing the crisp time of day into a single byte (of the program data storage). The following definitions show some creative usage of the definition of linguistic variables.

The crisp hours is a wraparound number system that resets at midnight. The linguistic variable day is conventional: it starts being "day" at 5:30 am and is truly "day" at 6:30 am; "day" continues to 5:30 pm, where it declines until 6:30 pm. The linguistic variable night avoids this complication by being defined as a fuzzy function hours IS NOT day. A new linguistic variable

can be defined in terms of other defined linguistic variables. The night setback time (nightsb) is
night but no evening.

```
// hours 0.1 resolution hour 0 .. 240 for a day
LINGUISTIC hours TYPE char MIN 0 MAX 240
  {
    MEMBER day    { 55, 65, 175, 185 }
    MEMBER night  { FUZZY { hours IS NOT day }}
    MEMBER morning { 50, 60, 190, 200 }
    MEMBER evening { 160, 170, 190, 200 }
    MEMBER nightsb {FUZZY { hours IS night AND
                    hours IS NOT evening }}
  {
```

## 10.6.2 Linguistic Comparisons

Many crisp equality comparisons are replaced with a fuzzy comparison that accounts for a range
of data. This comparison is based on three data values: the comparison pint, range until the comparison has failed (delta), and the current variable value. Delta is the distance to a value where
the current comparison ceases to be important. Consider for a moment the following definitions;
in each case the delta value returns a fuzzy zero or fuzzy one, and any further deviation from the
center point will not change the result. The following definition of Fuzzy EQUAL (Figure 10.9)
shows a definition that is easily implemented.

**FIGURE 10.9**   F_EQ (Fuzzy
Equal)

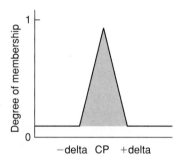

The arguments for F_EQ are: v for the crisp variable under test, cp for the center point, and
delta, which is the significant distance from the center point in the fuzzy comparisons. The F_EQ
function can be used in any expression that accepts linguistic variables as shown in this C code
example.

```
DOMtype F_EQ(v, cp, delta)
  {
    long m = ABS(cp-v);
    if (m > delta) return (F_ZERO);
      return ( (m/delta) * (F_ONE-F_ZERO));
  }
```

In essence, cp and delta help declare an anonymous linguistic variable. This technique can
be extended to all of the normal arithmetic comparisons.

Linguistic variables provide a normalized number system whose resolution is dependent
on the consequence requirements of the application. Linguistic variables provide a natural
smooth transition between competing rules describing different strategies. Linguistic rules focus
on problem solution, not problem analysis.

The implementation of linguistic variables and their use work well on conventional embedded microcontrollers and are generally not as computationally intensive as alternative application

implementations. The reduction of computation requirements is almost entirely due to the normal-ization of the data of interest to the application. Linguistic variables can easily be combined with conventional application software.

## 10.7      FUZZY LOGIC APPLICATION

Fuzzy logic requires some numerical parameters to operate, such as what is considered signifi-cant error and significant rate-of-change-error, but exact values of these numbers are usually not critical unless very responsive performance is required, in which case empirical tuning would de-termine them. For example, a simple temperature control system could use a single temperature feedback sensor whose data is subtracted from the command signal to compute "error" and then time-differentiated to yield the error slope rate-of-change-of-error, hereafter called "error-dot."

The controller might have units of degrees Fahrenheit (F) and a small error considered to be 2°F while a large error is 5°F. The "error-dot" might then have units of degrees/minute with a small error-dot being 5°F/min and a large one being 15°F/min. These values don't have to be sym-metrical and can be "tweaked" once the system is operating to optimize performance. Generally, fuzzy logic is so forgiving that the system will probably work the first time without any tweaking.

Fuzzy logic offers several unique features that make it a particularly good choice for many control problems.

1. It is inherently robust because it does not require precise, noise-free inputs and can be pro-grammed to fail safely if a feedback sensor quits or is destroyed. The output control is a smooth control function despite a wide range of input variations.
2. Because the fuzzy logic controller processes user defined rules governing the target control system, it can be modified and tweaked easily to improve or drastically alter system per-formance. New sensors can easily be incorporated into the system simply by generating ap-propriate governing rules.
3. Fuzzy logic is not limited to a few feedback inputs and one or two control outputs, nor is it necessary to measure or compute rate-of-change parameters for it to be implemented. Any sensor data that produces some indication of a system's actions and reactions is sufficient. This allows the sensors to be inexpensive and imprecise, thus keeping the overall system cost and complexity low.
4. Because of the rule-based operation, any reasonable number of inputs (1–8 or more) can be processed and numerous outputs (1–4 or more) generated, although defining the rule base quickly becomes complex if too many inputs and outputs are chosen for a single implemen-tation because rules defining their interrelations must also be defined. It would be better to break the control system into smaller chunks and use several smaller fuzzy logic controllers distributed on the system, each with more limited responsibilities.
5. Fuzzy logic can control nonlinear systems that would be difficult or impossible to model mathematically. This opens doors for control systems that would normally be deemed unfea-sible for automation.

### 10.7.1   How Fuzzy Logic Is Used

1. Define the control objectives and criteria: What am I trying to control? What do I have to do to control the system? What kind of response do I need? What are the possible (probable) system failure modes?
2. Determine the input and output relationships and choose a minimum number of variables for input to the fuzzy logic engine (typically error and rate-of-change-of-error).

3. Using the rule-based structure of fuzzy logic, break the control problem down into a series of IP X AND Y THEN Z rules that define the desired system output response for given system input conditions. The number and complexity of rules depends on the number of input parameters that are to be processed and the number of fuzzy variables associated with each parameter. If possible, use at least one variable and its time derivative. Although it is possible to use a single, instantaneous error parameter without knowing its rate of change, this cripples the system's ability to minimize overshoot and for a step inputs.
4. Create fuzzy logic membership functions that define the meaning (values) of input/output terms used in the rules.
5. Create the necessary pre- and postprocessing fuzzy logic routines if implementing in software, otherwise program the rules into the fuzzy logic hardware engine.
6. Test the system, evaluate the results, tune the rules and membership functions, and retest until satisfactory results are obtained.

Think of fuzzy variables as linguistic objects or words, rather than numbers. The sensor input is a noun (e.g., "temperature," "displacement," or "velocity"). Because error is just the difference, it can be thought of the same way. The fuzzy variables themselves are adjectives that modify the variable. At a minimum, one could simply have "positive," "zero," and "negative" variables for each of the parameters. Additional ranges, such as "very large" and "very small," could also be added to extend the responsiveness to exceptional or very nonlinear conditions, but aren't necessary in a basic system.

## 10.8    THE RULE MATRIX

The fuzzy parameters of error (command-feedback) and error-dot (rate-of-change-of-error) are modified by the adjective "negative," "zero," and "positive." To picture this, imagine the simplest practical implementation, a 3-by-3 matrix. The columns represent "negative error," "zero error," and "positive error" inputs from left to right. The rows represent "negative," "zero," and "positive," "error-dot" input from top to bottom. This planar construct is called a rule matrix. It has two input conditions, "error" and "error-dot," and one output response conclusion (at the intersection of each row and column). In this case there are nine possible logical product (AND) output response conclusions.

**FIGURE 10.10**   3 × 3 Rule Matrix

| "error-dot" \ "error" | "negative" | "zero" | "positive" |
|---|---|---|---|
| "negative" | N * N | N * Z | N * P |
| "zero" | Z * N | Z * Z | Z * P |
| "positive" | P * N | P * Z | P * P |

Although not absolutely necessary, rule matrices usually have an odd number of rows and columns to accommodate a "zero" center row and column region. This may not be needed as long as the functions on either side of the center overlap somewhat, and continuous dithering of the output is acceptable. Because the "zero" regions correspond to "no change" output, responses to a lack of this region will cause the system to continually hunt for "zero."

It is also possible to have a different number of rows than columns. This occurs when numerous degrees of inputs are needed. The maximum number of possible rules is simply the product of the number of rows and columns, but definition of all these rules may not be necessary because some input conditions may never occur in practical operation. The primary objective of this construct is to map out the universe of possible inputs while keeping the system sufficiently under control.

## 10.8.1 Fuzzy Logic Implementation

The first step in implementing fuzzy logic is to decide exactly what is to be controlled and how. For example, suppose there is a design for a simple proportional temperature controller (Figure 10.11) with an electric heating element and a variable-speed cooling fan. A positive signal output calls for 0 to 100% heat while a negative signal output calls for 0 to 100% cooling. Control is achieved through proper balance and control of these two active devices.

**FIGURE 10.11** Simple Fuzzy Logic Control System

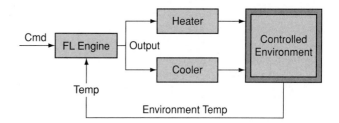

The variables are defined as follows:

```
Cmd:        Target temperature
Temp:       Feedback sensor in controlled environment
Error:      Cmd-Temp (+ = too cold, − = too hot)
Error.dot:  Time derivative or Error (+ = getting hotter, − = getting
            cooler)
Output:     HEAT or NO CHANGE or COOL
```

It is necessary to establish a meaningful system for representing the linguistic variables in the matrix. For this example, the following will be used:

```
"N" = "negative" error or error-dot input level
"Z" = "zero" error or error-dot input level
"P" = "positive" error or error-dot input level
"H" = "Heat" output response
"-" = "No Change" to current output
"C" = "Cool" output response
```

Define the minimum number of possible input product combinations and corresponding output response conclusions using these terms. For a three by three matrix with heating and cooling output responses, all nine rules will need to be defined. The conclusions to the rules with the linguistic variables associated with the output response for each rule are transferred to the matrix. Figure 10.12 shows what command and error look like in a typical control system relative to the command setpoint as the system hunts for stability. Definitions for this example are shown in Figure 10.18.

Example rule definitions for this control system are as follows:

```
INPUT#1: ("Error", positive(P), zero(Z), negative(N))
INPUT#2: ("Error.dot", positive(P), zero(Z), negative(N))
```

**FIGURE 10.12**   Typical Control
System Response

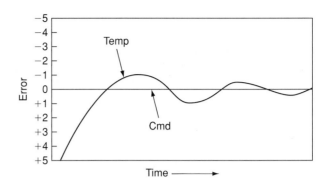

```
CONCLUSION: ("Output", Heat(H), No Change(-), Cool(C)
INPUT#1 System Status
Error = Command-Feedback
P = Too Cold, Z = Just right, N = Too hot
INPUT#2 System Status
Error-dot = d(Error/dt)
P = Getting hotter, Z = Not changing, N = Getting colder
OUTPUT Conclusion and System Response
Output H = Call for heating, - = Don't change anything, C = Call for cooling
```

   Linguistic rules describing the control system consist of two parts: an antecedent block
(between the IF and THEN) and a consequent block (following THEN). Depending on the system, it may not be necessary to evaluate every possible input combination (for 5 by 5 and up matrices) because some may rarely, if ever, occur. By making this type of evaluation, usually done
by an experienced operator, fewer rules can be evaluated, thus simplifying the processing logic
and perhaps even improving the fuzzy logic system performance.

   After transferring the conclusions from the nine rules to the matrix, there is a noticeable
symmetry to the matrix. This suggests (but does not guarantee) a reasonably well behaved (linear) system. This implementation may prove to be too simplistic for some control problems;
however it does illustrate the process. Additional degrees of error and error-dot may be included
if the desired system response calls for this. This will increase the rule base size and complexity,
but may also increase the quality of the control, as shown in Figure 10.13.

## 10.8.2  Membership Functions

The membership function is a graphical representation of the magnitude of participation of
each input. It associates a weighting with each of the inputs that are processed, defines functional overlap between inputs, and ultimately determines an output response. The rules use
the input membership values as weighting factors to determine their influence on the fuzzy
output sets of the final output conclusion. Once the functions are inferred, scaled, and combined, they are defuzzified into a crisp output that drives the system. There are different
membership functions associated with each input and output response. Some features to
note are

**SHAPE**                    triangular is common, but bell, trapezoidal, haversine, and
                             exponential have been used. More complex functions are possible
                             but require greater computing overhead to implement.

**SHOULDERING**              (Locks height at maximum if an outer function. Shouldered
                             functions evaluate as 1.0 past their center), CENTER points (center
                             of the member function shape), and OVERLAP (N&Z, Z&P,
                             typically about 50% of width but can be less).

**FIGURE 10.13**   Rules Structure
and Rule Matrix

|  | Antecedent Block | | | Consequent Block |
|---|---|---|---|---|

1. IF Cmd-Temp=N AND d(Cmd-Temp)/dt=N THEN Output=C
2. IF Cmd-Temp=Z AND d(Cmd-Temp)/dt=N THEN Output=H
3. IF Cmd-Temp=P AND d(Cmd-Temp)/dt=N THEN Output=H
4. IF Cmd-Temp=N AND d(Cmd-Temp)/dt=Z THEN Output=C
5. IF Cmd-Temp=Z AND d(Cmd-Temp)/dt=Z THEN Output=NC
6. IF Cmd-Temp=P AND d(Cmd-Temp)/dt=Z THEN Output=H
7. IF Cmd-Temp=N AND d(Cmd-Temp)/dt=P THEN Output=C
8. IF Cmd-Temp=Z AND d(Cmd-Temp)/dt=P THEN Output=C
9. IF Cmd-Temp=P AND d(Cmd-Temp)/dt=P THEN Output=H

Error-(Cmd-Temp)

Error-dot-(dcmd-Temp)(dt)

|  | N | Z | P |
|---|---|---|---|
| **N** | 1   C | 2   H | 3   H |
| **Z** | 4   C | 5   NC | 6   H |
| **P** | 7   C | 8   C | 9   H |

Figure 10.14 illustrates the features of the triangular membership function, which is used in this example because of its mathematical simplicity. Other shapes can be used, but the triangular shape lends itself to this illustration.

**FIGURE 10.14**   Membership
Functions

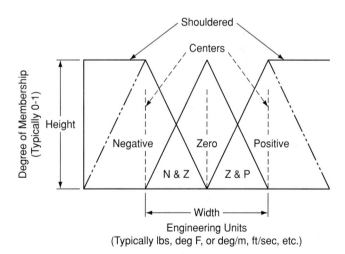

The degree of membership (DOM) is determined by plugging the selected input parameter (error or error-dot) into the horizontal axis and projecting vertically to the upper boundary of the membership function(s). In Figures 10.15(a) and (b), consider an "error" of −1.0 and an "error-dot" (rate of error change) of +2.5. These particular input conditions indicate that the feedback has exceeded the command and is still increasing.

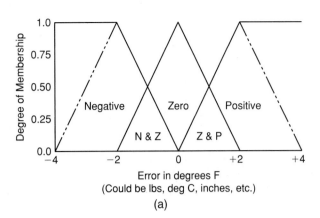 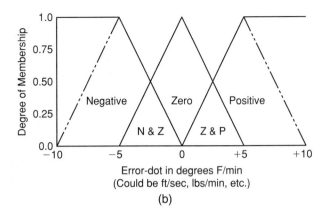

**FIGURE 10.15(a)**   Example Error Membership Function; **(b)** Example Error-Dot Membership Function

The degree of membership for an "error" of $-1.0$ projects up to the middle of the overlapping part of the "negative" and "zero" function so the result is "negative" membership $= 0.5$ and "zero" membership $= 0.5$. Only rules associated with "negative" and "zero" error will actually apply to the output response. This selects only the left and middle columns of the rule matrix.

For an "error-dot" of $+2.5$, a "zero" and "positive" membership of 0.5 is indicated. This selects the middle and bottom rows of the rule matrix. By overlaying the two regions of the rule matrix, it can be seen that only the rules in the 2-by-2 square in the lower left corner (rules 4, 5, 7, 8) of the rules matrix will generate nonzero output conclusions. The others have a zero weighting due to the logical AND in the rules.

There is a unique membership function associated with each input parameter. The membership functions associate a weighting factor with values of each input and the effective rules. These weighting factors determine the degree of influence or DOM each active rule has. By computing the logical product of the membership weights for each active rule, a set of fuzzy output response magnitudes are produced. All that remains is to combine and defuzzify these output responses. The fuzzy output response magnitudes for each of the effective rules must be processed and combined in some manner to produce a single, crisp (defuzzified) output.

As inputs are received by the system, the rule base is evaluated. The antecedent (IF X AND Y) blocks test the inputs and produce conclusions. The consequent (THEN Z) blocks of some rules are satisfied, whereas others are not. The conclusions are combined to form logical sums. These conclusions feed into the inference process where each response output member function's firing strength (0 to 1) is determined.

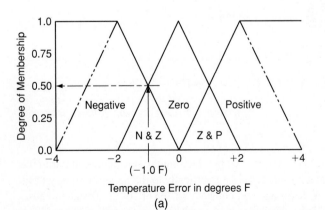 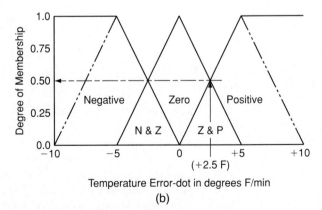

**FIGURE 10.16(a)**   Error Membership Function; **(b)** Error-Dot Membership Function

### 10.8.3 Input Degree of Membership

Referring back to the rules, plug in the membership function weights shown in Figure 10.17 "Error" selects rules 1, 2, 4, 5, 7, and 8, whereas "error-dot" selects rules 4 through 9. "Error" and "error-dot" for all rules are combined to a logical product (LP or AND, that is the minimum of either term).

**FIGURE 10.17** Membership Function Weights

"error" = −1.0: "negative" = 0.5 and "zero" = 0.5
"error-dot" = +2.5: "zero" = 0.5 and "positive" = 0.5
ANTECEDENT & CONSEQUENT BLOCKS (e = error, er = error-dot or error-rate)

Of the nine rules selected, only four (rules 4, 5, 7, and 8) fire or have nonzero results. This leaves fuzzy output response magnitudes for only "Cooling" and "No Change," which must be inferred, combined, and defuzzified to return the actual crisp output. In the following rule list, the following definitions apply: (e) = error, (er) = error-dot.

```
1. If (e < 0) AND (er < 0) then Cool 0.5 & 0.0 = 0.0
2. If (e = 0) AND (er < 0) then Heat 0.5 & 0.0 = 0.0
3. If (e > 0) AND (er < 0) then Heat 0.0 & 0.0 = 0.0
4. If (e < 0) AND (er = 0) then Cool 0.5 & 0.5 = 0.5
5. If (e = 0) AND (er = 0) then No_Change 0.5 & 0.5 = 0.5
6. If (e > 0) AND (er = 0) then Heat 0.0 & 0.5 = 0.0
7. If (e < 0) AND (er > 0) then Cool 0.5 & 0.5 = 0.5
8. If (e = 0) AND (er > 0) then Cool 0.5 & 0.5 = 0.5
9. If (e > 0) AND (er > 0) then Heat 0.0 & 0.5 = 0.0
```

The inputs are combined logically using the AND operator to produce output response values for all expected inputs. The active conclusions are then combined into a logical sum for each membership function. A firing strength for each output membership function is computed. All that remains is to combine these logical sums in a defuzzification process to produce the crisp output.

### 10.8.4 Inferencing

The last step completed in the example was to determine the firing strength of each rule. It turned out that rules 4, 5, 7, and 8 each fired at 50%, or 0.5, whereas rules 1, 2, 3, 6, and 9 did not fire at all (0%, or 0.0). The logical products for each rule must be combined or inferred (max-min'd, max-dot'd, averaged, root-sum-squared, etc.) before being passed on to the defuzzification process for crisp output generation. Several inference methods exist that were mentioned previously.

The MAX-MIN method tests the magnitudes of each rule and selects the highest one. The horizontal coordinate of the "fuzzy centroid" of the area under that function is taken as the output. This method does not combine the effects of all applicable rules but does produce a continuous output function and is easy to implement.

The MAX-DOT or MAX-PRODUCT method scales each member function to fit under its respective peak value and takes the horizontal coordinate of the "fuzzy" centroid of the composite area under the function(s) as the output. Essentially, the member function(s) are shrunk so that their peak equals the magnitude of their respective function ("negative," "zero," and "positive"). This method combines the influence of all active rules and produces a smooth, continuous output.

The AVERAGING method is another approach that works, though it fails to give increased weighting to more rule votes per output member function. For example, if three "negative" rules fire, but only one "zero" rule does, averaging will not reflect this difference because both averages will equal 0.5. Each function is clipped at the average, and the "fuzzy" centroid of the composite area is computed.

The ROOT-SUM-SQUARE (RSS) method combines the effects of all applicable rules, scales the functions at their respective magnitudes, and computes the "fuzzy" centroid of the composite area. This method is more complicated mathematically than other methods, but was selected for this example because it seemed to give the best weighted influence to all firing rules.

## 10.9          DEFUZZIFICATION

The RSS method was chosen to include all contributing rules because there are so few member functions associated with the inputs and outputs. For the ongoing example, an error of $-1.0$ and an error-dot of $+2.5$ selects regions of the "negative" and "zero" output membership functions. The respective output membership function strengths (range: 0–1) from the possible rules (R1–R9) are

```
"negative" = (R1² + R4² + R7² + R8²)(Cooling) = (0.00² + 0.50²+
             0.50² + 0.50²)².5 = 0.866
"zero" = (R5²).5 = (0.50²).5(No Change) = 0.500
"positive" = (R2² + R3² + R6² + R9²)(Heating) = (0.00² + 0.00² +
             0.00² + 0.00²).5 = 0.000
```

### 10.9.1 Fuzzy Centroid Algorithm

The defuzzification of the data into a crisp output is accomplished by combining the results of the inference process and then computing the "fuzzy centroid" of the area. The weighted strengths of each output member function are multiplied by their respective output membership function center points and summed. Finally, this area is divided by the sum of the weighted member function strengths, and the result is taken as the crisp output. One feature to note is that because the zero center is at zero, any zero strength will automatically compute to zero. If the center of the zero function happened to be offset from zero (which is likely in a real system where heating and cooling effects are not perfectly equal), then this factor would have an influence.

$$\frac{(\text{neg\_center} * \text{neg\_strength} + \text{zero\_center} * \text{zero\_strength} + \text{pos\_center} * \text{pos\_strength})}{(\text{neg\_strength} + \text{zero\_strength} + \text{pos\_strength})} = \text{OUTPUT}$$

With the appropriate substitutions the following output is generated:

$$\frac{(-100 * 0.866 + 0 * 0.500 + 100 * 0.000)}{(0.866 + 0.500 + 0.000)} = 63.4\%$$

The horizontal coordinate of the centroid of the area marked in Figure 10.18 is taken as the normalized, crisp output. This value of $-63.4\%$ (63.4% Cooling) seems logical because the particular input conditions (Error = $-1$, Error-dot = $+2.5$) indicate that the feedback has exceeded the command and is still increasing, therefore cooling is the expected and required system response.

**FIGURE 10.18** Output Membership Function

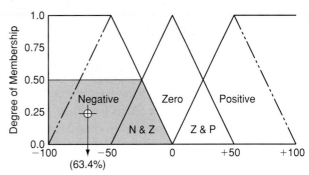

Percent Output-(−100 to 0 = Cooling, 0 to +100 = Heating)

## 10.10        TUNING AND SYSTEM ENHANCEMENT

Tuning the system can be done by changing the rule antecedents or conclusions, changing the centers of the input and/or output membership functions, or adding additional degrees to the input and/or output functions such as "low," "medium," and "high" levels of "error," "error-dot," and output response. These new levels would generate additional rules and membership functions, which would overlap with adjacent functions forming longer "mountain ranges" of functions and responses. The techniques for doing this systematically are a subject unto itself.

The logical product of each rule is inferred to arrive at a combined magnitude for each output membership function. This can be done by max-min, max-dot, averaging, RSS, or other methods. Once inferred, the magnitudes are mapped into their respective output membership functions, delineating all or part of them. The "fuzzy centroid" of the composite area of the member functions is computed and the final result taken as the crisp output. Tuning the system amounts to "tweaking" the rules and membership function definition parameters to achieve acceptable system response.

## QUESTIONS AND PROBLEMS

1. Why does fuzzy logic prevent overshoot?
2. Describe the three basic components of a fuzzy logic system.
3. Give five examples of nonfuzzy data.
4. Give three examples of fuzzy sets.
5. Explain how fuzzy logic makes use of common sense.
6. What is meant by degree of membership?
7. Give three examples of linguistic variables.
8. Write five fuzzy logic rules.
9. What is defuzzification?
10. Select a linguistic variable and diagram it.
11. Describe a PID controller primary function.
12. Give two examples of significant error and rate-of-change error.
13. What is the value of a zero center row and column in a rule matrix?
14. Diagram a fuzzy logic system to control a pizza oven.
15. Give an example of defuzzification.

## SOURCES

Banks, Walter. *Linguistic variables: Clear thinking with fuzzy logic.* Byte Craft Ltd., Waterloo, Ontario, Canada.

Brubaker, David. 1992. *Fuzzy logic overview.* Menlo Park, CA: The Huntington Group.

HCS12X Microcontrollers. 2005. *S12XCPUV1 reference manual,* document S12CPU1, v01.01. Freescale Semiconductors.

Ibrahim, Ahmad M. 2004. *Fuzzy logic for embedded system applications.* Burlington, MA: Elsever Science.

Kaehler, Steven D. "Fuzzy logic tutorial, Parts 1–6. Encoder," *Newsletter of the Seattle Robotics Society.*

Miller, Byron. 2004, June. "Fuzzy logic does real time on the DSP." *Embedded Systems Design.* CMP Media, Inc., San Francisco, CA.

Pack, Daniel J. and Steven F. Barrett. 2002. *68HC12 Microcontroller theory and applications.* Upper Saddle River, NJ: Prentice Hall.

Petriu, Emil M. *Fuzzy systems for control applications.* School of Information Technology and Engineering, University of Ottawa, Canada.

Sowell, Thomas. *Fuzzy logic for just plain folks.* Retrieved from http://www.fuzzy-logic.com/.

Wikipedia. *Fuzzy logic.* Retrieved from http://en.wikipedia.org/wiki/Fuzzy_logic.

# CHAPTER 11

## 8-Bit Microcontrollers

**OBJECTIVE: AN INTRODUCTION TO TWO POPULAR "OFF-THE-SHELF" 8-BIT MICROCONTROLLERS**

The reader will learn:

1. The PIC18F4520 8-bit microcontroller from MicroChip.
2. The eZ8 Encore XP F0830 microcontroller from ZiLOG.

## 11.0 GENERAL-PURPOSE MICROCONTROLLERS

With tried-and-true architectures, general-purpose microcontollers (GPMs) have stood the test of time. One of the key features of the devices is their mature architectures. This provides an established product development tool market that further decreases the cost of the design.

A drawback of general-purpose microcontrollers is their fixed architecture, which does not have a minimized set of functions for a specific design implementation. In response, manufacturers design in peripheral functions that cover a broad range of applications. This keeps manufacturing costs down by increasing volume and spreading it over a wide range of applications.

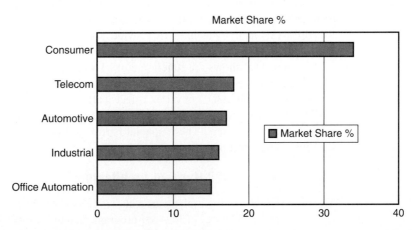

**FIGURE 11.1** Applications by Market Segment for PIC Microcontrollers (Copyright 2004 Microchip Technology Incorporated. Reprinted with permission.)

Using a general-purpose microcontroller for a high-volume application is often dependent on design costs as much as the price per unit. Although a custom chip may have a lower per unit cost over the life of the product, a general-purpose microcontroller can have significantly lower front-end costs and correspondingly less financial risk, as shown in Figure 11.2. These factors have to be taken together to determine whether to go with a custom, semicustom, or general-purpose microcontroller.

**FIGURE 11.2** Cost Comparison SoC to GPM

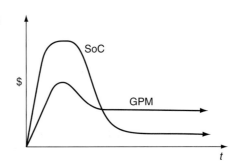

Other key factors for a decision to go with a GPM include time-to-market and performance. A distinct advantage that systems on a chip (SoCs) have over GPMs is in the performance category. An SoC can be designed for a specific application and have its performance optimized in hardware. Often a GPM has to rely on software implementations of critical functions that degrade performance.

As can be seen, many factors affect the decision to go with a GPM besides low volume and low design costs. Two architectures are covered that are based on well-established market success with wide application focus and correspondingly large-volume usage.

## 11.1          MICROCHIP PIC18F4520

This is an improved version of the successful PIC18F452 product family. It is an 8-bit Harvard architecture device utilizing proven design technologies with a wide application range. Established FLASH memory technology provides for high volume and low price points, which makes it applicable to a wider range of designs.

Figure 11.3 shows the functional block diagram of the PIC18F4520. In addition to standard I/O ports, the device contains a number of additional functions. These include multiple timer/counters, an A/D converter, electrically erasable programmable read-only memory (EEPROM), universal asynchronous received transmitter (UART), and capture and compare functions, among others. These special hardware implemented functions interface externally via multiplexed port pins.

### 11.1.1  PIC 18F4520 Harvard Architecture

The PIC18F4520 family implements a classic Harvard architecture. There are 1536 bytes of data memory and 32,758 bytes of FLASH program memory. Figure 11.4 shows the separate memories and their associated buses. With the Harvard architecture of separate data and program buses, a data byte can be accessed during each instruction fetch.

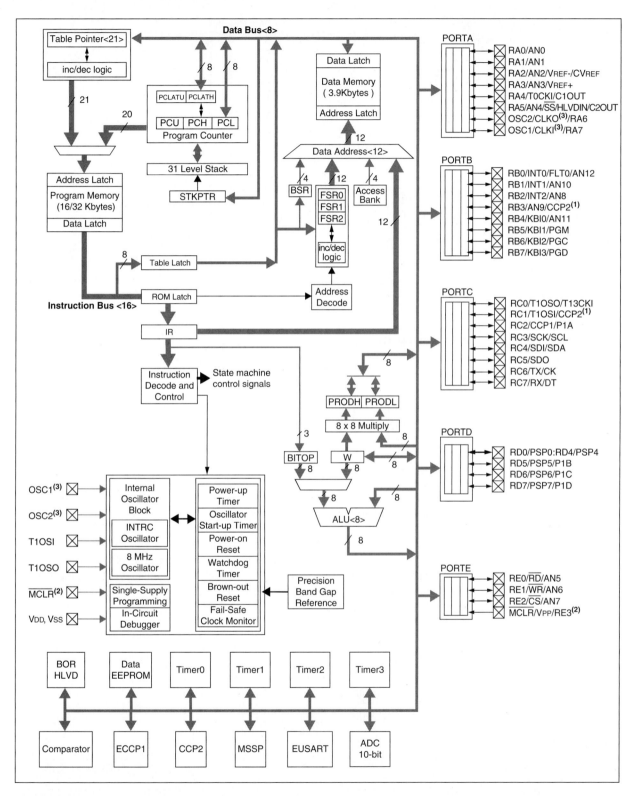

**FIGURE 11.3** PIC18F4520 Functional Block Diagram (Copyright 2004 Microchip Technology Incorporated. Reprinted with permission.)

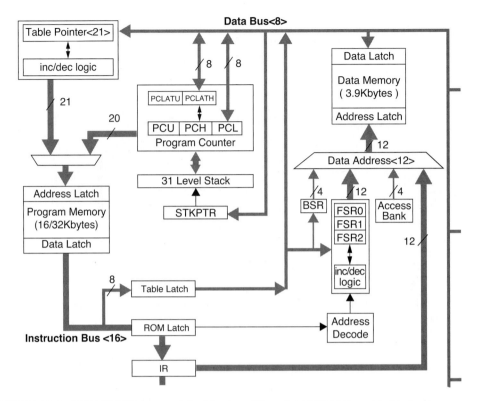

**FIGURE 11.4**   PIC18F4520 Harvard Architecture (Copyright 2004 Microchip Technology Incorporated. Reprinted with permission.)

## 11.1.2  Instruction Pipeline

The PIC18F4520 implements a classic instruction pipeline with the overlap of instruction fetch with instruction decode. This two-stage pipeline is simple to implement (e.g., small die footprint) and yet provides a doubling of instruction throughput. With only two stages, the control logic is reasonably simple and the overhead minimal compared to five- or seven-stage pipelines.

Figure 11.5 shows how the pipeline is implemented with the overlap of instruction fetch and execution. As with all pipelines without look ahead logic, when a change in program flow (any branch type instruction) is encountered, the pipeline will flush, and extra cycles are required. An example of "BRA" (branch) instruction requiring two instruction cycles is shown in Figure 11.5.

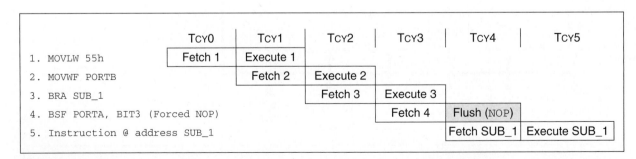

**FIGURE 11.5**   PIC18F4520 Two-Stage Pipeline (Copyright 2004 Microchip Technology Incorporated. Reprinted with permission.)

### 11.1.3 Special Features

The PIC18F4520 incorporates functional blocks to reduce the requirements for external circuitry. This simplifies the design while providing high-end enhancements. This is particularly valuable when designing the production PCB without the overhead of external components. In essence, the prototype PCB becomes the production PCB.

Nine major features are configured via a set of 11 configuration registers. They are located at program memory addresses 0x300000 through 0-x3FFFFF, which are out of range of the program address counter. This prevents them from being inadvertently changed via in-circuit programming. The special features include

| Oscillator Selection | Resets |
|---|---|
| Power-on Reset (POR) | Watchdog Timer (WDT) |
| Power-up Timer (PWRT) | Fail-Safe Clock Monitor |
| Power-up Timer (PWRT) | Two-Speed Start-up |
| Oscillator Start-up Timer (OST) | Code Protection |
| Brown-out Reset (BOR) | ID Locations |
| Interrupts | |
| In-Circuit Programming | |

### 11.1.4 Power Management Modes

Managing the power consumption of a device can be crucial. An example would be for battery-powered applications where power consumption affects the utility of the product. A sensor in the field may be solar powered and have limited capability for power.

There are three major power management modes in the PIC18F4520:

Run modes
Idle modes
Sleep mode

These categories define which portions of the device are clocked and depending on mode, at what speed. The RUN and IDLE modes can use any of the three input clock sources, whereas the SLEEP mode does not use a clock source. The table in Figure 11.6 shows the different combinations of mode, clock source, and speeds.

In the RUN modes, the clocks to the core and peripherals remain active. The difference between the modes is the clock source. In the PRI_RUN mode, the clock source is the main oscillator input, whereas in the SEC_RUN mode, the clocking is from Timer1 external oscillator. This provides a lower power option while still using a high-accuracy clock source.

In RC_RUN mode, the CPU and peripherals are clocked from the internal RC oscillator. This mode provides for program execution with the least power consumption. The trade-off is lower clock accuracy, which is fine for many battery backup type of applications.

In Sleep mode, the device does not execute program instructions. The clock sources a shutdown for maximum power savings because no instructions are being executed. A wake-up event occurs when an interrupt, RESET, or WDT time-out occurs.

Idle mode is a way for the CPU to be shut down while keeping the peripherals active. This saves power when an interrupt is expected after a long period of time (relatively speaking). The

| Mode | OSCCON Bits | | Module Clocking | | Available Clock and Oscillator Source |
| | IDLEN[1] <7> | SCS1:SCS0 <1:0> | CPU | Peripherals | |
|------|------|------|------|------|------|
| Sleep | 0 | N/A | Off | Off | None—All clocks are disabled |
| PRI_RUN | N/A | 00 | Clocked | Clocked | Primary—LP, XT, HS, HSPLL, RC, EC and Internal Oscillator Block[2], This is the normal full power execution mode. |
| SEC_RUN | N/A | 01 | Clocked | Clocked | Secondary—Timer1 Oscillator |
| RC_RUN | N/A | 1x | Clocked | Clocked | Internal Oscillator Block[2] |
| PRI_IDLE | 1 | 00 | Off | Clocked | Primary—LP, XT, HS, HSPLL, RC, EC |
| SEC_IDLE | 1 | 01 | Off | Clocked | Secondary—Timer1 Oscillator |
| RC_IDLE | 1 | 1x | Off | Clocked | Internal Oscillator Block[2] |

Note 1:  IDLEN reflects its value when the SLEEP instruction is executed.
     2:  Includes INTOSC and INTOSC postscaler, as well as the INTRC source.

**FIGURE 11.6**    PIC18F4520 power Management Modes (Copyright 2004 Microchip Technology Incorporated. Reprinted with permission.)

CPU is not executing instructions and simply waits for an interrupt, WDT time-out, or RESET to resume instruction execution.

### 11.1.5  Oscillator Configurations

There are ten different oscillator modes for the PIC18F4520, which are configured through bits FOSC3:FOSC0 in the 0x01 configuration register. They are

| | | |
|---|---|---|
| 1 | LP | Low-Power Crystal |
| 2 | XT | Crystal/Resonator |
| 3 | HS | High-Speed Crystal/Resonator |
| 4 | HSPLL | High-Speed Crystal/Resonator with PLL enabled |
| 5 | RC | External Resistor/Capacitor with Fosc/4 output on RA6 |
| 6 | RCIO | External Resistor/Capacitor with I/O on RA6 |
| 7 | INTIO1 | Internal Oscillator with Fosc/4 output on RA6 and I/O on RA7 |
| 8 | INTIO2 | Internal Oscillator with I/O on RA6 and RA7 |
| 9 | EC | External Clock with Fosc/4 Output |
| 10 | ECIO | External Clock with I/O on RA6 |

Oscillator configuration may be selected based on a combination of cost, clock accuracy, and application. For minimal cost for applications where timing accuracy is not critical, the external RC and RCIO modes can be used.

Additionally the internal oscillator may be used for minimal parts count. An oscillator "tuning" register is included to allow the timing to be adjusted by the user application code. The factory setting is 8Mhz; however, this can drift as Vdd or temperature (of the die) change.

### 11.1.6 Reset

The PIC18F4520 includes a number of oscillator configurations to fit a combination of accuracy and cost effectiveness. They include

Power-on-Reset (POR)
MCLR/ Reset during normal operation
MCRL/ Reset during power managed modes
Watchdog Timer (WDT) Reset (during execution)
Programmable Brown-out Reset (BOR)
RESET Instruction
Stack Full Reset
Stack Underflow Reset

Figure 11.7 shows the simplified version of the RESET logic functional block diagram. Most registers are unaffected by a RESET. If the processor is in a sleep mode, then Wake-up via the WDT or an Interrupt will not affect most registers. The exceptions would be the registers that contain bits directly pertinent to the Wake-up condition.

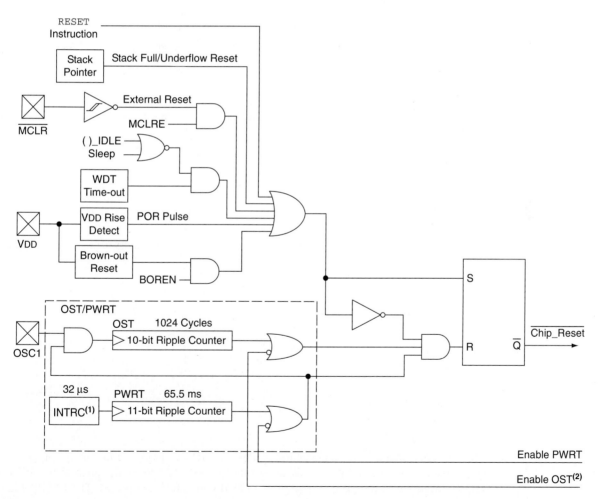

**FIGURE 11.7** PIC18F4520 On-chip Reset Circuit Block Diagram (Copyright 2004 Microchip Technology Incorporated. Reprinted with permission.)

MCLR/, WDT Reset, RESET Instruction, and the Stack Resets basically leave the control registers unchanged. They reset the registers involved with program flow to the initialization state. POR and BOR result in complete reset of all registers to their initial states.

### 11.1.7 Memory Organization

The PIC18F4520 has three memory spaces: program memory, data memory, and peripheral EEPROM. In the Harvard style, both the program and data memories can be accessed during a single instruction cycle. The EEPROM memory is treated as a peripheral function and lies outside the data or program memory spaces. The basic program memory map is shown in Figure 11.8.

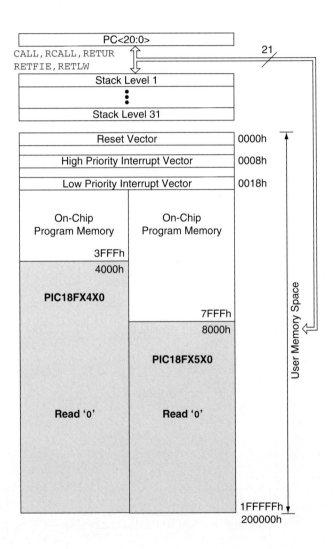

**FIGURE 11.8**   PIC18F4520 Program Memory Map (Copyright 2004 Microchip Technology Incorporated. Reprinted with permission.)

Program memory is implemented using FLASH technology and can be programmed externally as well as internally. This allows the PIC18F4520 to be designed into a PCB, be shipped from the factory, and still be programmed remotely with updates. The PIC18F4520 is implemented with 32k bytes (16 k words) of FLASH out of an address space of 2 Mb.

Data memory is organized into register banks of 256 bytes, as shown in Figure 11.9. Although the BSR (Bank Select Register) supports addressing 16 banks, only 6 are used. The

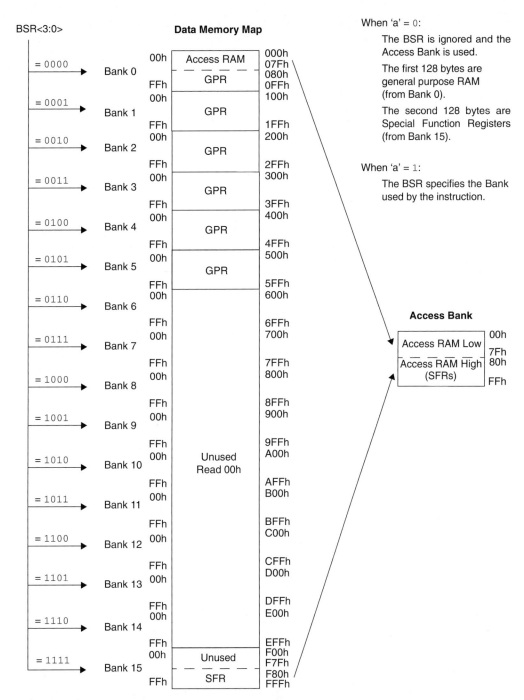

**FIGURE 11.9**  PIC18F4520 Data Memory Map (Copyright 2004 Microchip Technology Incorporated. Reprinted with permission.)

programmer must keep track of the bank being selected to ensure the proper data location is written to or read from.

An Access Bank is defined so that the first 128 bytes of data memory and the high-order 128 bytes of register bank 15 are mapped into a single bank. The Access Bank does not use the BSR. Instructions are defined with the a = 0 setting, "forcing" the use of the Access Bank and bypassing the BSR.

EEPROM memory is separately addressed and outside the program/data memory address space. It is accessed via the Special Function Registers (SFRs). The EEADR register is used to address one of 256 possible memory locations 0x00 through 0xFF. EEDATA register holds the read or write byte, depending on the operation. EECON1 and EECON2 are the control registers required for the read and write operation.

An application example would be for a data logger. As data is inputted to the system, it is stored in the EEPROM memory. Over a period of time, the power could be cycled to reduce consumption while the critical data is retained.

## 11.1.8 Interrupt Structure

The PIC18F4520 supports a sophisticated interrupt structure for peripheral functions as well as external interrupts. Two levels of interrupt priority are defined, as shown in Figure 11.10. All interrupts are assignable to either level, except for INT0, which is high priority only. The low-level interrupt vector is at address 0x0018 and the high level at 0x0008. INT0 (RB<0>) is defined to be a high-priority interrupt only for backward compatibility.

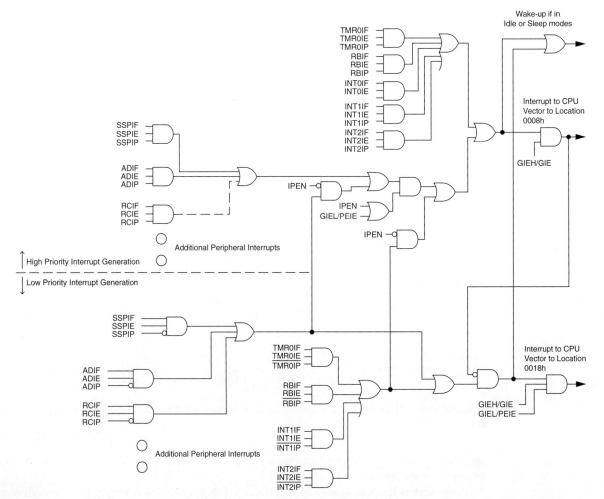

**FIGURE 11.10** PIC18F4520 Interrupt Structure (Copyright 2004 Microchip Technology Incorporated. Reprinted with permission.)

Other than INT0, each interrupt has an associated enable, flag, and priority bit. When an interrupt occurs, the associated flag bit is set; if the enable bit is also set, then an interrupt will vector to the appropriate address. Interrupts can be polled by testing for the interrupt flag bit.

During interrupts, the return PC address is saved on the stack. Additionally the working register (WREG), Status, and BSR registers are saved on the fast return stack. If a fast return from interrupt is not used, the user may need to save the WREG, Status and BSR register on entry to the Interrupt Service Routine. Depending on the application, other registers may also need to be saved.

### 11.1.9 Input/Output Ports

There are five I/O ports defined for the PIC18F4520. Most pins are multiplexed with peripheral functions to reduce the overall package pin count. In general, all pins are available for bit I/O operations except when used with a peripheral function. Figure 11.11 shows the diagram of a generic I/O port.

**FIGURE 11.11** PIC18F4520 Generic I/O Port (Copyright 2004 Microchip Technology Incorporated. Reprinted with permission.)

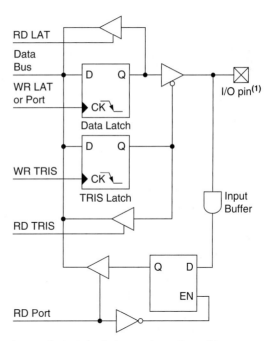

**Note1:** I/O pins have diode protection to V<sub>DD</sub> and V<sub>SS</sub>.

Each I/O pin has three registers associated with it: TRIS for data direction, PORT register, which reads the levels on the pins, and LAT register, which is the output latch. It is important that the TRIS register bit be set for input or output as appropriate when used with a peripheral function.

### 11.1.10 Timer-Related Functions

Four timer modules are included as peripherals in the PIC18F4520. They are also used in combination with the Capture/Compare function and pulse-width modulation (PWM) function. Not all functions may be used simultaneously, depending on which ones are selected.

## 11.1.11  Timer Modules

There are four 16-bit timer modules in the PIC18F4520. Timer0 can be configured in either 8- or 16-bit mode. The 8-bit mode is a compatibility mode with previous PIC devices. The functional block diagram of Timer0 is shown in Figure 11.12(a) and (b). Note that it supports an 8-bit prescalar with values from 2 through 256 in powers of 2. In 16-bit mode that yields a maximum count number of $2^8 \times 2^{16} = 2^{24}$.

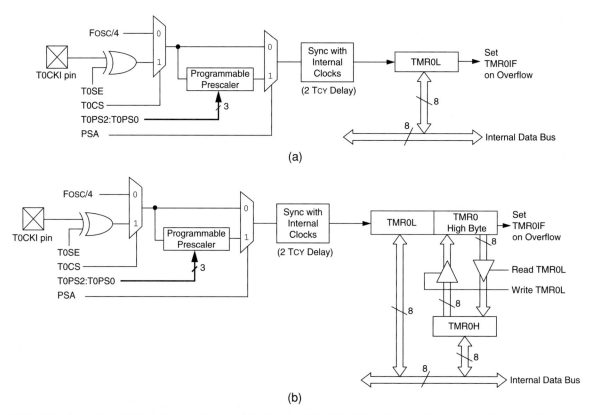

**FIGURE 11.12 (a)**   PIC18F4520 Timer0 Module 8-Bit Mode; **(b)** PIC18F4520 Timer0 Module 16-Bit Mode (Copyright 2004 Microchip Technology Incorporated. Reprinted with permission.)

The Timer0 interrupt flag is set on overflow of the counter. In 8-bit mode, to count upward to 100, you would set Timer0 to 156. If the Timer0 Interrupt Enable (TMR0IE) bit is set, then the overflow will create a vectored interrupt to the location specified by the priority setting. When used in 16-bit mode, TMR0H (high byte) is first loaded and then TMR0L (low byte). Moving a value to TMR0L will load the value from TMR0H to TMR0 so that the 16-bit value is loaded in one instruction cycle.

Timer1 is shown in Figures 11.13(a) and (b). It can be configured for 8- or 16-bit operation similar to Timer0, however, with a more limited prescalar option. Three major modes of operation are supported.

Timer
Synchronous counter
Asynchronous counter

Timer1 can be configured to support an external crystal. This is valuable when using Timer1 as a clock source in power managed modes. Timer1 can also be used as a real-time clock.

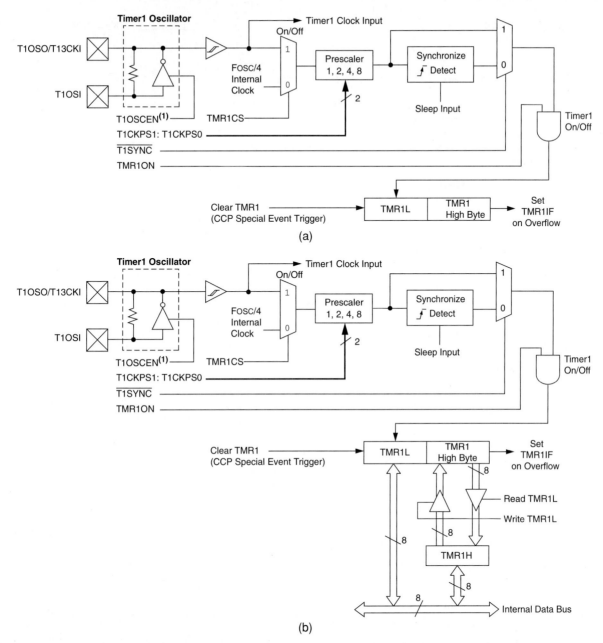

**FIGURE 11.13 (a)** PIC18F4520 Timer1 Block Diagram (8-bit Mode); **(b)** PIC18F4520 Timer1 Block Diagram (16-Bit Mode) (Copyright 2004 Microchip Technology Incorporated. Reprinted with permission.)

Using a 32.768 kHz crystal in 16-bit Read/Write mode will take 2 seconds to overflow. Shorter intervals can be programmed by appropriately setting the TMR1H and TMR1L register pair.

The 8-bit Timer2 with pre- and postscaler capability is shown in Figure 11.14. It incorporates both a prescaler and postscaler function. The value of TMR2 register can be compared to the Period Register (PR) and used as output to the PWM or master synchronous serial port (MSSP) modules.

Timer3 has both 8- and 16-bit modes with prescaler only capability. The 8-bit mode of Timer3 is shown in Figure 11.15(a) and the 16-bit mode in Figure 11.15(b). As with Timer1, it

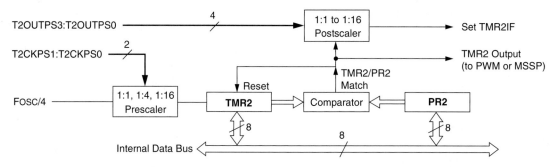

**FIGURE 11.14**  PIC18F4520 Timer2 Block Diagram (8-Bit Mode) (Copyright 2004 Microchip Technology Incorporated. Reprinted with permission.)

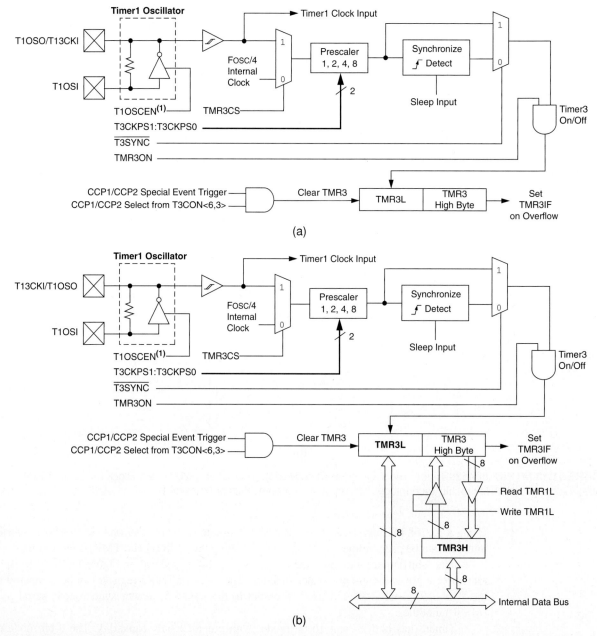

**FIGURE 11.15 (a)** PIC18F4520 Timer3 Block Diagram; **(b)** PIC18F4520 Timer3 Block Diagram (16-Bit Read/Write Mode) (Copyright 2004 Microchip Technology Incorporated. Reprinted with permission.)

can be configured for use in counter, synchronous and asynchronous modes. It is also used in conjunction with CCP module.

## 11.1.12 Capture/Compare/PWM Functions

Three additional functions utilize the Timers. They are Capture (Figure 11.16), Compare (Figure 11.17), and Pulse-Width Modulation (Figure 11.18). Not all the Timers can be active at the same

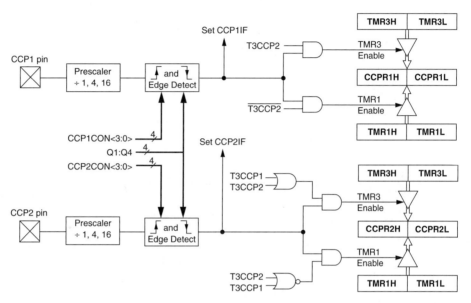

**FIGURE 11.16** PIC18F4520 Capture Function Block Diagram (Copyright 2004 Microchip Technology Incorporated. Reprinted with permission.)

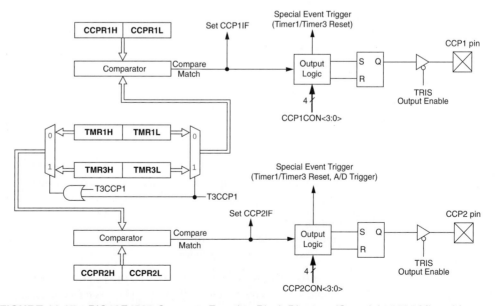

**FIGURE 11.17** PIC18F4520 Compare Function Block Diagram (Copyright 2004 Microchip Technology Incorporated. Reprinted with permission.)

216 CHAPTER 11

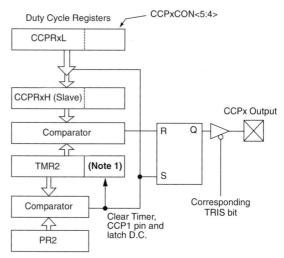

**FIGURE 11.18** PIC18F4520 PWM Block Diagram (Copyright 2004 Microchip Technology Incorporated. Reprinted with permission.)

**Note 1:** The 8-bit TMR2 value is concatenated with the 2-bit internal Q clock, or 2 bits of the prescaler, to create the 10-bit time base.

time. This limits the number of options that can be incorporated in any single design. It depends on which combination is selected for use as a Timer, CCP, or PWM function.

Pulse-Width Modulation is supported in both normal and enhanced modes. Figure 11.18 shows a simplified block diagram of the PWM function. It supports up to 10 bits of resolution. This standard PWM capability functions only in the single output mode.

Figure 11.19 shows the typical output waveform produced by the single-mode PWM. It shows that period register 2 (PR2) and timer register 2 (TMR2) of the Timer2 function block are used for computing the waveform. The period is specified by writing to the PR2 register, and the duty cycle is specified by writing to the CCPRxL register and associated CCPxCON<5:4> bits.

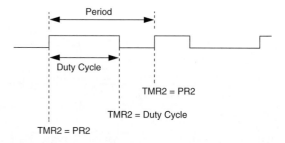

**FIGURE 11.19** PIC18F4520 PWM Output (Copyright 2004 Microchip Technology Incorporated. Reprinted with permission.)

An enhanced PWM mode provides additional PWM output options (Figure 11.20) for a broader range of control applications. It supports up to four outputs. There are three modes of operation:

Half-Bridge Output
Full-Bridge Output, Forward Mode
Full-Bridge Output, Reverse Mode

In the Half-Bridge output mode, two pins are used as outputs to drive push-pull loads. The PWM output signal is output on the P1A pin, whereas the complementary PWM output signal is output on the P1B pin. This mode can be used for half-bridge or full-bridge applications, as shown in Figure 11.21(a) and (b).

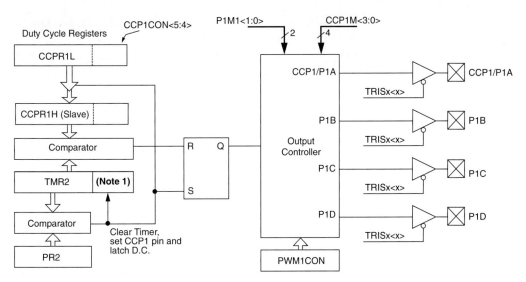

**FIGURE 11.20**   PIC18F4520 Enhanced PWM Block Diagram (Copyright 2004 Microchip Technology Incorporated. Reprinted with permission.)

**FIGURE 11.21 (a)** PIC18F4520 Standard Half-Bridge Circuit ("Push-Pull"); **(b)** PIC18F4520 Half-Bridge Output Driving a Full-B (Copyright 2004 Microchip Technology Incorporated. Reprinted with permission.)

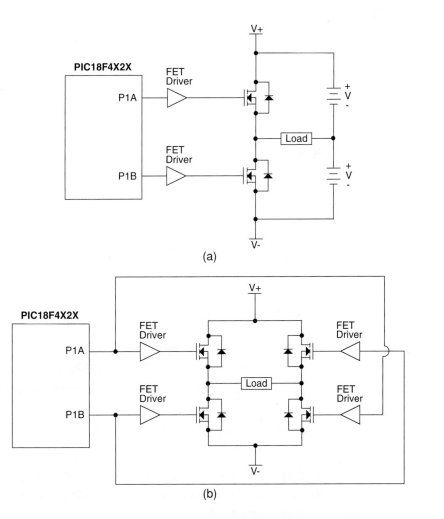

In Full-Bridge output mode, four pins are used as output; however, only two outputs are active at a time. In the forward mode, pin P1A is continuously active, and P1D is modulated. In the reverse mode, pin P1C is continuously active, and pin P1B is modulated. An application of Full-Bridge mode is shown in Figure 11.22.

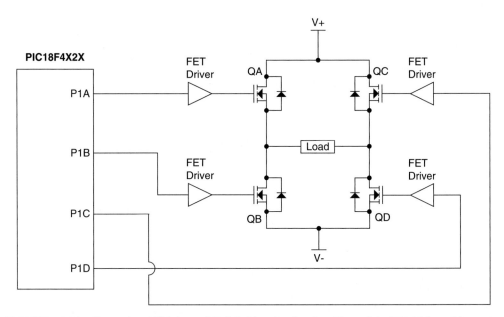

**FIGURE 11.22**    Example of PIC-based Full-Bridge Application (Copyright 2004 Microchip Technology Incorporated. Reprinted with permission.)

## 11.1.13  Serial Communication Interface

The PIC18F4520 supports four major serial I/O functions: MSSP with both SPI and $I^2C$ modes and an enhanced USART that also supports LIN 1.2 functionality. This gives broad support for a spectrum of applications requiring high-performance serial I/O functionality.

**11.1.13.1  MSSP**    The MSSP module is a serial interface, useful for communicating with other peripheral or microcontroller devices. These devices may be, for example, EEPROMs, shift registers, display drivers, or A/D converters. The MSSP module can operate in one of two modes:

> Serial Peripheral Interface (SPI)
> Interintegrated Circuit ($I^2C$)
>> Full Master mode
>> Slave mode (with general address call)

The I2C interface supports the following modes in hardware:

> Master Mode
> Multi-Master mode
> Slave mode

**11.1.13.2  SPI**    SPI is a synchronous serial interface that can transfer data in 8-bit byte format a single bit at a time. It can be used to interface to peripheral devices and another microcontroller

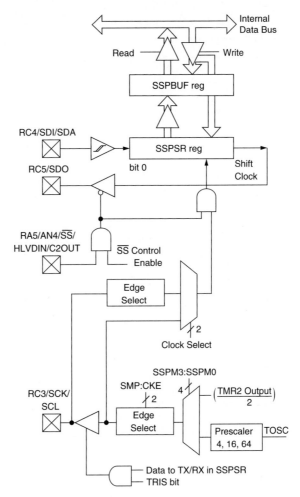

**FIGURE 11.23**   PIC18F4520 MSSP Block Diagram (SPI Mode) (Copyright 2004 Microchip Technology Incorporated. Reprinted with permission.)

supporting the SPI interface. This allows for the possibility of multimicrocontroller system designs. Figure 11.23 shows the basic SPI interface logic.

A typical connection between two microcontrollers is shown in Figure 11.24. The master controller initiates the data transfer by sending the Serial Clock (SCK) signal. Data is shifted out both shift registers on their programmed clock edge and latched on the opposite edge of the clock.

In Master mode, the master can initiate the data transfer at any time because it controls the SCK. The master determines when the slave (Processor 2 in Figure 11.24) is to broadcast data by the software protocol. In Slave mode, the data is transmitted and received as the external clock pulses appear on SCK.

**11.1.13.3  $I^2C$**   The MSSP module in $I^2C$ mode fully implements all master and slave functions (including general call support) and provides interrupts on Start and Stop bits in hardware to determine a free bus (multimaster function). The MSSP module implements the standard mode specifications as well as 7-bit and 10-bit addressing. The serial clock (SCK) and serial data (SDA) pins are configured through the port pins RC3 and RC4.

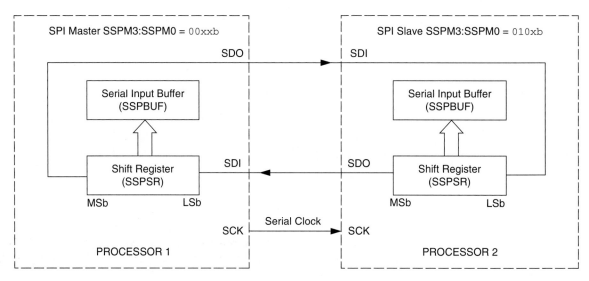

**FIGURE 11.24**   PIC18F4520 SPI Master/Slave Connection (Copyright 2004 Microchip Technology Incorporated. Reprinted with permission.)

**FIGURE 11.25**   PIC18F4520 I²C Functional Diagram (Copyright 2004 Microchip Technology Incorporated. Reprinted with permission.)

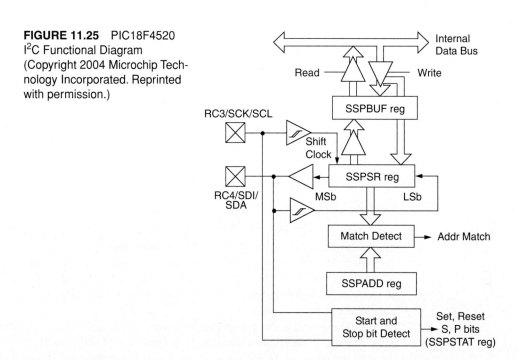

**11.1.13.4   *EUSART***   The PIC18F4520 supports an enhanced USART module, which is generically referred to as a Serial Communications Interface or SCI. The EUSART can be configured as a full-duplex asynchronous system that can communicate with peripheral devices, such as CRT terminals and personal computers. It can also be configured as a half-duplex synchronous system that can communicate with peripheral devices, such as A/D or D/A converters and serial EEPROMS.

The EUSART can be configured in the following modes:

Asynchronous (full duplex) with:
    Auto-wake-up on character reception
    Auto-baud calibration
    12-bit Break character transmission
Synchronous—Master (half duplex) with selectable clock polarity
Synchronous—Slave (half duplex) with selectable clock polarity

The EUSART includes the ability for automatic baud rate detection on reception. The input to the baud rate generator (BRG) clocking is reversed so that the RX signal does the clocking.

In the asynchronous mode the EUSART includes the following major functions:

Baud Rate Generator (BRG)
Sampling Circuit
Asynchronous Transmitter
Asynchronous Receiver
Auto-Wake-Up on Sync Break Character
12-bit Break Character Transmit
Auto-Baud Rate Detection

The EUSART transmitter block diagram is shown in Figure 11.26. Once the appropriate control registers are configured, transmitting a character is accomplished by loading the TX holding register (TXREG). The value in TXREG is then moved to the TSR, where it is transmitted. Thus, once the control register values are set, a simple MOV-type instruction will create the transfer.

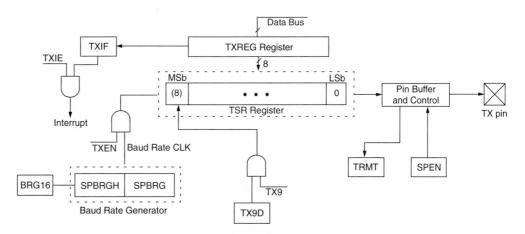

**FIGURE 11.26**   PIC18F4520 EUSART Transmit Function (Copyright 2004 Microchip Technology Incorporated. Reprinted with permission.)

The receiver block diagram is shown in Figure 11.27. It incorporates a data recovery block that operates at 16 times the baud rate. For RS-232 reception, the main receive serial shifter operates at the bit rate of Fose.

Receiving can be done by interrupt or polling. Checking for the receive flag (RCIF) can be done by polling, for example, and the received byte can be simply moved (MOV) from the RCREG to a file location.

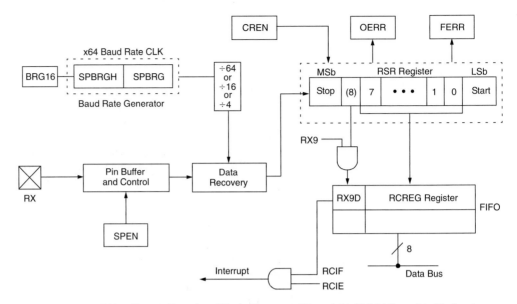

**FIGURE 11.27**   PIC18F4520 Receiver Block Diagram (Copyright 2004 Microchip Technology Incorporated. Reprinted with permission.)

## 11.1.14 Analog-to-Digital Converter

The analog-to-digital converter function shown in Figure 11.28, has 10 inputs for the smaller 28-pin devices and 13 inputs for the 40/44-pin devices. The module supports conversion of an analog input signal to the corresponding 10-bit digital number. The output is generated via the successive approximation method.

The analog reference voltage is software selectable to either the device's positive and negative supply voltage (Vdd and Vss) or the voltage level on the RA3/AN3/Vref+ and RA2/AN2/Vref−/Cvref pins. The A/D converter also has the unique feature to operate in Sleep mode. To operate in Sleep, the A/D conversion clock must be derived from the A/D's internal RC oscillator.

Using the A/D is a straightforward process of setting the proper configuration bits in the ADCON0, ADCON1, and ADCON2 control registers. The following steps can be followed to perform a conversion:

1. Configure the A/D module:
   Configuzre analog pins, voltage reference, and digital I/O (ADCON1)
   Select A/D input channel (ADCON0)
   Select A/D input channel (ADCON0)
   Select A/D acquisition time (ADCON2)
   Select A/D conversion clock (ADCON2)
   Turn on A/D module (ADCON0)
2. Configure A/D interrupt (if desired):
   Clear ADIF bit
   Set ADIE bit
   Set GIE bit
3. Wait the required acquisition time (if required).
4. Start conversion:
   Set GO/DONE/ bit (ADCON0 register)

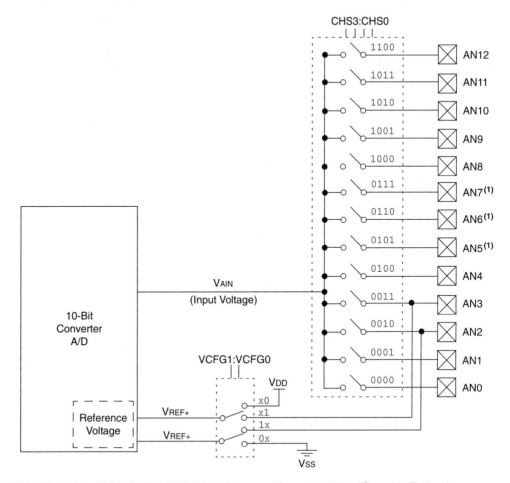

**FIGURE 11.28** PIC18F4520 ADC Block Diagram (Copyright 2004 Microchip Technology Incorporated. Reprinted with permission.)

5. Wait for A/D conversion to complete, by either
   Polling for the GO/DONE/ bit to be cleared or
   Waiting for the A/D interrupt.
6. Read A/D Result registers (ADRESH:ADRESL); clear bit ADIF, if required.
7. For next conversion, got to step 1 or step 2, as required. The A/D conversion time per bit is defined as $T_{AD}$. A minimum wait of 2 $T_{AD}$ is required before the next acquisition starts.

## 11.1.15 Analog Comparator

A single comparator is shown in Figure 11.29, along with the relationship between the analog input levels and the digital output. When the analog input Vin+ is less than the analog input Vin−, the output of the comparator is a digital low level. When the analog input at Vin+ is greater than the analog input Vin−, the output of the comparator is a digital high level. The shaded areas of the output of the comparator in Figure 11.29 represent the uncertainty due to input offsets and response time.

Depending on the comparator operating mode, either an external or internal voltage reference may be used. The analog signal present at Vin− is compared to the signal at Vin+, and the digital output of the comparator is adjusted accordingly (Figure 11.29).

**FIGURE 11.29** PIC18F4520 Single Comparator (Copyright 2004 Microchip Technology Incorporated. Reprinted with permission.)

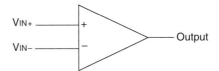

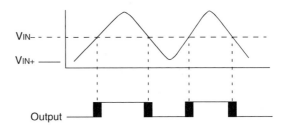

The comparator outputs are read through the CMCON register. These bits are read-only. The comparator outputs may also be directly output to the RA4 and RA5 I/O pins. When enabled, multiplexers in the output path of the RA4 and RA5 pins will switch, and the output of each pin will be the unsynchronized output of the comparator. The uncertainty of each of the comparators is related to the input offset voltage and response time given in the specifications.

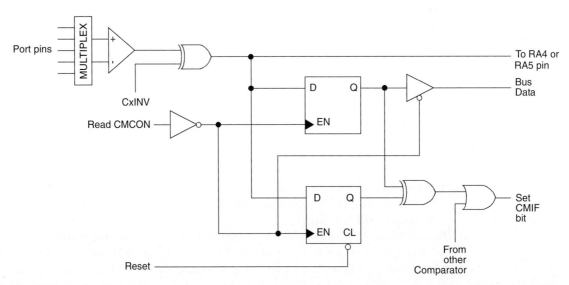

**FIGURE 11.30** PIC18F4520 Comparator Block Diagram (Copyright 2004 Microchip Technology Incorporated. Reprinted with permission.)

## 11.1.16 Special Features of the CPU

The PIC18F4520 includes several features intended to maximize reliability and minimize cost through elimination of external components. These are

Oscillator Selection
Resets:
    Power-on Reset (POR)
    Power-up Timer (PWRT)

Oscillator Start-up Timer (OST)
Brown-out Reset (BOR)
Interrupts
Watchdog Timer (WDT)
Fail-Safe Clock Monitor
Two-Speed Start-up
Code Protection
ID Locations
In-Circuit Serial Programming

The oscillator can be configured for the application, depending on frequency, power, accuracy, and costs. In addition to the Power-up and Oscillator Start-up Timers provided for Resets, the PIC18FD4520 has a Watchdog Timer, which is either permanently enabled via the configuration bits or software controlled (if configuration is disabled).

## 11.1.17 Instruction Set

The PIC18F4520 implements the Reduced Instruction Set Architecture (RISC) concept in that most instructions (nonbranching) can execute in a single instruction cycle. The PIC18F4520 incorporates the standard set of 75 PIC18 core instructions, as well as an extended set of eight new instructions, for the optimization of code that is recursive or utilizes a software stack.

The standard PIC18 instruction set adds many enhancements to the PICmicro instruction sets, while maintaining an easy migration from these PICmicro instruction sets. Most instructions are a single program memory word (16 bits), but four instructions require two program memory locations. They are the MOVFF, CALL, GOTO, and LSFR.

Each single-word instruction is a 16-bit word divided into an opcode, which specifies the instruction type and one or more operands, which further specify the operation of the instruction.

The instruction set is highly orthogonal and is grouped into four basic categories:

Byte-oriented operations
Bit-oriented operations
Control operations
Literal operations

All instructions are a single word, except for the four double-word instructions. These instructions are double word to contain the required address information in 32 bits. In the second word, the four MSBs are 1s. If this second word is executed as an instruction (by itself), it will execute as an NOP.

All single-word instructions are executed in a single instruction cycle, unless a conditional test is true of the program counter and is changed as a result of the instruction. In these cases, the execution takes two instruction cycles with the additional instruction cycle(s) executed as an NOP. The double-word instructions execute in two instruction cycles.

One instruction cycle consists of four oscillator periods. Thus, for an oscillator frequency of 4 MHz, the normal instruction execution time is 1 μs. If a conditional test is true, or the program counter is modified as a result of an instruction, the instruction execution time is 2 μs. Two-word branch instructions (if true) would take 3 μs.

## 11.1.18 Electrical Characteristics

The primary electrical specifications are given in Figure 11.31. Note the limitations on total current sourced or sunk by all ports simultaneously. This can be a limiting factor in the design.

## Absolute Maximum Ratings[t]

| | |
|---|---|
| Ambient temperature under bias | −40°C to +125°C |
| Storage temperature | −65°C to +150°C |
| Voltage on any pin with respect to VSS (except VDD, $\overline{MCLR}$ and RA4) | −0.3V to (VDD + 0.3V) |
| Voltage on VDD with respect to VSS | −0.3V to +7.5V |
| Voltage on $\overline{MCLR}$ with respect to VSS (Note 2) | 0V to +13.25V |
| Total power dissipation (Note 1) | 1.0W |
| Maximum current cut of VSS pin | 300 mA |
| Maximum current into VDD pin | 250 mA |
| Input damp current, IIk (Vi < 0 or Vi > VDD) | ±20 mA |
| Output damp current, Ick (Vo < 0 or Vo > VDD) | ±20 mA |
| Maximum output current sunk by any I/O pin | 25 mA |
| Maximum output current sourced by any I/O pin | 25 mA |
| Maximum current sunk by all ports | 200 mA |
| Maximum current sourced by all ports | 200 mA |

**FIGURE 11.31**   PIC18F4520 Electrical Characteristics (Copyright 2004 Microchip Technology Incorporated. Reprinted with permission.)

## 11.2          ZiLOG Z8 ENCORE! XP F0830 SERIES

The Z8 Encore! XP F0830 Series devices are Flash microcontrollers based on the ZiLOG eZ8 CPU. The MCU is part of Z8 Encore! XP family of devices. The instruction set architecture has ties back to the original Z8 design.

The ZiLOG Z8 Encore! XP F0830 Series is designed as a general-purpose 8-bit microcontroller. Its flexible hardware design allows it to be configured to a variety of applications in a cost-effective manner. It supports a basic pipelined architecture. Instruction fetch and execution are optimized for multibyte, multicycle instructions.

A block diagram of the MCU is shown in Figure 11.32. It supports up to 12 KB flash program memory and 256 bytes register RAM. It also incorporates peripheral functions such as PWM and WDT, as well as up to eight channels of fast analog-to-digital conversion (11.9 µs) with a SAR converter.

Z8 Encore! XP F0830 Series MCU includes the following key features:

20 MHz eZ8 CPU
Up to 12 KB flash memory with in-circuit programming capability
Up to 256 B register RAM
64 B Nonvolatile data storage (NVDS)
Up to 25 I/O pins, depending on package
Internal precision oscillator (IPO)
External crystal oscillator
Two enhanced 16-bit timers with capture and compare
Watchdog Timer (WDT) with dedicated internal RC oscillator
Single-pin, on-chip debugger (OCD)
Optional 8-channel, 10-bit analog-to-digital converter (ADC)
On-chip analog comparator
Up to 17 interrupt sources
Voltage brownout protection (VBO)
Power-on reset (POR)
2.7 V to 3.6 V operating voltage
Up to 13 5-V tolerant input pins
20- and 28-pin packages

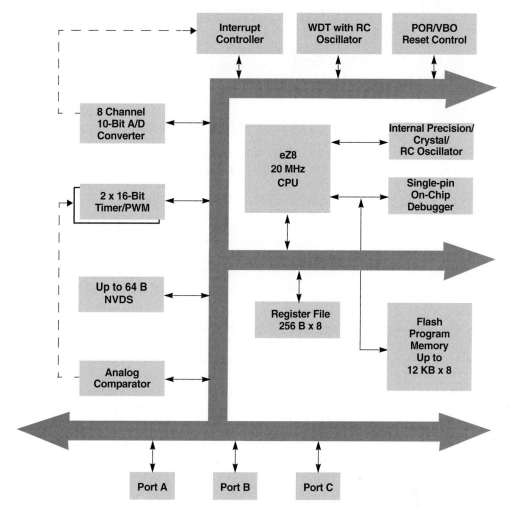

**FIGURE 11.32**   Z8 Encore! XP F0830 Series Block Diagram (Reprinted by permission of ZiLOG Corporation.)

## 11.2.1  eZ8 CPU Description

The ZiLOG Z8 Encore! XP F0830 eZ8 CPU meets the continuing demand for faster and more code-efficient microcontrollers. It executes a superset of the original Z8 instruction set in a more optimized format that permits minimum byte count ensuring tighter code design. This also provides for faster execution as the fewest number of instruction bytes flow through the execution unit.

Key features of the eZ8 CPU include the following:

Direct register-to-register architecture allows each register to function as an accumulator, improving execution time and decreasing the required program memory.

Software stack allows much greater depth in subroutine calls and interrupts than hardware stacks.

Compatible with existing Z8 code.

Expanded internal register file allows access up to 4 KB.

New instructions improve execution efficiency for code developed using higher-level programming languages, including C.

Pipelined instruction fetch and execution.

New instructions for improved performance including BIT, BSWAP, BTJ, CPC, LDC, LDCI, LEA, MULT, and SRL.

New instructions support 12-bit linear addressing of the register file.

Up to 10 MIPS operation.

C-Compiler friendly.

Two to nine clock cycles per instruction.

## 11.2.2 The Z8 Encore! CPU Architecture

The Z8 Encore! CPU contains two major functional blocks: the fetch unit and the execution unit. Each of these units can operate independently overlapping fetch and execution of instructions. Each unit contains functional blocks that implement a pipelined design and are optimized to support the variable number of clock cycles per instruction.

**11.2.2.1 Fetch Unit**    The fetch unit controls the memory interface. Its primary function is to fetch opcodes and operands from memory. The fetch unit is pipelined and operates semi-independently from the rest of the eZ8 CPU. It also fetches interrupt vectors or reads and writes memory in the program or data memory. The fetch unit performs a partial decoding of the Op code to determine the number of bytes to fetch for the operation. The fetch unit operation sequence is as follows:

Step 1. Fetch the Op code.
Step 2. Determine the operand size (number of bytes).
Step 3. Fetch the operands.
Step 4. Present the Op Code and operands to the instruction state machine.

**11.2.2.2   Execution Unit**    The execution unit is further subdivided into the instruction state machine, program counter, CPU control registers, and arithmetic logic unit (ALU). Figure 11.33 illustrates the functional design of the execution unit.

**FIGURE 11.33**   Z8 Encore! CPU Block Diagram (Reprinted by permission of ZiLOG Corporation.)

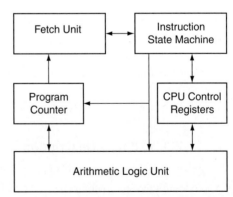

The eZ8 CPU execution unit is controlled by the instruction state machine. After the initial operation decode by the fetch unit, the instruction state machine takes control and completes the instruction. The instruction state machine performs register read and write operations and generates addresses. The instruction cycle times vary from instruction to instruction, allowing higher performance given a specific clock speed.

Minimum instruction execution time for standard CPU instructions is two clock cycles (only the BRK instruction executes in a single cycle). Because of the variation in the number of bytes required for different instructions, delay cycles can occur between instructions. Delay cycles are added any time the number of bytes in the next instruction exceeds the number of clock cycles the current instruction takes to execute. For example, if the eZ8 CPU executes a two-cycle instruction while fetching a 3-byte instruction, a delay cycle occurs because the fetch unit has only two cycles to fetch the three bytes. The execution unit is idle during a delay cycle.

The program counter contains a 16-bit counter and a 16-bit adder. It monitors the address of the current memory address and calculates the next memory address. It is also incremented automatically according to the number of bytes fetched by the fetch unit. A 16-bit adder increments and handles program counter jumps for relative addressing.

## 11.2.3 Address Space

The eZ8 CPU can access three distinct address spaces as follows:

> The register file contains addresses for the general-purpose registers and the eZ8 CPU, peripheral, and general-purpose I/O port control registers.
> The program memory contains addresses for all memory locations having executable code and/or data.
> The data memory contains addresses for all memory locations that contain data only.

***11.2.3.1 Register File***   The register file address space in the Z8 Encore! MCU is 4 KB (4096 bytes). The register file is composed of two sections: control registers and general-purpose registers. When instructions are executed, registers defined as sources are read and registers defined as destinations are written. The architecture of the eZ8 CPU allows all general-purpose registers to function as accumulators, address pointers, index registers, stack areas, or scratch pad memory in an orthogonal manner.

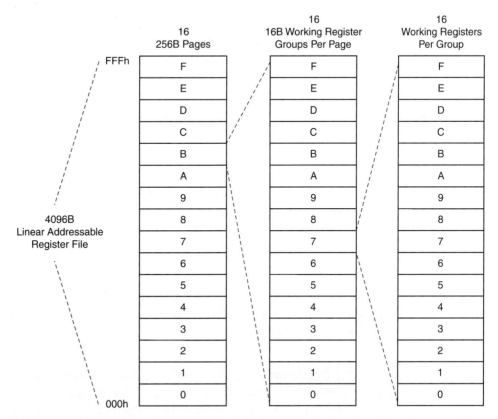

**FIGURE 11.34**   Z8 Encore! XP F0830 Register File Organization (Reprinted by permission of ZiLOG Corporation.)

The upper 256 bytes of the 4 KB register file address space is reserved for eZ8 CPU control, the on-chip peripherals, and the I/O ports. These registers are located at addresses from F00H to FFFH. Some of the addresses within the 256-byte control register section are reserved (unavailable). Reading from a reserved register file address returns an undefined value. Writing to reserved register file addresses is not recommended, and it produces unpredictable results.

The on-chip RAM always begins at address 000H in the register file address space. Reading from register file addresses outside the available RAM addresses (and not within the control register address space) returns an undefined value. Writing to these register file addresses produces no effect.

### 11.2.3.2 Program Memory

The eZ8 CPU supports up to 64 KB of program memory address space. The Z8 Encore! XP F0830 series devices contain up to 12 KB of on-chip flash memory in the program memory address space, depending on the device. Reading from program memory addresses outside the available flash memory addresses returns FFh. Writing to these unimplemented program memory addresses produces no effect. The following table (Figure 11.35) describes the program memory maps.

| Program Memory Address (Hex) Function | |
|---|---|
| **Z8F0830 and Z8F0831 Products** | |
| 0000–0001 | Flash Option Bits |
| 0002–0003 | Reset Vector |
| 0004–003D | Interrupt Vectors* |
| 003E–1FFF | Program Memory |
| **Z8F0430 and Z8F0431 Products** | |
| 0000–0001 | Flash Option Bits |
| 0002–0003 | Reset Vector |
| 0004–003D | Interrupt Vectors* |
| 003E–0FFF | Program Memory |
| **Z8F0130 and Z8F0131 Products** | |
| 0000–0001 | Flash Option Bits |
| 0002–0003 | Reset Vector |
| 0004–003D | Interrupt Vectors* |
| 003E–03FF | Program Memory |
| **Z8F0230 and Z8F0231 Products** | |
| 0000–0001 | Flash Option Bits |
| 0002–0003 | Reset Vector |
| 0004–003D | Interrupt Vectors* |
| 003E–07FF | Program Memory |

**FIGURE 11.35**    Z8 Encore! XP F0830 Program Memory Map (Reprinted by permission of ZiLOG Corporation.)

### 11.2.3.3 Data Memory

In addition to the register file and the program memory, the eZ8 CPU also supports a maximum of 64 KB (65,536 bytes) of data memory. The data memory space provides data storage only. Op code and operand fetches cannot be executed out of this space. Access is obtained by the use of the LDE and LDEI instructions. Valid addresses for the data memory are from 0000h to FFFFh.

Each Z8 Encore! device can have different amounts of data memory available. Doing this allows for smaller die sizes and reduces costs for the target application. Attempts to read unavailable data memory addresses returns FFh. Attempts to write to unavailable data memory addresses produce no effect.

## 11.2.4 Peripherals Overview

Figure 11.36 shows the basic functional block diagram of the Z8 Encore XP F0830 series. Different devices of the family will have various combinations of peripheral functional blocks depending on their application. Smaller pin count devices will have correspondingly fewer functions available. This allows the engineer to balance functionality with product cost.

The Z8 Encore! XP F0830 series can be configured with up to 17 or 25 port pins (Ports A–D) for general-purpose input/output (GPIO). The number of GPIO pins available is a function of package size, as shown in Figures 11.37(a) and (b). Each pin is individually programmable.

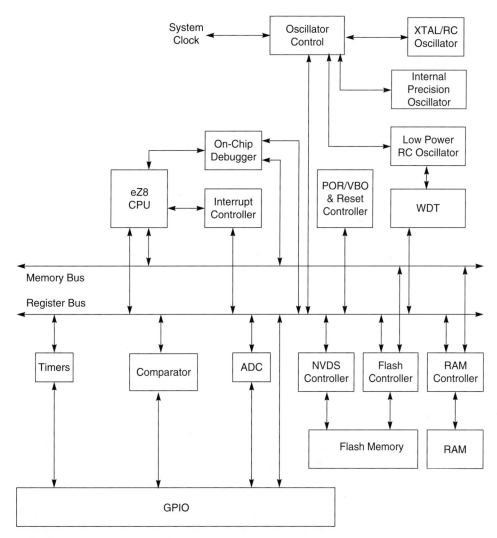

**FIGURE 11.36** Z8 Encore! XP F0830 Block Diagram (Reprinted by permission of ZiLOG Corporation.)

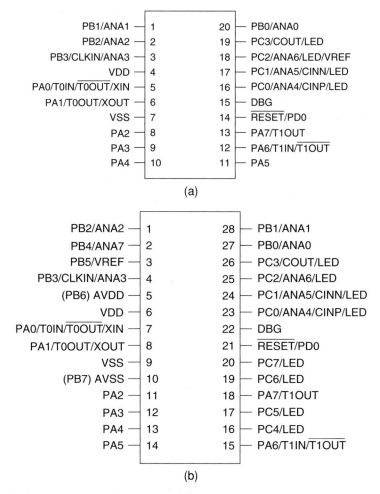

**FIGURE 11.37 (a)** Z8 Encore! 20-Pin SOIC, SSOP or PDIP Package **(b)** Z8 Encore! 28-Pin SOIC, SSOP or PDIP package (Reprinted by permission of ZiLOG Corporation.)

The flash controller programs and erases flash memory. It supports protection against accidental program and erasure. Flash can be erased in-circuit, which provides the ability to update the program after the device has been soldered to the PCB. This provides a cost-effective way to get the product to market quickly, yet allow for last-minute program changes.

The internal precision oscillator (IPO) is a trimmable clock source that requires no external components, thus keeping product costs to a minimum. The crystal oscillator circuit provides highly accurate clock frequencies with the use of an external crystal, ceramic resonator, or RC network.

The optional ADC converts an analog input signal to a 10-bit binary number. The ADC accepts inputs from eight different analog input pins in both single-ended and differential modes. The ADC block includes a transimpedance amplifier for current measurements.

The analog comparator compares the signal at an input pin with either an internal programmable voltage reference or a second input pin. The comparator output is used either to drive an output pin or to generate an interrupt.

Two enhanced 16-bit reloadable timers are used for timing/counting events or for motor control operations. These timers provide a 16-bit programmable reload counter and operate in ONE-SHOT, COMPARE, CONTINUOUS, CAPTURE and COMPARE, GATED, PWM SINGLE OUTPUT, CAPTURE, PWM DUAL OUTPUT modes, and CAPTURE RESTART modes.

Up to 17 interrupts sources are supported. These interrupts consist of eight internal peripheral interrupts and 12 general-purpose I/O pin interrupt sources. The interrupts have three levels of programmable interrupt priority.

Resetting the processor can be done using any one of the following: the RESET pin, Power-on reset, Watchdog Timer (WDT) time-out, STOP mode exit, or Voltage Brownout (VBO) warning signal. The RESET pin is bidirectional; that is, it functions as reset source as well as a reset indicator.

An integrated on-chip debugger (OCD) is supported. The OCD provides a rich set of debugging capabilities, such as reading and writing registers, programming flash memory, setting breakpoints, and executing code. The OCD uses one single-pin interface for communication with an external host.

## 11.2.5  Reset Controller and Stop Mode Recovery

The reset controller controls reset and Stop Mode Recovery operations and provides indication of low-supply voltage conditions. In typical operation, the following events cause a reset:

Power-on reset (POR)

Voltage Brownout (VBO)

Watchdog Timer (WDT) time-out (when configured by the WDT_RES flash option bit to initiate a reset)

External RESET pin assertion (when the Alternate Reset function is enabled by the GPIO register)

On-chip debugger (OCD) initiated Reset (OCDCTL[0] set to 1)

When the device is in STOP mode, a Stop Mode Recovery is initiated by either of the following: WDT time-out or GPIO port input pin transition on an enabled Stop Mode Recovery source. The VBO circuitry on the device generates the VBO reset when the supply voltage drops below a minimum safe level.

## 11.2.6  Low-Power Modes

Important power-saving features are incorporated into the design. The highest level of power reduction is provided by the STOP mode. The next lower level of power reduction is provided by the HALT mode. Further power saving is implemented by disabling individual peripheral blocks while operating in NORMAL mode.

The device enters STOP mode after executing the eZ8 CPU's STOP instruction. In STOP mode, the operating characteristics are

Primary crystal oscillator and internal precision oscillator are stopped; XIN and XOUT (if previously enabled) are disabled, and pins PA0/PA1 revert to the states programmed by the GPIO registers.

System clock is stopped.

eZ8 CPU is stopped.

Program counter (PC) stops incrementing.

Watchdog timer (WDT) internal RC oscillator continues to operate if enabled by the oscillator control register.

If enabled, the WDT logic continues to operate.

If enabled for operation in STOP mode by the associated flash option bit, the voltage brownout (VBO) protection circuit continues to operate.

All other on-chip peripherals are nonactive.

To minimize the current in STOP mode, all GPIO pins that are configured as digital inputs must be driven to $V_{DD}$ when the pull-up register bit is enabled or to one of the power rails ($V_{DD}$ or GND) when the pull-up register bit is disabled. The device can be brought out of STOP mode using Stop Mode Recovery.

The device enters HALT mode after executing the eZ8 CPU HALT instruction. In HALT mode, the operating characteristics are

Primary oscillator is enabled and continues to operate.
System clock is enabled and continues to operate.
eZ8 CPU is stopped.
Program counter (PC) stops incrementing.
WDT's internal RC oscillator continues to operate.
If enabled, the WDT continues to operate.
All other on-chip peripherals continue to operate.

The eZ8 CPU is brought out of HALT mode by any of the following operations:

Interrupt
WDT time-out (interrupt or reset)
Power-on reset (POR)
VBO reset
External RESET pin assertion

To minimize current in HALT mode, all GPIO pins that are configured as inputs must be driven to one of the supply rails ($V_{DD}$ or $V_{SS}$).

## 11.2.7  General-Purpose Input/Output

A maximum of 25 port pins (Ports A–D) for general-purpose input/output (GPIO) operations are supported. Each port contains control and data registers. The GPIO control registers determine data direction, open-drain, output drive current, programmable pull-ups, Stop Mode Recovery functionality, and alternate pin functions. Each port pin is individually programmable. In addition, the Port C pins are capable of direct LED drive at programmable drive strengths.

***11.2.7.1  GPIO Architecture***    Figure 11.38 illustrates a simplified block diagram of a GPIO port pin. In this figure, the ability to accommodate alternate functions and variable port current drive strength are not illustrated.

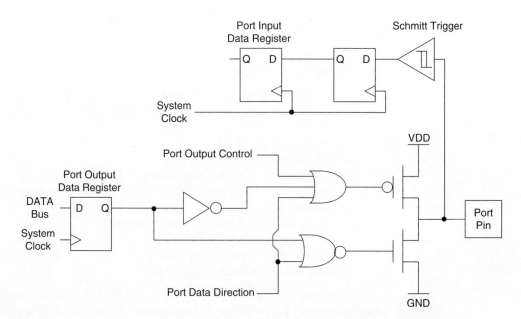

**FIGURE 11.38**    GPIO Block Diagram (Reprinted by permission of ZiLOG Corporation.)

***11.2.7.2 GPIO Alternate Functions***   Many of the GPIO port pins are used for GPIO and access to on-chip peripheral functions, such as the timers and serial communication devices. The Port A–D alternate function subregisters configure these pins for either GPIO or alternate function operation. When a pin is configured for alternate function, control of the port pin direction (input/output) is passed from the Port A–D data direction registers to the alternate function assigned to this pin.

The crystal oscillator functionality is not controlled by the GPIO block. When the crystal oscillator is enabled in the oscillator control block, the GPIO functionality of PA0 and PA1 is overridden. In that case, those pins function as input and output for the crystal oscillator.

PA0 and PA6 contain two different timer functions, a timer input and a complementary timer output. Both these functions require the same GPIO configuration. The selection between the two functions is based on the timer mode.

***11.2.7.3 GPIO Interrupts***   Many of the GPIO port pins are used as interrupt sources. Some port pins are configured to generate an interrupt request on either the rising edge or falling edge of the pin input signal. Other port pin interrupt sources generate an interrupt when any edge occurs (both rising and falling).

## 11.2.8  Interrupt Controller

The interrupt controller prioritizes the interrupt requests from the on-chip peripherals and the GPIO port pins. The interrupt controller architecture is show in Figure 11.39. Key features of the interrupt controller include the following:

Seventeen interrupt sources using 16 unique interrupt vectors:
  Twelve GPIO port pin interrupt sources.
  Five on-chip peripheral interrupt sources.
Flexible GPIO interrupts:
  Eight selectable rising and falling edge GPIO interrupts.
  Four dual-edge interrupts.

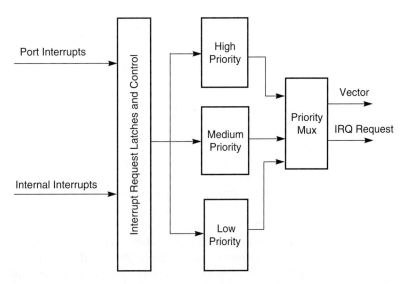

**FIGURE 11.39**   Z8 Encore! XP F0830 Interrupt Controller Block Diagram (Reprinted by permission of ZiLOG Corporation.)

Three levels of individually programmable interrupt priority.

Watchdog timer (WDT) can be configured to generate an interrupt.

Interrupt requests (IRQs) allow peripheral devices to suspend CPU operation in an orderly manner and force the CPU to start an interrupt service routine (ISR). Usually this ISR is involved with the exchange of data, status information, or control information between the CPU and the interrupting peripheral. When the service routine is completed, the CPU returns to the operation from which it was interrupted.

The eZ8 CPU supports both vectored and polled interrupt handling. For polled interrupts, the interrupt controller has no effect on operation. The interrupt vector is stored with the most significant byte (MSB) at the even program memory address and the least significant byte (LSB) at the following odd program memory address. Some port interrupts are not available on the 20-pin and 28-pin packages. The ADC interrupt is unavailable on devices not containing an ADC.

### 11.2.8.1 *Master Interrupt Enable*

The master interrupt enable bit (IRQE) in the interrupt control register globally enables and disables interrupts. Interrupts are globally enabled by any of the following actions:

Execution of an Enable Interrupt (EI) instruction.

Execution of an Return from Interrupt (IRET) instruction.

Writing a 1 to the IRQE bit in the interrupt control register.

Interrupts are globally disabled by any of the following actions:

Execution of a Disable Interrupt (DI) instruction.

eZ8 CPU acknowledgment of an interrupt service request from the interrupt controller

Writing a 0 to the IRQE bit in the interrupt control register

Reset

Execution of a Trap instruction

Illegal instruction Trap

Primary Oscillator Fail Trap

Watchdog Oscillator Fail Trap

### 11.2.8.2 *Interrupt Vectors and Priority*

The interrupt controller supports three levels of interrupt priority. Level 3 is the highest priority, Level 2 is the second highest priority, and Level 1 is the lowest priority. If all the interrupts are enabled with identical interrupt priority (for example, all interrupts as Level 2 interrupts), the interrupt priority is assigned from highest to lowest. Level 3 interrupts are always assigned higher priority than Level 2 interrupts, which in turn always are assigned higher priority than Level 1 interrupts. Within each interrupt priority level (Level 1, Level 2, or Level 3), priority is assigned as specified by that priority level. Reset, Watchdog Timer interrupt (if enabled), Primary Oscillator Fail Trap, Watchdog Oscillator Fail Trap, and Illegal Instruction Trap always have highest (Level 3) priority.

Interrupt sources assert their interrupt requests for only a single system clock period (single pulse). When the interrupt request is acknowledged by the eZ8 CPU, the corresponding bit in the Interrupt Request register is cleared until the next interrupt occurs. Writing a 0 to the corresponding bit in the Interrupt Request register clears the interrupt request.

Program code generates interrupts directly. Writing 1 to the correct bit in the Interrupt Request register triggers an interrupt (assuming that interrupt is enabled). When the interrupt request is acknowledged by the eZ8 CPU, the bit in the Interrupt Request register is automatically cleared to 0.

## 11.2.9          TIMERS

Their are up to two 16-bit reloadable timers in the Z8 Encore! XP F0830 that can be used for timing, event counting, or generation of pulse width modulated (PWM) signals. These two timers as shown in Figure 11.40, include the following functions:

16-bit reload counter

Programmable prescaler with prescale values ranging from 1 to 128

PWM output generation

Capture and compare capability

External input pin for timer input, clock gating, or capture signal

External input pin signal frequency is limited to a maximum of one-fourth the system clock frequency

Timer output pin

Timer interrupt

In PWM SINGLE/DUAL mode, the comparator output can disable the timer, when the option bit OMPB is set to 0

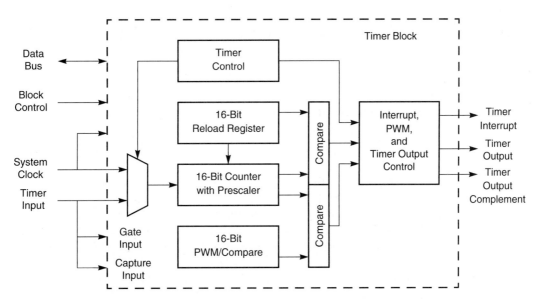

**FIGURE 11.40**   Z8 Encore! XP F0830 Timer Block Diagram (Reprinted by permission of ZiLOG Corporation.)

Both timers are 16-bit up-counters. Minimum time-out delay is set by loading the value 0001H into the Timer Reload High and Low Byte registers and setting the prescalar value to 1. Maximum time-out delay is set by loading the value 0000H into the Timer Reload High and Low Byte registers and setting the prescalar value to 128. If the timer reaches FFFFH, the timer rolls over to 0000H and continues counting.

***11.2.9.1 ONE-SHOT Mode***   In ONE-SHOT mode, the timer counts up to the 16-bit Reload value stored in the Timer Reload High and Low Byte registers. The timer input is the system clock. On reaching the Reload value, the timer generates an interrupt, and the count value in the Timer High and Low Byte registers is reset to 0001H. The timer is automatically disabled and stops counting.

Also, if the Timer Output Alternate function is enabled, the timer output pin changes state for one system clock cycle (from Low to High or from High to Low) on timer Reload.

$$\text{ONE-SHOT Mode Time-Out Period (s)} = \frac{(\text{Reload Value} - \text{Start Value}) \times \text{Prescale}}{\text{System Clock Frequency (Hz)}}$$

### 11.2.9.2 CONTINUOUS Mode

In CONTINUOUS mode, the timer counts up to the 16-bit Reload value stored in the Timer Reload High and Low Byte registers. The timer input is the system clock. When the timer count reaches the Reload value, it generates an interrupt, and the count value in the Timer High and Low Byte registers is reset to 0001H and counting resumes. Also, if the Timer Output alternate function is enabled, the Timer Output pin changes state (from Low to High or from High to Low) at timer Reload.

$$\text{CONTINUOUS Mode Time-Out Period (s)} = \frac{\text{Reload Value} - \text{Prescale}}{\text{System Clock Frequency (Hz)}}$$

### 11.2.9.3 COMPARATOR COUNTER Mode

In COMPARATOR COUNTER mode, the timer counts input transitions from the analog comparator output. The TPOL bit in the Timer Control register selects whether the count occurs on the rising edge or the falling edge of the comparator output signal. In COMPARATOR COUNTER mode, the prescaler is disabled. The frequency of the comparator output signal must not exceed one-fourth the system clock frequency. After reaching the Reload value stored in the Timer Reload High and Low Byte registers, the timer generates an interrupt, the count value in the Timer High and Low Byte registers is reset to 0001H, and counting resumes. Also, if the Timer Output alternate function is enabled, the Timer Output pin changes state (from Low to High or from High to Low) at timer Reload.

In COMPARATOR COUNTER mode, the number of comparator output transitions since the timer start is given by the following equation:

$$\text{Comparator Output Transitions} = \text{Current Count Value} - \text{Start Value}$$

### 11.2.9.4 PWM SINGLE OUTPUT Mode

In PWM SINGLE OUTPUT mode, the timer outputs a PWM output signal through a GPIO port pin. The timer input is the system clock. The timer first counts up to the 16-bit PWM match value stored in the Timer PWM High Byte and Low Byte registers. When the timer count value matches the PWM value, the Timer Output toggles. The timer continues counting until it reaches the Reload value stored in the Timer Reload High and Low Byte registers. On reaching the Reload value, the timer generates an interrupt, the count value in the Timer High and Low Byte registers is reset to 0001H, and counting resumes.

If the TPOL bit in the Timer Control register is set to 1, the Timer Output signal begins as a High (1) and transitions to a Low (0) when the timer value matches the PWM value. The Timer Output signal returns to a High (1) after the timer reaches the Reload value and is reset to 0001H. If the TPOL bit in the Timer Control register is set to 0, the Timer Output signal begins as a Low (0) and transitions to a High (1) when the timer value matches the PWM value. The Timer Output signal returns to a Low (0) after the timer reaches the Reload value and is reset to 0001H.

$$\text{PWM Period (s)} = \frac{\text{Reload Value} - \text{Prescale}}{\text{System Clock Frequency (Hz)}}$$

### 11.2.9.5 PWM DUAL OUTPUT Mode

In PWM DUAL OUTPUT mode, the timer outputs a PWM output signal pair (basic PWM signal and its complement) through two GPIO port pins. The timer input is the system clock. The timer first counts up to the 16-bit PWM match value stored in the Timer PWM High and Low Byte registers. When the timer count value matches the PWM value, the timer output toggles. The timer continues counting until it reaches the Reload value stored in the Timer Reload High and Low Byte registers. On reaching the Reload value, the timer generates an interrupt, the count value in the Timer High and Low Byte registers is reset to 0001H, and counting resumes.

The timer also generates a second PWM output signal, Timer Output Complement. The Timer Output Complement is the complement of the Timer Output PWM signal. A programmable deadband delay can be configured to time delay (0 to 128 system clock cycles). PWM output transitions on these two pins from a Low to a High (inactive to active). This ensures a time gap between the deassertion of one PWM output to the assertion of its complement.

The PWM period is represented by the following equation:

$$\text{PWM Period (s)} = \frac{\text{Reload Value} - \text{Prescale}}{\text{System Clock Frequency (Hz)}}$$

### 11.2.9.6 CAPTURE Mode

In CAPTURE mode, the current timer count value is recorded when the appropriate external Timer Input transition occurs. The Capture count value is written to the Timer PWM High Byte and Timer PWM Low Byte registers. The timer input is the system clock. The TPOL bit in the Timer Control register determines if the Capture occurs on a rising edge or a falling edge of the Timer Input signal. When the Capture event occurs, an interrupt is generated and the timer continues counting. The INPCAP bit in TxCTL1 register is set to indicate the timer interrupt is resulted by an input capture event.

The timer continues counting up to the 16-bit Reload value stored in the Timer Reload High and Low Byte registers. On reaching the Reload value, the timer generates an interrupt and continues counting. The INPCAP bit in TxCTL1 register clears, indicating that the timer interrupt is not because of an input capture event.

### 11.2.9.7 CAPTURE RESTART Mode

In CAPTURE RESTART mode, the current timer count value is recorded when the acceptable external Timer Input transition occurs. The Capture count value is written to the Timer PWM High and Low Byte registers. The timer input is the system clock. The TPOL bit in the Timer Control register determines if the Capture occurs on a rising edge or a falling edge of the Timer Input signal. When the Capture event occurs, an interrupt is generated, and the count value in the Timer High and Low Byte registers is reset to 0001H and counting resumes. The INPCAP bit in TxCTL1 register is set to indicate the timer interrupt results from an input capture event.

If no Capture event occurs, the timer counts up to the 16-bit Compare value stored in the Timer Reload High Byte and Timer Reload Low Byte registers. On reaching the Reload value, the timer generates an interrupt, the count value in the Timer High Byte and Timer Low Byte registers is reset to 0001H, and counting resumes. The INPCAP bit in TxCTL1 register is cleared to indicate the timer interrupt is not caused by an input capture event.

In CAPTURE mode, the elapsed time from timer start to Capture event is calculated using the following equation:

$$\text{Capture Elapsed Time (s)} = \frac{(\text{Capture Value} - \text{Start Value}) \times \text{Prescale}}{\text{System Clock Frequency (Hz)}}$$

### 11.2.9.8 COMPARE Mode

In COMPARE mode, the timer counts up to the 16-bit maximum Compare value stored in the Timer Reload High and Low Byte registers. The timer input is the system clock. On reaching the Compare value, the timer generates an interrupt and continues counting (the timer value is not reset to 0001H). Also, if the Timer Output alternate function is enabled, the Timer Output pin changes state (from Low to High or from High to Low) on Compare. If the Timer reaches FFFFH, the timer takes value 0000H and continues counting.

In COMPARE mode, the system clock always provides the timer input. The Compare time is calculated using the following equation:

$$\text{Compare Mode Time (s)} = \frac{(\text{Compare Value} - \text{Start Value}) \times \text{Prescale}}{\text{System Clock Frequency (Hz)}}$$

**11.2.9.9 *GATED Mode***   In GATED mode, the timer counts only when the Timer Input signal is in its active state (asserted), as determined by the TPOL bit in the Timer Control register. When the Timer Input signal is asserted, counting begins. A timer interrupt is generated when the Timer Input signal is deasserted or a timer reload occurs. To determine if a Timer Input signal deassertion generated the interrupt, read the associated GPIO input value and compare to the value stored in the TPOL bit. The timer counts up to the 16-bit Reload value stored in the Timer Reload High and Low Byte registers. The timer input is the system clock. When reaching the Reload value, the timer generates an interrupt, the count value in the Timer High and Low Byte registers is reset to 0001H and counting resumes (assuming the Timer Input signal remains asserted).

Also, if the Timer Output alternate function is enabled, the Timer Output pin changes state (from Low to High or from High to Low) at timer reset.

**11.2.9.10 *CAPTURE/COMPARE Mode***   In CAPTURE/COMPARE mode, the timer begins counting on the first external Timer Input transition. The acceptable transition (rising edge or falling edge) is set by the TPOL bit in the Timer Control register. The timer input is the system clock. Every subsequent acceptable transition (after the first) of the Timer Input signal captures the current count value. The Capture value is written to the Timer PWM High and Low Byte registers. When the Capture event occurs, an interrupt is generated, the count value in the Timer High and Low Byte registers is reset to 0001H, and counting resumes. The INPCAP bit in TxCTL1 register is set to indicate the timer interrupt is caused by an input capture event.

If no Capture event occurs, the timer counts up to the 16-bit Compare value stored in the Timer Reload High and Low Byte registers. On reaching the Compare value, the timer generates an interrupt, the count value in the Timer High and Low Byte registers is reset to 0001H, and counting resumes. The INPCAP bit in TxCTL1 register is cleared to indicate the timer interrupt is not because of an input capture event.

In CAPTURE/COMPARE mode, the elapsed time from timer start to Capture event is calculated using the following equation:

$$\text{Capture Elapsed Time (s)} = \frac{(\text{Capture Value} - \text{Start Value}) \times \text{Prescale}}{\text{System Clock Frequency (Hz)}}$$

## 11.2.10  Watchdog Timer

The Watchdog Timer (WDT) protects against corrupt or unreliable software, power faults, and other system-level problems, which places hardware into unsuitable operating states. The program counter can be reset and the program regains normal execution. The WDT includes the following features:

On-chip RC oscillator
A selectable time-out response: reset or interrupt
24-bit programmable time-out value

The WDT is a retriggerable one-shot timer that resets or interrupts when the WDT reaches its terminal count. The WDT uses a dedicated on-chip RC oscillator as its clock source. The WDT operates in only two modes: ON and OFF.

Once enabled, it always counts and must be refreshed to prevent a time-out. Perform an enable by executing the WDT instruction or by setting the WDT_AO Flash Option Bit. The WDT_AO bit forces the WDT to operate immediately on reset, even if a WDT instruction has not been executed. The WDT is a 24-bit reloadable down counter that uses three 8-bit registers in the eZ8 CPU register space to set the reload value. The nominal WDT time-out period is described by the following equation:

$$\text{WDT Time-out Period (ms)} = \frac{\text{WDT Relaod Value}}{10}$$

where the WDT reload value is the decimal value of the 24-bit value given by {WDTU[7:0], WDTH[7:0], WDTL[7:0]}, and the typical Watchdog Timer RC oscillator frequency is 10 KHz. The WDT cannot be refreshed after it reaches the value 000002H.

## 11.2.11 Analog-to-Digital Converter

An eight-channel successive approximation register (SAR) analog-to-digital converter (ADC) is part of the peripheral function blocks. The ADC converts an analog input signal to a 10-bit binary number. The features of the SAR ADC include

Eight analog input sources multiplexed with general-purpose I/O ports
Fast conversion time, less than 11.9 μs
Programmable timing controls
Interrupt on conversion complete
Internal voltage reference generator
Ability to select external reference voltage
When configuring ADC using external Vref, PB5 is used as Vref in 28-pin package, otherwise PC2(ANA6) is used as Vref in 20-pin package

The ADC architecture, as shown in Figure 11.41, consists of an 8-input multiplexer, sample and hold amplifier, and 10-bit SAR ADC. The ADC digitizes the signal on a selected

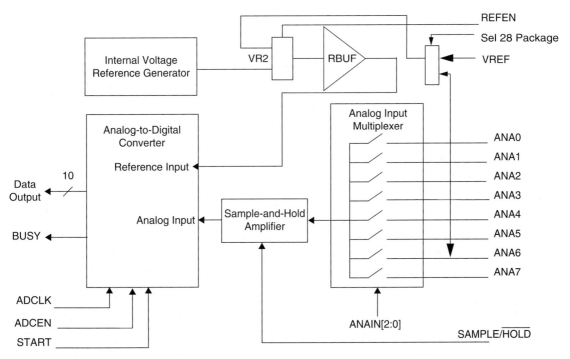

**FIGURE 11.41**   Z8 Encore! XP F0830 Analog-to-Digital Converter Block Diagram (Reprinted by permission of ZiLOG Corporation.)

channel and stores the digitized data in the ADC data registers. In an environment with high electrical noise, an external RC filter must be added at the input pins to reduce high-frequency noise.

**11.2.11.1 *ADC Operation***   The ADC converts the analog input, ANAX, to a 10-bit digital representation. The equation for calculating the digital value is represented by

$$ADCOutput = 1024 \times (ANA_X \div V_{REF})$$

Assuming zero gain and offset errors, any voltage outside the ADC input limits of $AV_{SS}$ and $V_{REF}$ returns all 0s or 1s, respectively. A new conversion can be initiated by a software write to the ADC control register's start bit. Initiating a new conversion stops any conversion currently in progress and begins a new conversion. To avoid disrupting a conversion already in progress, the START bit can be read to determine ADC operation status (busy or available).

**11.2.11.2 *ADC Timing***   Each ADC measurement consists of three steps:

Step 1. Input sampling (programmable, minimum of 1.0 µs)
Step 2. Sample-and-hold amplifier settling (programmable, minimum of 0.5 µs)
Step 3. Conversion is 13 ADCLK cycles. Figures 11.42(a) and (b) illustrate the timing of an ADC conversion.

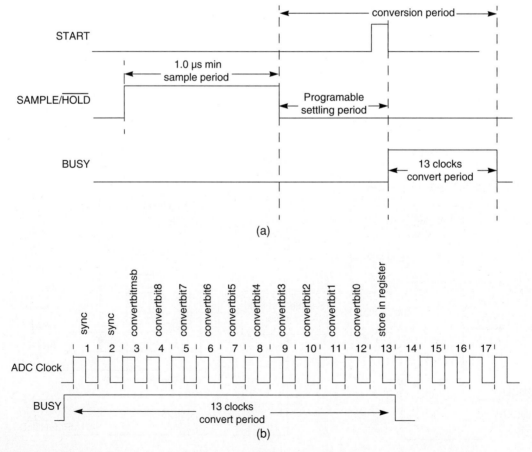

**FIGURE 11.42 (a)** ADC Timing Diagram; **(b)** ADC Convert Timing (Reprinted by permission of ZiLOG Corporation.)

## 11.2.12 Comparator

The Z8 Encore! XP F0830 Series devices feature a general-purpose comparator that compares two analog input signals. A GPIO (CINP) pin provides the positive comparator input. The negative input (CINN) can be taken from either an external GPIO pin or from an internal reference. The output is available as an interrupt source or can be routed to an external pin using the GPIO multiplexor.

The comparator includes the following features:

Positive input is connected to a GPIO pin.
Negative input can be connected to either a GPIO pin or a programmable internal reference.
Output can be either an interrupt source or an output to an external pin.

One of the comparator inputs is connected to an internal reference, which is a user-selectable reference, and it is programmable with 200 mV resolution. The comparator is powered down to save on supply current. Because of the propagation delay of the comparator, it is not recommended to enable the comparator without disabling the interrupts first and waiting for the comparator output to settle. This delay prevents spurious interrupts after comparator enabling.

## 11.2.13 Flash Memory

The products in the Z8 Encore! XP F0830 Series feature either 1 KB, 2 KB, 4 KB, 8 KB (with NVDS), or 12 KB with (no NVDS) of nonvolatile flash memory with read/write/erase capability. The flash memory can be programmed and erased in-circuit by either user code or through the on-chip debugger.

The flash memory array is arranged in pages with 512 bytes per page. The 512-byte page is the minimum flash block size that can be erased. Each page is divided into eight rows of 64 bytes. For program/data protection, the flash memory is also divided into sectors. Each sector maps to one page (for 1 KB, 2 KB, and 4 KB devices), two pages (8 KB device) or three pages (12 KB device).

Figure 11.43 illustrates the FLASH memory arrangement for 8KB devices.

Programmable flash option bits allow the user hard configuration of certain aspects of device operations. The feature configuration data is stored in the flash program memory and read during reset. The features available for control through the flash option bits are

Watchdog Timer (WDT) time-out response selection–interrupt or system reset.
Watch Dog Timer enabled at reset.
The ability to prevent unwanted read access to user code in program memory.
The ability to prevent accidental programming and erasure of all or a portion of the user code in Program Memory.
Voltage Brownout (VBO) configuration always enabled or disabled during STOP mode to reduce STOP mode power consumption.
Oscillator mode selection-for high, medium, and low power crystal oscillators, or external RC oscillator.
Factory trimming information for the Internal Precision Oscillator and VBO voltage.

## 11.2.14 Nonvolatile Data Storage

A nonvolatile data storage (NVDS) function of up to 64 bytes (except in flash 12-KB mode) is part of the device. This memory can perform over 100,000 write cycles. It can be used for constants storage through power cycling.

The NVDS is implemented by special-purpose ZiLOG software stored in areas of program memory not accessible to the designer. These special-purpose routines use the flash memory to store the data. The routines incorporate a dynamic addressing scheme to maximize the write/erase endurance of the flash.

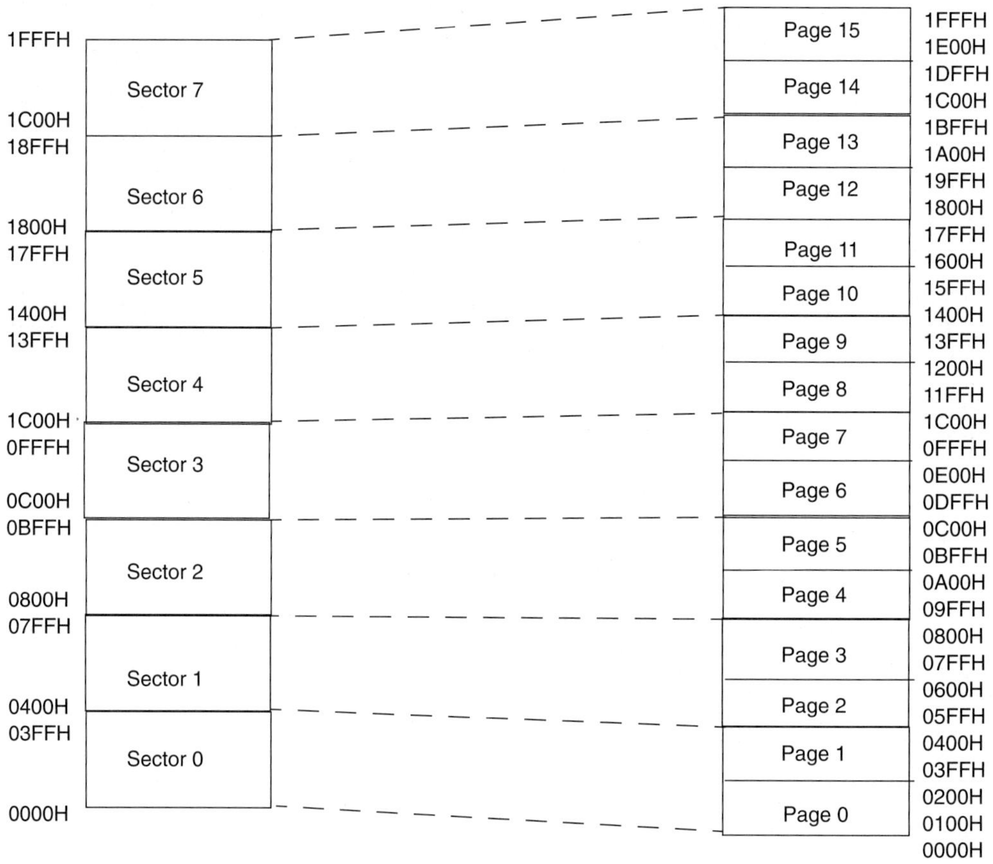

**FIGURE 11.43**   Z8 Encore! XP F0830 8K Flash with NVDS (Reprinted by permission of ZiLOG Corporation.)

Two subroutines are required to access the NVDS, a write routine and a read routine. Both of these routines are accessed with a CALL instruction to a predefined address outside the program memory space. Both the NVDS address and data are single-byte values. In order NOT to disturb the user code, these routines save the working register set before using it, so 16 bytes of stack space are needed to preserve the site. After finishing the call to these routines, the working register set of the user code is recovered.

During both read and write accesses to the NVDS, interrupt service is NOT disabled. Any interrupts that occur during the NVDS execution must not disturb the working register and existing stack contents, or else the array becomes corrupted. Disabling interrupts before executing NVDS operations is recommended. Use of the NVDS requires 16 bytes of available stack space. The contents of the working register set are saved before calling NVDS read or write routines.

## 11.2.15  On-Chip Debugger

A key feature is the integrated on-chip debugger (OCD), as shown in Figure 11.44. It provides advanced debugging features, including

- Reading and writing of the register file
- Reading and writing of program memory and data memory
- Setting of breakpoints and watchpoints
- Executing eZ8 CPU instructions

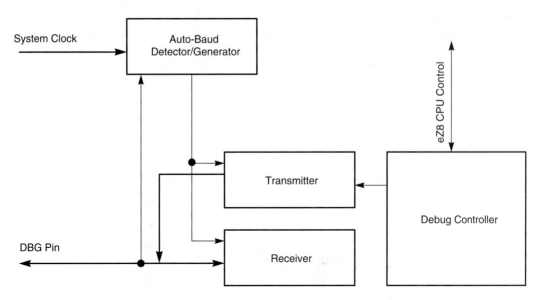

**FIGURE 11.44**   Z8 Encore! XP F0830 On-Chip Debugger Block Diagram (Reprinted by permission of ZiLOG Corporation.)

The OCD consists of four primary functional blocks: transmitter, receiver, auto-baud detector/generator, and debug controller.

The OCD uses the DBG pin for communication with an external host. This one-pin interface is a bidirectional open-drain interface that transmits and receives data. Data transmission is half-duplex, in that transmit and receive cannot occur simultaneously. The serial data on the DBG pin is sent using the standard asynchronous data format defined in RS-232. This pin creates an interface to the serial port of a host PC using minimal external hardware.

## 11.2.16  Oscillator Control

Support is provided for five possible clocking schemes, each user-selectable:

> On-chip precision trimmed RC oscillator
> On-chip oscillator using off-chip crystal or resonator
> On-chip oscillator using external RC network
> External clock drive
> On-chip low precision Watchdog Timer oscillator

In addition, clock failure detection and recovery circuitry is provided, allowing continued operation despite a failure of the primary oscillator. The table in Figure 11.45 gives the oscillator configurations and selection.

***11.2.16.1  Crystal Oscillator***   A multifunctional crystal oscillator is provided on-chip for use with external crystals with 32-KHz to 20-MHz frequencies. In addition, the oscillator supports external RC networks with oscillation frequencies up to 4 MHz or ceramic resonators with frequencies up to 8 MHz. The on-chip crystal oscillator can be used to generate the primary system clock for the internal eZ8 CPU and the majority of the on-chip peripherals.

Alternatively, the $X_{IN}$ input pin can also accept a CMOS-level clock input signal (32 KHz–20 MHz). If an external clock generator is used, the $X_{OUT}$ pin must be left unconnected. The on-chip crystal oscillator also contains a clock filter function. But by default, this clock filter is disabled and no divide to the input clock, namely the frequency of the signal on the $X_{IN}$ input pin, determines the frequency of the system clock in default settings.

| Clock Source | Characteristics | Required Setup |
|---|---|---|
| Internal Precision RC Oacllator | • 32.8 KHz or 5.53 MHz<br>• ± 4% accuracy when trimmed<br>• No external componants required | • Unlock and write Oscillator Control Register (OSCCTL) to enable and select oscillator at either 5.53 MHz or 32.8 KHz. |
| Extemal Crystal Resonator | • 32 KHz to 20 MHz<br>• Very high accuracy (dependent on crystal or resonator used)<br>• Requires external components | • Configure Flash option bits for correct external oscillator mode.<br>• Unlock and write OSCCTL to enable crystal oscillator, wait for it to stabilize and select as system clock (if the XTLDIS option bit has been deasserted, no waiting is required). |
| External RC Oscillator | • 32 KHz to 4 MHz<br>• Acuracy dependent on external componants | • Configure Flash option bits for correct external osillator mode.<br>• Unlock and write OSCCTL to enable crystal oscillator and select as system clock. |
| External Clock Drive | • 0 to 20 MHz<br>• Accuracy dependent on external clock source | • Write GPIO registers to configure PB3 pin for external clock function.<br>• Unlock and write OSCCTL to select external system clock.<br>• Apply external clock signal to GPIO. |
| Internal Watchdog Timer Oscillator | • 10 KHz nominal<br>• ± 40% accuracy; no external components required<br>• Low power consumption | • Enable WDT if not enabled and wait until WDT Oscillator is operating.<br>• Unlock and write Oscillator Control register (OSCCTL) to enable and select oscillator. |

**FIGURE 11.45**  Z8 Encore! XP F0830 Oscillator Configuration and Selection (Reprinted by permission of ZiLOG Corporation.)

The OSCILLATOR mode is selected using user-programmable flash option bits. The four OSCILLATOR modes supported are

Minimum power for use with very low frequency crystals (32 KHz–1 MHz)
Medium power for use with medium frequency crystals or ceramic resonators (0.5 MHz to 8 MHz)
Maximum power for use with high-frequency crystals (8 MHz to 20 MHz)
On-chip oscillator configured for use with external RC networks (< 4 MHz)

***11.2.16.2 Internal Precision Oscillator***    The internal precision oscillator (IPO) is designed for use without external components to reduce part count and reduce design costs. The oscillator for a nonstandard frequency can either be manually trimmed or use the automatic factory trimmed version to achieve a 5.53 MHz frequency with ± 4% accuracy and 45%~55% duty cycle over the operating temperature and supply voltage of the device. Maximum startup time of IPO is 25 µs.
IPO features include

On-chip RC oscillator that does not require external components
Output frequency of either 5.53 MHz or 32.8 KHz (contains both a FAST and a SLOW mode)

Trimming possible through flash option bits with user override

Elimination of crystals or ceramic resonators in applications where high timing accuracy is not required

The internal oscillator is an RC relaxation oscillator with minimized sensitivity to power supply variation. By using ratio tracking thresholds, the effect of power supply voltage is cancelled out. The dominant source of oscillator error is the absolute variance of chip level fabricated components, such as capacitors. An 8-bit trimming register, incorporated into the design, compensates for absolute variation of oscillator frequency. Once trimmed, the oscillator frequency is stable and does not require subsequent calibration. Trimming is performed during manufacturing and is not necessary for the user to repeat unless a frequency other than 5.53 MHz (FAST mode) or 32.8 KHz (SLOW mode) is required.

## 11.2.17 eZ8 CPU Instructions and Programming

This compact instruction set provides sufficient flexibility to support programming across a broad range of applications. It is evolved from the original Z8 and executes a superset of the instructions.

The eZ8 CPU executes all Z8 assembly language instructions except for the Watchdog Timer Enable During HALT Mode instruction (WDh, at Op Code 4Fh). Users with existing Z8 assembly code can easily compile their code to use the eZ8 CPU. The assembler for the eZ8 CPU is available for download at www.zilog.com.

When compared to the Z8 CPU instruction set, the eZ8 CPU features many new instructions that increase processor efficiency and allow access to the expanded 4 KB register file. A total of 83 instructions can be grouped functionally into the following eight groups:

Arithmetic
Bit manipulation
Block transfer
CPU control
Load
Logical
Program control
Rotate and shift

Six addressing modes are supported to provide greater flexibility and efficiency in program design. This can result in more compact code and faster execution speed. The six modes are

Register (R)
Indirect register (IR)
Indexed (X)
Direct (DA)
Relative (RA)
Immediate data (IM)
Extended register (ER)

***11.2.17.1 Program Stack*** A stack programming architecture is implemented. Stack operations occur in the general-purpose registers of the register file. The eZ8 CPU allows the user to relocate the stack within the register file. The stack can be located at addresses from 000h to EFFh. The 12-bit stack pointer (SP) value is provided by {SPH[3:0], SPL[7:0]}. The SP has a 12-bit increment/decrement capability for stack operations, allowing the SP to operate over more than one page.

Stack operations occur in the general-purpose registers of the register file. The register pair FFEh and FFFh form the 16-bit SP used for all stack operations. The SP holds the current stack address. The SP must be always be set to point to a section of the register file that does not cause user program data to be overwritten. Even for linear program code that may not employ the stack for Call and/or Interrupt routines, the SP must be set to prepare for possible illegal instruction traps.

The stack address decrements prior to a PUSH operation and increments after a POP operation. The stack address always points to the data stored at the top of the stack. The stack is a return stack for interrupts and CALL and TRAP instructions. It can also be employed as a data stack. During a CALL instruction, the contents of the program counter are saved on the stack. The program counter is restored during execution of a Return (RET).

Interrupts and traps (either the TRAP instruction or an illegal instruction trap) save the contents of the program counter and the flags register on the stack. The Interrupt Return (IRET) instruction restores them. Figure 11.46 illustrates the contents of the stack and the location of the SP following Call, Interrupt and Trap operations

**FIGURE 11.46**   Stack Pointer Registers (Reprinted by permission of ZiLOG Corporation.)

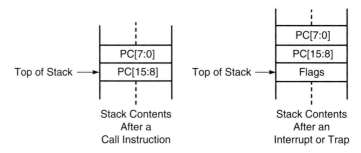

An overflow or underflow can occur when the stack address is incremented or decremented beyond the available address space. The programmer must prevent this occurrence, or unpredictable operation will result.

## QUESTIONS AND PROBLEMS

1. Explain the advantages of GPMs over SoCs.
2. List the primary functional blocks of the PIC18F4520.
3. Describe the instruction pipeline in the PIC18F4520.
4. What does the PIC18F4520 configuration register do?
5. Describe the difference in the three power modes of the PIC18F4520.
6. What are the key advantages of using flash program memory in the PIC18F4520?
7. Why use two interrupt vectors in the PIC18F4520?
8. What is meant by a Tri-State I/O port?
9. What are three advantages to using timers in the PIC18F4520?
10. Diagram an application using the Capture/Compare function.
11. How might you use pulse-width modulation in an application?
12. Why have multiple serial I/O modes in the PIC18F4520?
13. What is the value of having an on-chip A/D converter?
14. List the PIC18F4520's five major types of instructions.
15. List the key functional blocks of the Zilog Z8 Encore! CPU.
16. What are key architectural differences between the PIC18F4520 and the Zilog Z8 Encore!?
17. When do delay cycles occur in the eZ8 CPU?

18. List the major addressing modes of the eZ8 CPU.
19. What is the advantage of the three level interrupt scheme of the eZ8 CPU?
20. Why have three power saving modes for the Z8 Encore! XP F0830?
21. Describe an application of using Z8 Encore! XP F0830's timer in CONTINUOUS mode.
22. What is the most important use of the WDT in the Z8 Encore! XP F0830?
23. What are the five clocking schemes of the Z8 Encore! XP F0830?

## SOURCES

Customer presentation. 2006. Microchip Technology, Marketing Department.

Fischer, Richard L. 2003. *Using the PICmicro MSSP module for master I²C communications, application note*. Microchip Technology.

Garbutt, Mike. 2003. *Asynchronous communications with the PICmicro USART, application note*. Microchip Technology.

Palmer, Mark. 1997. *Using the capture module, application note*. Microchip Technology.

Palmer, Mark. 1997. Using the CCP module(s), application note. Microchip Technology.

Palmer, Mark. 1997. *Using the PWM, application note*. Microchip Technology.

Palmer, Mark. 1997. *Using Timer1 in asynchronous clock mode, application note*. Microchip Technology.

PIC18F4520 data sheet, 44-Pin Enhanced Flash Microcontrollers with 10-Bit A/D and nanoWatt Technology. 2004. Microchip Technology.

Z8 Encore! Microcontrollers eZ8 CPU Core. *User Manual UM012815-0606*. June 2006. ZiLOG, Inc.

Z8 Encore! XP F0830 Series with Extended Peripherals. *Product brief, PB016107-0706*. July 2006. ZiLOG Inc.

Z8 Encore! XP F0830 Series with Extended Peripherals. *Product specification, PS024707-0606*. June 2006. ZiLOG Inc.

# CHAPTER 12

# 16-Bit Microcontroller

**OBJECTIVE: UNDERSTANDING A GENERAL-PURPOSE 16-BIT MICROCONTROLLER**

The reader will learn about:

1. Freescale S12XD family of embedded microcontrollers.
2. Texas Instruments MSP430™ series of mixed-signal microcontrollers.

## 12.0 16-BIT PROCESSOR OVERVIEW

The 16-bit microcontrollers offer a step up in performance over their 8-bit counterparts. They can be found in a variety of applications where increased processing power is required, yet fall in a lower price point than full 32-bit microcontrollers. As with their 8-bit counterparts, they incorporate a broad range of flexible peripheral modules.

The Freescale S12XD family of microcontrollers represents an evolution from the previous generation of devices to address the challenge that system engineers face in practical designs. The S12X family advances the concepts of low cost, low power, low EMC (electromagnetic compatibility), high code-size efficiency, and high flexibility, which has particular usage in automotive applications. Based on an enhanced HCS12 core, the S12XD family is designed to deliver two to five times the performance of a 25-MHz HCS12 while still retaining a high degree of pin and code compatibility with the HCS12.

Texas Instruments MSP430 device-based solutions efficiently use a 16-bit RISC (reduced instruction set computing) CPU to achieve high performance. The family is designed for ultra-low-power applications via five different operating modes. They take into account three specific needs: ultra-low-power, speed and data throughput, and minimization of individual peripheral current consumption. Through an appropriate interrupt-driven design, chip-powered uptime can be kept to a minimum, saving power, while still achieving high system-level performance.

## 12.1 FREESCALE S12XD PROCESSOR OVERVIEW

At the level of the core blocks, the Freescale MC9S12XD families of microcontrollers share the same architecture. Some devices incorporate more functions and options than others. Differentiation also depends on package size with its corresponding pin count. Figure 12.1 shows the basic functional block diagram of the Freescale MC9S12XD family.

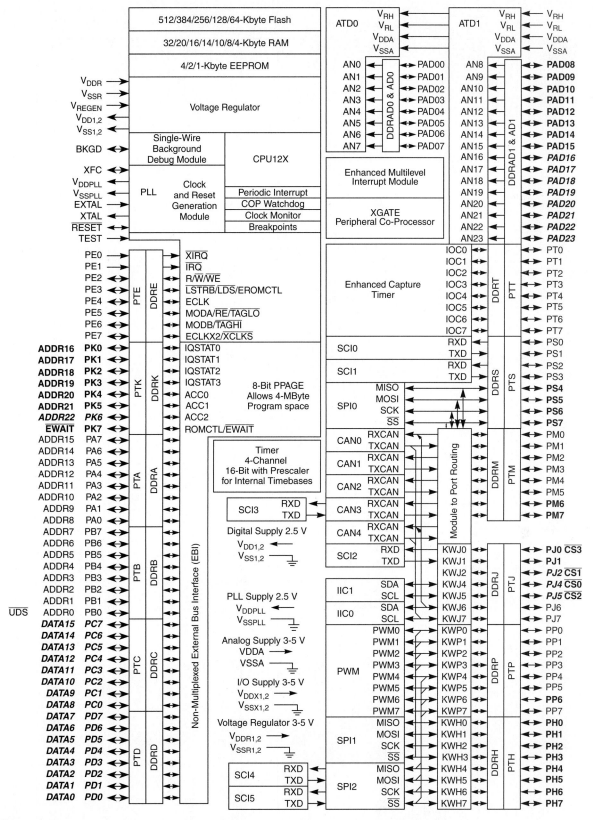

**FIGURE 12.1** S12XD Family Functional Block Diagram (Copyright of Freescale Semiconductor, Inc. 2007, used by permission.)

251

| Attribute | S12X Cores | | S12XS Cores | S12 Core |
|---|---|---|---|---|
| | S12XD -- New! (XB and XD Families) | S12XE -- New! (XE Family) | S12XS -- New! (XS Family) | S12 -- Legacy (C, D, H, Q, and R Families) |
| Bus Speed | 40 MHz | 50 MHz | 40 MHz | 25 MHz |
| CPU | 16-bit CPU12XV1 | 16-bit CPU12XV2 | 16-bit CPU12XV2 | 16-bit CPU12 |
| Debug | - BDM (enhanced to support global paging accesses) - DBG Debugger monitor CPU - XGATE busses and four comparators | - BDM (enhanced to support global paging accesses) - DBG Debugger monitor CPU - XGATE busses and four comparators | - BDM (enhanced to support global paging accesses) - DBG Debugger monitor CPU | - BDM (Single-wire background debug) |
| System Protection Features | - Low-voltage detect/interrupt | - Memory Protection Unit (MPU) to protect undesired accesses - Low-voltage detect/interrupt on all devices in family - Error Correction Code | - Low voltage detect/interrupt on all devices in family - Error Correction Code | - Low voltage detect/interrupt |
| Oscillator Choices | - Loop controlled or full swing Pierce (XOSC) circuitry enhanced to dynamically control gain of the output amplitude for low harmonic distortion - Low power and good noise immunity - Eliminates bias resistor | - Loop controlled or full swing Pierce (XOSC) circuitry enhanced to dynamically control gain of the output amplitude for low harmonic distortion - Low power and good noise immunity - Eliminates bias resistor | - Loop controlled or full swing Pierce (XOSC) circuitry enhanced to dynamically control gain of the output amplitude for low harmonic distortion - Low power and good noise immunity - Eliminates bias resistor | - Colpitts of Full swing Pierce |
| Clock | - Phased Lock Loop (PLL) | - Phase Lock Loop (PLL) enhanced circuitry for elimination of external components | - Phase Lock Loop (PLL) enhanced circuitry for elimination of external components | - Phase Lock Loop (PLL) |
| Analog to Digital Converter (ADC) | - 8/10 bit resolution, 7μs conversion time | - 8/10/12 bit resolution, conversion time as low as 2.12μs | - 8/10/12 bit resolution, conversion time as low as 2.12μs | - 8/10 bit resolution, 7μs conversion time |
| EEPROM | - Small sector Flash to emulate EEPROM | - Emulated EEPROM with Data Flash and RAM buffer | - Small sector Flash to emulate EEPROM | - Small sector Flash to emulate EEPROM |
| Timer | Enhanced Capture Time (ECT) - Timer (TIM) - Periodic Interrupt Timer (PIT) | Enhanced Capture Time (ECT) - Timer (TIM) - Periodic Interrupt Timer (PIT) | Enhanced Capture Time (ECT) - Timer (TIM) - Periodic Interrupt Timer (PIT) | - Enhanced Capture Timer (ECT) - Timer (TIM) |
| XGATE | - XGATE programmable high performance I/O co-processor module (up to 80 MIPS RISC performance) | - XGATE programmable high performance I/O co-processor module (up to 80 MIPS RISC performance) | Not applicable | Not applicable |
| Interrupt Nesting | - Enhanced Interrupt Module with eight levels | - Enhanced Interrupt Module with eight levels | - Enhanced Interrupt Module with eight levels | - Standard Interrupt module |

**FIGURE 12.2** S12XD Family Comparison Chart (Copyright of Freescale Semiconductor, Inc. 2007, used by permission.)

The S12XD family incorporates the performance-boosting XGATE coprocessor module. Using enhanced direct memory access (DMA) functionality, this parallel processing RISC-based module offloads the CPU by providing high-speed data processing and transfer between peripheral modules, RAM, and I/O ports. Providing up to 80 MIPS of performance additional to the CPU, the XGATE enables innovation and efficiency of designs.

Key features of the S12XD family include

HCS12X core
16-bit HCS12X CPU
Upward compatible with HCS12 instruction set
Interrupt stacking and programmer's model identical to HCS12
Instruction queue
Enhanced indexed addressing
Enhanced instruction set
EBI (external bus interface)
MMC (module mapping control)
INT (interrupt controller)
DBG (Debug module to monitor HCS12X CPU and XGATE bus activity)
BDM (background debug mode)

### 12.1.1  XGATE Overview

The XGATE module on the advanced HCS12X family of 16-bit microcontrollers (MCUs) is a highly flexible, high-performance and cost-sensitive parallel processing solution. The XGATE module is a peripheral coprocessor that allows autonomous high-speed data processing and transfers between the MCU's peripherals and the internal RAM and I/O ports. It has a built-in RISC core that is able to preprocess the transferred data and perform complex communication protocols.

The XGATE module is intended to increase the MCU's data throughput by lowering the S12X CPU's interrupt load. Running at up to 80 MHz in parallel to the S12X CPU, the XGATE is easily programmable in C code. The XGATE processor function is fully integrated with the S12X CPU as shown in Figure 12.3.

**FIGURE 12.3**  XGATE Integration (Copyright of Freescale Semiconductor, Inc. 2007, used by permission.)

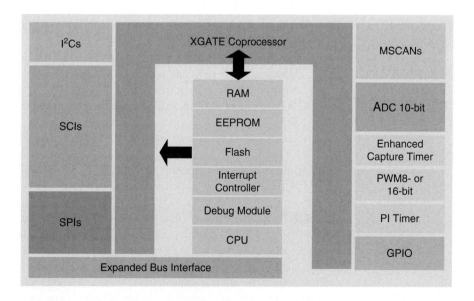

Features of the XGATE function include

Optimized 16-bit RISC core for data manipulation
Hardware semaphores for secure data sharing between cores
Interrupt-driven operations
Up to 112 XGATE channels
104 hardware-triggered channels
Eight software-triggered channels
Able to trigger S12X_CPU interrupts on completion of an XGATE transfer
Barrel shifter for fast data manipulation
Smart memory access protection avoiding conflicting CPU12 and XGATE accesses

Figure 12.4 shows the interface of the XGATE function with the CAN subsystem. It shows the memory mapping of the CAN code to be executed by the XGATE RISC processor. This provides for the independent execution of XGATE instructions with the S12X_CPU.

**FIGURE 12.4** XGATE with CAN Interface (Copyright of Freescale Semiconductor, Inc. 2007, used by permission.)

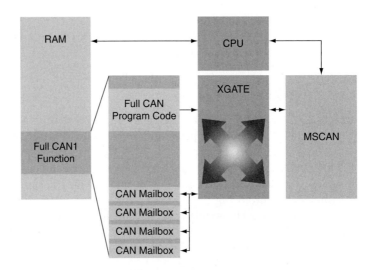

Key benefits of the XGATE function include

Advanced, easy-to-use interrupt processing peripheral
Programmable "Smart" DMA
Cost-effective CPU boost of up to 80 MHz parallel processing
Data transfer between all peripherals and RAM
Access to 30 K of flash (read only)
Very flexible full CAN mailbox and filtering management when used with MSCAN
Efficient multi-LIN master capability
Easily programmable in standard C, compiler and debugger are integrated into the S12X tool suite
Single-background debug mode (BDM) for debugging both S12X_CPU and XGATE simultaneously while running at full speed

**12.1.1.1   *XGATE Module*** The core of the XGATE module is a RISC processor, which is able to access the MCU's internal memories and peripherals. The RISC processor always remains in an idle state until it is triggered by an XGATE request. Then it executes a code sequence that is associated with the request and optionally triggers an interrupt to the S12X_CPU on completion. Code sequences are not interruptible. A new XGATE request can only be serviced when the previous sequence is finished and the RISC core becomes idle.

The XGATE module also provides a set of hardware semaphores that are necessary to ensure data consistency whenever RAM locations or peripherals are shared with the S12X_CPU. The following sections describe the components of the XGATE module in further detail.

***12.1.1.2 XGATE RISC Core*** Figure 12.5 is a functional block diagram of the XGATE module. The RISC core is a 16-bit processor with an instruction set that is well suited for data transfers, bit manipulations, and simple arithmetic operations. It is able to access the MCU's internal memories and peripherals without blocking these resources from the S12X_CPU1. Whenever the S12X_CPU and the RISC core access the same resource, the RISC core will be stalled until the resource becomes available again.

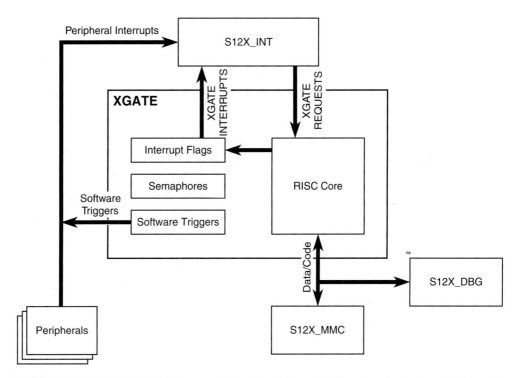

**FIGURE 12.5** XGATE Block Diagram (Copyright of Freescale Semiconductor, Inc. 2007, used by permission.)

The XGATE offers a high access rate to the MCU's internal RAM. Depending on the bus load, the RISC core can perform up to two RAM accesses per S12X_CPU bus cycle. Bus accesses to peripheral registers or flash are slower. A transfer rate of one bus access per S12X_CPU cycle cannot be exceeded. The XGATE module is intended to execute short interrupt service routines that are triggered by peripheral modules or by software.

***12.1.1.3 XGATE Programmer's Model*** The programmer's model of the XGATE RISC core is shown in Figure 12.6. The processor offers a set of seven general-purpose registers (R1–R7), which serve as accumulators and index registers. An additional eighth register (R0) is tied to the value "$0000." Register R1 has an additional functionality. It is preloaded with the initial variable pointer of the channel's service request vector (see Figures 12.6). The initial content of the remaining general-purpose registers is undefined.

The 16-bit program counter allows the addressing of a 64 K byte address space. The condition code register contains 4 bits: the sign bit (S), the zero flag (Z), the overflow flag (V), and the carry bit (C). The initial content of the condition code register is undefined.

**FIGURE 12.6** XGATE Programmer's Model (Copyright of Freescale Semiconductor, Inc. 2007, used by permission.)

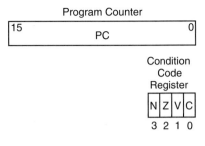

***12.1.1.4 XGATE Memory Map*** The XGATE's RISC core is able to access an address space of 64 K bytes. The allocation of memory blocks within this address space is determined on chip level. The XGATE vector block assigns a start address and a variable pointer to each XGATE channel. Its position in the XGATE memory map can be adjusted through the XGVBR register.

Figure 12.7 shows the layout of the vector block. Each vector consists of two 16-bit words. The first contains the start address of the service routine. This value will be loaded into the program counter before a service routine is executed. The second word is a pointer to the service routine's variable space. This value will be loaded into register R1 before a service routine is executed.

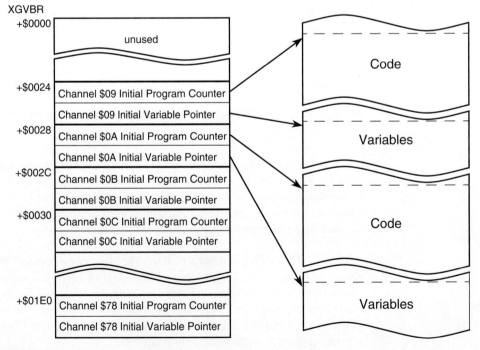

**FIGURE 12.7** XGATE Vector Block (Copyright of Freescale Semiconductor, Inc. 2007, used by permission.)

**12.1.1.5 XGATE Semaphores** The XGATE module offers a set of eight hardware sema-phores. These semaphores provide a mechanism to protect system resources that are shared be-tween two concurrent threads of program execution: one thread running on the S12X_CPU and one running on the XGATE RISC core. Each semaphore can only be in one of the three states: "Unlocked," "Locked by S12X_CPU," and "Locked by XGATE."

The S12X_CPU can check and change a semaphore's state through the XGATE sema-phore register. The RISC core does this through its SSEM and CSEM instructions. Figure 12.8 illustrates the valid state transitions.

Figure 12.9 gives an example of the typical usage of the XGATE hardware semaphores. Two concurrent threads are running on the system. One is running on the S12X_CPU, and the other is running on the RISC core. They both have a critical section of code that accesses the same system resource. To guarantee that the system resource is only accessed by one thread at a time, the critical code sequence must be embedded in a semaphore lock/release sequence, as shown.

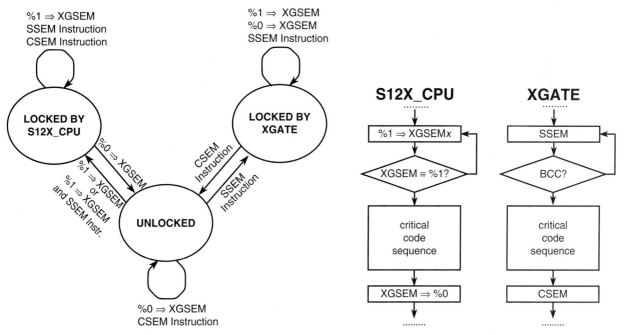

**FIGURE 12.8** XGATE Semaphore State Transitions (Copyright of Freescale Semiconductor, Inc. 2007, used by permission.)

**FIGURE 12.9** XGATE Algorithm for Locking and Releasing Semaphores (Copyright of Freescale Semiconductor, Inc. 2007, used by permission.)

**12.1.1.6 XGATE Modes of Operation** There are four run modes on S12X devices. They are the Run mode, Wait mode, Stop mode, and Freeze mode (with BDM active). The XGATE is able to operate in Run, Stop, and Freeze system modes. Clock activity will be automatically stopped when the XGATE module is idle. In Freeze mode, all clocks of the XGATE module may be stopped, depending on the module configuration.

## 12.1.2 Clocking

There are two clock functions: the clock and reset generator functional block, and the Pierce oscillator functional block.

**12.1.2.1  *Clock and Reset Generator (CRG)*** The CRG is composed of five major features. They are the Phase locked loop (PLL) frequency multiplier, System clock generator, Computer operating properly (COP) watchdog timer with time-out, System reset generation, and Real-time interrupt (RTI).

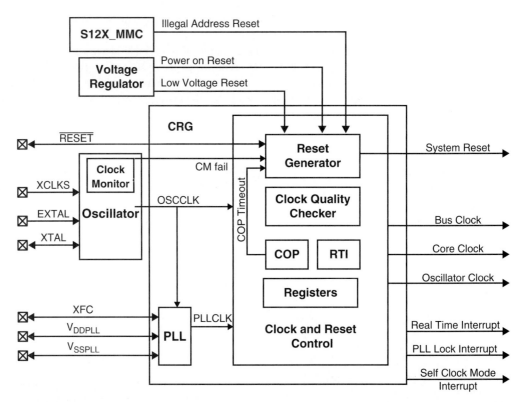

**FIGURE 12.10** CRG Block Diagram (Copyright of Freescale Semiconductor, Inc. 2007, used by permission.)

**12.1.2.2  *Pierce Oscillator (XOSC)*** The Pierce oscillator module provides a robust, low-noise and low-power clock source. The module will be operated from the VDDPLL supply rail (2.5 V nominal) and require the minimum number of external components. It is designed for optimal start-up margin with typical crystal oscillators.

The XOSC contains circuitry to dynamically control current gain in the output amplitude. This ensures a signal with low harmonic distortion, low power and good noise immunity. The basic features include

High noise immunity due to input hysteresis
Low RF emissions with peak-to-peak swing limited dynamically
Transconductance (gm) sized for optimum start-up margin for typical oscillators
Dynamic gain control eliminates the need for external current limiting resistor
Integrated resistor eliminates the need for external bias resistor
Low power consumption
Operates from 2.5 V (nominal) supply
Amplitude control limits power
Clock monitor

**FIGURE 12.11** Pierce Oscillator (XOSC) Block Diagram (Copyright of Freescale Semiconductor, Inc. 2007, used by permission.)

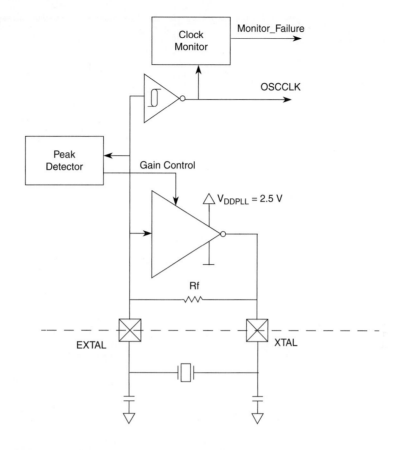

### 12.1.3 Analog-to-Digital Converter (ATD)

The ATD10B8C is an 8-channel, 10-bit, multiplexed input successive approximation analog-to-digital converter (Figure 12.12). The MC9S12XD also includes a 16-channel analog-to-digital converter (ATD10B16CV4). The key features of the 8-bit analog-to-digital converter are

8/10-bit resolution
7 μsec, 10-bit single conversion time
Sample buffer amplifier
Programmable sample time
Left/right justified, signed/unsigned result data
External trigger control
Conversion completion interrupt generation
Analog input multiplexer for 16 analog input channels
Analog/digital input pin multiplexing
1 to 16 conversion sequence lengths
Continuous conversion mode
Multiple channel scans
Configurable external trigger functionality on any AD channel or any of four additional external trigger inputs. The four additional trigger inputs can be chip external or internal.
Configurable location for channel wrap around (when converting multiple channels in a sequence).

There is software programmable selection between performing single or continuous conversion on a single channel or multiple channels.

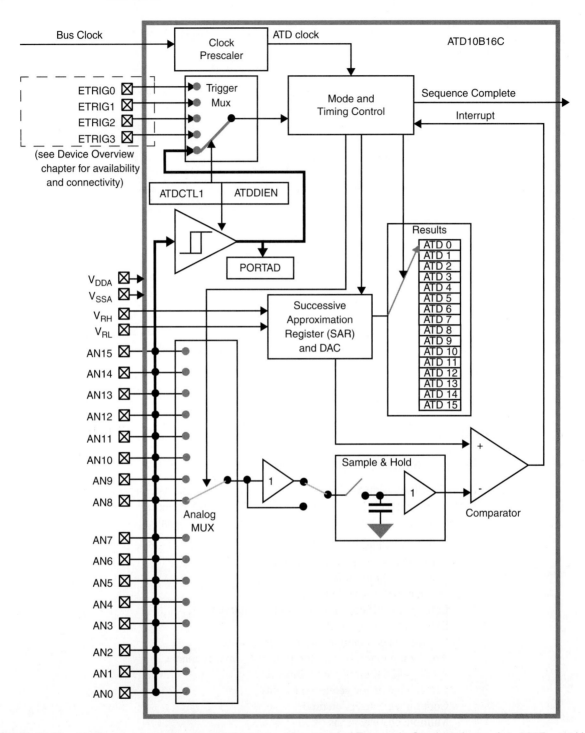

**FIGURE 12.12**   16-Channel Analog-to-Digital Converter (Copyright of Freescale Semiconductor, Inc. 2007, used by permission.)

## 12.1.4  Enhanced Capture Timer (ECT)

The HCS12 enhanced capture timer module has the features of the HCS12 standard timer module enhanced by additional features to enlarge the field of applications, in particular for automotive ABS applications (Figure 12.13). This design specification describes the standard timer as well as the additional features.

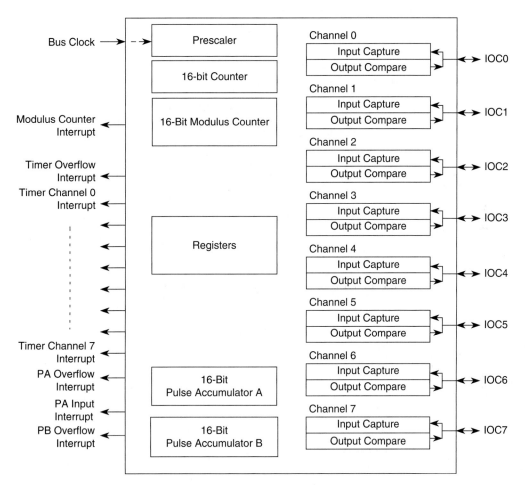

**FIGURE 12.13**  Enhanced Capture Timer (Copyright of Freescale Semiconductor, Inc. 2007, used by permission.)

The basic timer consists of a 16-bit, software-programmable counter driven by a prescaler. This timer can be used for many purposes, including input waveform measurements while simultaneously generating an output waveform. Pulse widths can vary from microseconds to many seconds. A full access for the counter registers or the input capture/output compare registers will take place in one clock cycle. Accessing high byte and low byte separately for all these registers will not yield the same result as accessing them in one word.

### 12.1.4.1  *Features*

16-bit buffer register for four input capture (IC) channels.
Four 8-bit pulse accumulators with 8-bit buffer registers associated with the four buffered IC channels. Configurable also as two 16-bit pulse accumulators.

16-bit modulus down-counter with 8-bit prescaler.
Four user-selectable delay counters for input noise immunity increase.

## 12.1.5  Pulse-Width Modulator (PWM)

The pulse-width modulator (PWM) contains the basic features of center-aligned output mode and four available clock sources (Figure 12.14). The PWM module has eight channels with independent control of left- and center-aligned outputs on each channel.

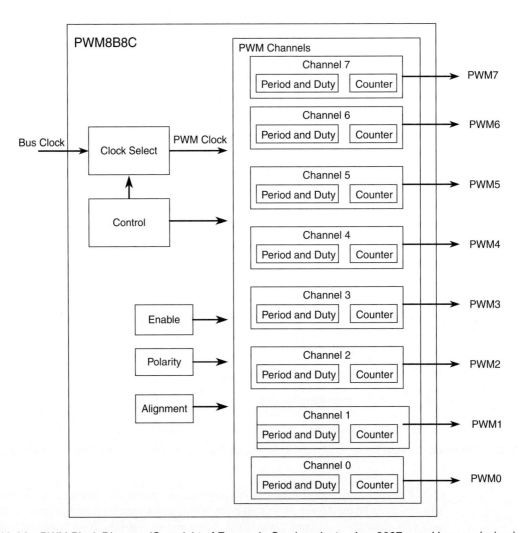

**FIGURE 12.14**   PWM Block Diagram (Copyright of Freescale Semiconductor, Inc. 2007, used by permission.)

Each of the eight channels has a programmable period and duty cycle as well as a dedicated counter. A flexible clock select scheme allows a total of four different clock sources to be used with the counters. Each of the modulators can create independent continuous waveforms with software-selectable duty rates from 0% to 100%. The PWM outputs can be programmed as left-aligned outputs or center-aligned outputs.

### *12.1.5.1 Features* The PWM block includes these distinctive features:

Eight independent PWM channels with programmable period and duty cycle
Dedicated counter for each PWM channel
Programmable PWM enable/disable for each channel
Software selection of PWM duty pulse polarity for each channel
Period and duty cycle are double buffered. Change takes effect when the end of the effective period is reached (PWM counter reaches zero) or when the channel is disabled.
Programmable center or left-aligned outputs on individual channels
Eight 8-bit channel or four 16-bit channel PWM resolution
Four clock sources (A, B, SA, and SB) provide for a wide range of frequencies
Programmable clock select logic
Emergency shutdown

## 12.1.6 Interintegrated Circuit (IIC)

The IIC bus is a two-wire, bidirectional serial bus that provides a simple, efficient method of data exchange between devices (Figure 12.15). Being a two-wire device, the IIC bus minimizes the need for large numbers of connections between devices and eliminates the need for an address decoder. This bus is suitable for applications requiring occasional communications over a short distance between a number of devices. It also provides flexibility, allowing additional devices to be connected to the bus for further expansion and system development.

**FIGURE 12.15** IIC Block Diagram (Copyright of Freescale Semiconductor, Inc. 2007, used by permission.)

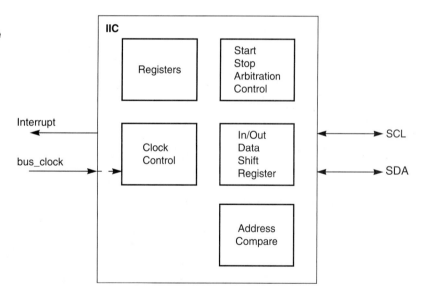

The IIC interface is designed to operate up to 100 kbps with maximum bus loading and timing. The device is capable of operating at higher baud rates, up to a maximum of clock/20, with reduced bus loading. The maximum communication length and the number of devices that can be connected are limited by a maximum bus capacitance of 400 pF.

### *12.1.6.1 Features* The IIC module has the following key features:

Compatible with $I^2C$ bus standard
Multimaster operation

Software programmable for one of 256 different serial clock frequencies
Software selectable acknowledge bit
Interrupt driven byte-by-byte data transfer
Arbitration lost interrupt with automatic mode switching from master to slave
Calling address identification interrupt
Start and stop signal generation/detection
Repeated start signal generation
Acknowledge bit generation/detection
Bus busy detection

## 12.1.7  Scalable Controller Area Network (CAN)

Freescale's scalable controller area network (CAN) definition is based on the MSCAN12 defini-tion, which is the specific implementation of the MSCAN concept targeted for the M68HC12 microcontroller family. The module (Figure 12.16) is a communication controller implementing the CAN 2.0A/B protocol, as defined in the Bosch specification dated September 1991.

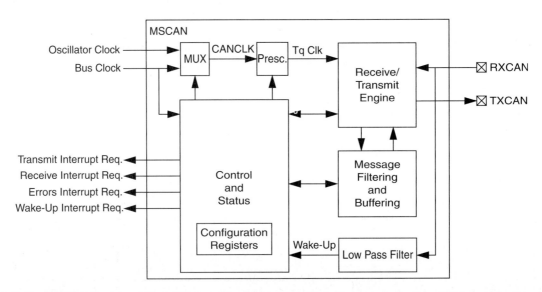

**FIGURE 12.16**    MSCAN Functional Block Diagram (Copyright of Freescale Semiconductor, Inc. 2007, used by permission.)

Though not exclusively intended for automotive applications, CAN protocol is designed to meet the specific requirements of a vehicle serial data bus: real-time processing, reliable opera-tion in the EMI environment of a vehicle, cost effectiveness, and required bandwidth. MSCAN uses an advanced buffer arrangement resulting in predictable real-time behavior and simplified application software.

### 12.1.7.1  Features    The basic features of the MSCAN are as follows:

Implementation of the CAN protocol–Version 2.0A/B
Standard and extended data frames
Zero to eight bytes data length
Programmable bit rate up to 1 Mbps1
Support for remote frames
Five receive buffers with FIFO storage scheme

Three transmit buffers with internal prioritization using a "local priority" concept

Flexible maskable identifier filter supports two full-size (32-bit) extended identifier filters, or four 16-bit filters, or eight 8-bit filters

Programmable wakeup functionality with integrated low-pass filter

Programmable loopback mode supports self-test operation

Programmable listen-only mode for monitoring of CAN bus

Programmable bus-off recovery functionality

Separate signaling and interrupt capabilities for all CAN receiver and transmitter error states (warning, error passive, bus-off)

Programmable MSCAN clock source either bus clock or oscillator clock

Internal timer for time-stamping of received and transmitted messages

Three low-power modes: sleep, power down, and MSCAN enable

Global initialization of configuration registers

***12.1.7.2   CAN System***   A typical CAN system with MSCAN is shown in Figure 12.17. Each CAN station is connected physically to the CAN bus lines through a transceiver device. The transceiver is capable of driving the large current needed for the CAN bus and has current protection against defective CAN or defective stations.

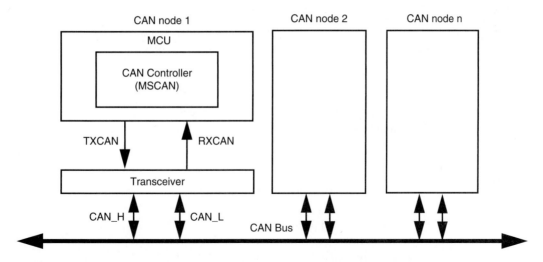

**FIGURE 12.17**   CAN System (Copyright of Freescale Semiconductor, Inc. 2007, used by permission.)

## 12.1.8  Serial Communication Interface (SCI)

The SCI allows asynchronous serial communications with peripheral devices and other CPUs. A block diagram is shown in Figure 12.18. The SCI is particularly flexible in its ability to be configured into many serial interfaces.

***12.1.8.1   Features***   The SCI includes these distinctive features:

Full-duplex or single-wire operation

Standard mark/space non-return-to-zero (NRZ) format

Selectable IrDA 1.4 return-to-zero-inverted (RZI) format with programmable pulse widths

13-bit baud rate selection

Programmable 8-bit or 9-bit data format

Separately enabled transmitter and receiver

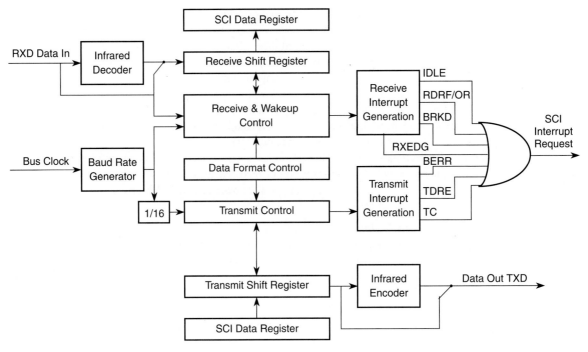

**FIGURE 12.18**  SCI Block Diagram (Copyright of Freescale Semiconductor, Inc. 2007, used by permission.)

Programmable polarity for transmitter and receiver
Programmable transmitter output parity
Two receiver wakeup methods:
    Idle line wakeup
    Address mark wakeup
Interrupt-driven operation with eight flags:
    Transmitter empty
    Transmission complete
    Receiver full
    Idle receiver input
    Receiver overrun
    Noise error
    Framing error
    Parity error
Receive wakeup on active edge
Transmit collision detect supporting LIN
Break detect supporting LIN
Receiver framing error detection
Hardware parity checking
1/16-bit-time noise detection

***12.1.8.2  Functional Description***  This section provides a complete functional description of the SCI block, detailing the operation of the design from the end user perspective in a number of subsections. Figure 12.19 shows the structure of the SCI module. The SCI allows full duplex, asynchronous, serial communication between the CPU and remote devices, including other

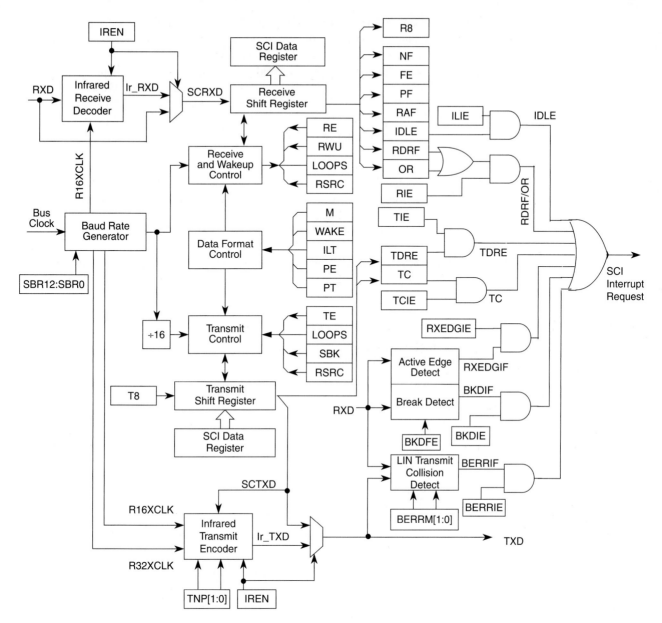

**FIGURE 12.19**  Detailed SCI Block Diagram (Copyright of Freescale Semiconductor, Inc. 2007, used by permission.)

CPUs. The SCI transmitter and receiver operate independently, although they use the same baud rate generator. The CPU monitors the status of the SCI, writes the data to be transmitted, and processes received data.

**12.1.8.2.1 Infrared Interface (IrDA)**     This module provides the capability of transmitting narrow pulses to an IR LED and receiving narrow pulses and transforming them to serial bits, which are sent to the SCI. The IrDA physical layer specification defines a half-duplex infrared communication link for exchange data. The full standard includes data rates up to 16 Mbits/s. This design covers only data rates between 2.4 Kbits/s and 115.2 Kbits/s.

The infrared submodule consists of two major blocks: the transmit encoder and the receive decoder. The SCI transmits serial bits of data, which are encoded by the infrared submodule to transmit a narrow pulse for every zero bit. No pulse is transmitted for every one bit.

When receiving data, the IR pulses should be detected using an IR photo diode and transformed to CMOS levels by the IR receive decoder (external from the MCU). The narrow pulses are then stretched by the infrared submodule to get back to a serial bit stream to be received by the SCI. The polarity of transmitted pulses and expected receive pulses can be inverted so that a direct connection can be made to external IrDA transceiver modules that use active low pulses.

**12.1.8.2.2 LIN Support**    This module provides some basic support for the local inter connect network (LIN) protocol. At first, this is a break detect circuitry, making it easier for the LIN software to distinguish a break character from an incoming data stream. As a further addition, it supports a collision detection at the bit level as well as canceling pending transmissions.

**12.1.8.3    *Data Formats***    The SCI uses the standard NRZ mark/space data format. When Infrared is enabled, the SCI uses RZI data format where zeroes are represented by light pulses and ones remain low (Figure 12.20).

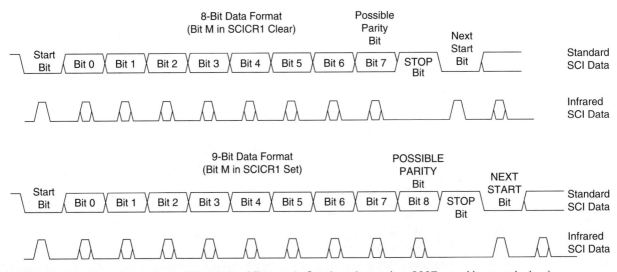

**FIGURE 12.20**    SCI Data Formats (Copyright of Freescale Semiconductor, Inc. 2007, used by permission.)

**12.1.8.4    *Receiver***    The SCI receiver functional block diagram is shown in Figure 12.21.

**12.1.8.5    *Transmitter***    The SCI transmitter functional block diagram is shown in Figure 12.22.

**12.1.8.6    *Baud Rate Generator***    A 13-bit modulus counter in the baud rate generator derives the baud rate for both the receiver and the transmitter. The value from 0 to 8191 written to the SBR12:SBR0 bits determines the bus clock divisor. The baud rate clock is synchronized with the bus clock and drives the receiver. The baud rate clock divided by 16 drives the transmitter. The receiver has an acquisition rate of 16 samples per bit time.

The table in Figure 12.23 lists some examples of achieving target baud rates with a bus clock frequency of 25 MHz.

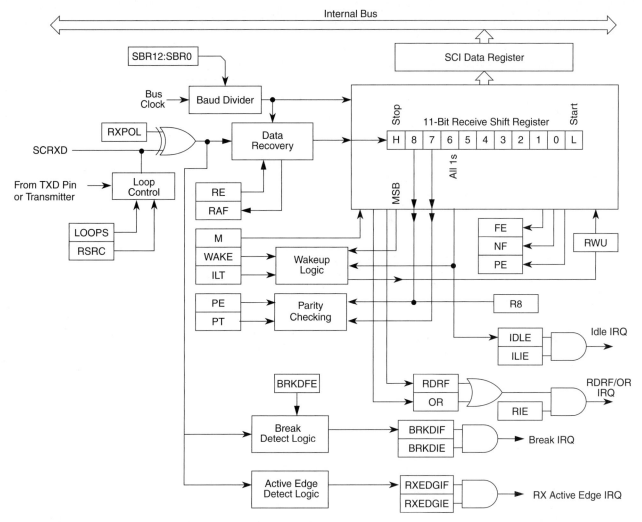

**FIGURE 12.21**    SCI Receiver Functional Block Diagram (Copyright of Freescale Semiconductor, Inc. 2007, used by permission.)

### 12.1.9  Serial Peripheral Interface (SPI)

The SPI module allows a duplex, synchronous, serial communication between the MCU and peripheral devices. Software can poll the SPI status flags, or the SPI operation can be interrupt driven.

*12.1.9.1    Features*    The SPI includes these distinctive features:

>   Master mode and slave mode
>   Bidirectional mode
>   Slave select output
>   Mode fault error flag with CPU interrupt capability
>   Double-buffered data register
>   Serial clock with programmable polarity and phase
>   Control of SPI operation during wait mode

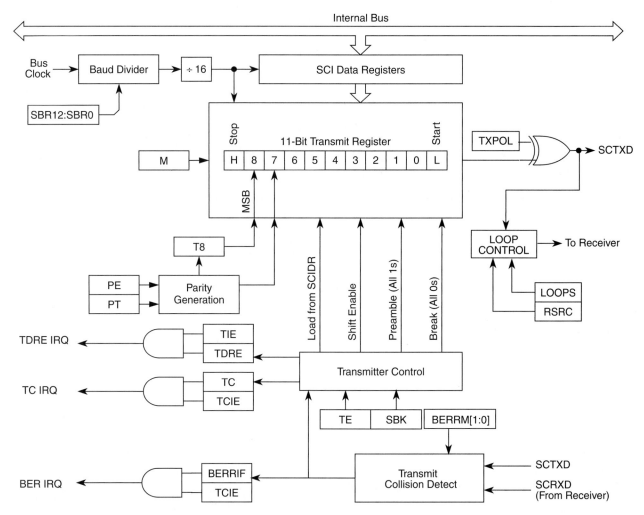

**FIGURE 12.22**    SCI Transmitter Functional Block Diagram(Copyright of Freescale Semiconductor, Inc. 2007, used by permission.)

**FIGURE 12.23**    Baud Rate Generator Table (Copyright of Freescale Semiconductor, Inc. 2007, used by permission.)

| Bits SBR[12:] | Receiver Clock (Hz) | Transmitter Clock (HZ) | Target Baud Rate | Error (%) |
|---|---|---|---|---|
| 41 | 609,756.1 | 38,109.8 | 38,400 | .76 |
| 81 | 308,642.0 | 19,290.1 | 19,200 | .47 |
| 163 | 153,374.2 | 9585.9 | 9,600 | .16 |
| 326 | 76,687.1 | 4792.9 | 4800 | .15 |
| 651 | 38,402.5 | 2400.2 | 2400 | .01 |
| 1302 | 19,201.2 | 1200.1 | 1200 | .01 |
| 2604 | 9600.6 | 600.0 | 600 | .00 |
| 5208 | 4800 | 300.0 | 300 | .00 |

Figure 12.24 gives an overview on the SPI architecture. The main parts of the SPI are status, control and data registers, shifter logic, baud rate generator, master/slave control logic, and port control logic.

The SPI module has the standard four external pins.

MOSI    Master Out/Slave In Pin
MISO    Master In/Slave Out Pin
SS    Slave Select Pin
SCK    Serial Clock Pin

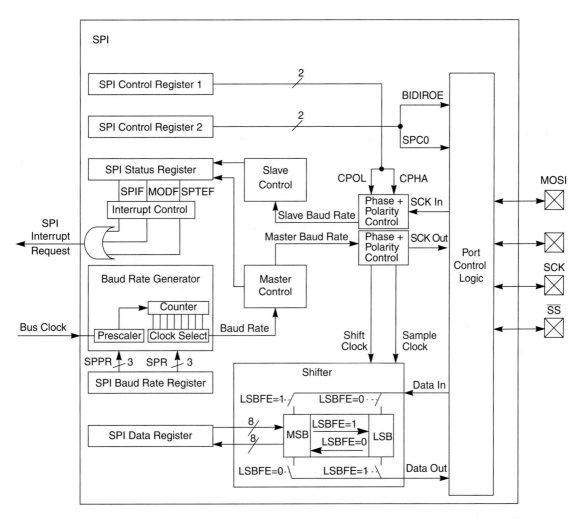

**FIGURE 12.24**    SPI Block Diagram (Copyright of Freescale Semiconductor, Inc. 2007, used by permission.)

***12.1.9.2 Functional Description***    The SPI module allows a duplex, synchronous, serial communication between the MCU and peripheral devices (Figure 12.25). Software can poll the SPI status flags, or SPI operation can be interrupt driven. The SPI system is enabled by setting the SPI enable (SPE) bit in SPI control register 1.

The main element of the SPI system is the SPI data register. The 8-bit data register in the master and the 8-bit data register in the slave are linked by the MOSI and MISO pins to form a distributed 16-bit register. When a data transfer operation is performed, this 16-bit register is serially

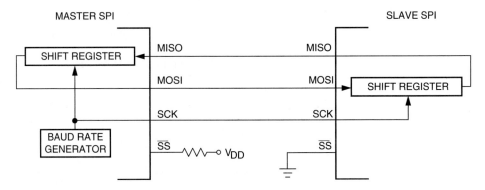

**FIGURE 12.25**    Master/Slave Transfer Block Diagram (Copyright of Freescale Semiconductor, Inc. 2007, used by permission.)

shifted 8 bit positions by the S-clock from the master, so data is exchanged between the master and the slave. Data written to the master SPI data register becomes the output data for the slave, and data read from the master SPI data register after a transfer operation is the input data from the slave.

**12.1.9.2.1 Master/Slave Modes**    The SPI operates in master mode when the MSTR bit is set. Only a master SPI module can initiate transmissions. A transmission begins by writing to the master SPI data register. If the shift register is empty, the byte immediately transfers to the shift register. The byte begins shifting out on the MOSI pin under the control of the serial clock.

Although the SPI is capable of duplex operation, some SPI peripherals are capable of only receiving SPI data in slave mode. For these simpler devices, there is no serial data out pin. As long as no more than one slave device drives the system slave's serial data output line, it is possible for several slaves to receive the same transmission from a master, although the master would not receive return information from all the receiving slaves.

During an SPI transmission, data is transmitted (shifted out serially) and received (shifted in serially) simultaneously. The serial clock (SCK) synchronizes shifting and sampling of the information on the two serial data lines. A slave select line allows selection of an individual slave SPI device; slave devices that are not selected do not interfere with SPI bus activities. Optionally, on a master SPI device, the slave select line can be used to indicate multiple-master bus contention.

## 12.1.10  Periodic Interrupt Timer (PIT)

The period interrupt timer is an array of 24-bit timers that can be used to trigger peripheral modules or raise periodic interrupts, as shown in Figure 12.26.

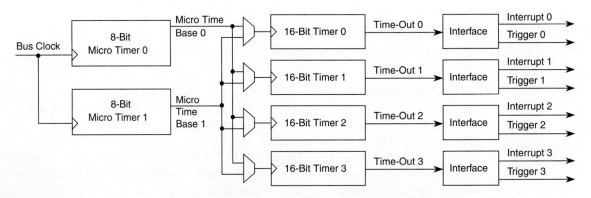

**FIGURE 12.26**    PIT Block Diagram (Copyright of Freescale Semiconductor, Inc. 2007, used by permission.)

***12.1.10.1 Features*** The PIT includes these features:

Four timers implemented as modulus down-counters with independent time-out periods
Time-out periods selectable between 1 and 224 bus clock cycles Time-out equals m*n bus clock cycles with $1 <= m <= 256$ and $1 <= n <= 65536$.
Timers that can be enabled individually
Four time-out interrupts
Four time-out trigger output signals available to trigger peripheral modules
Start of timer channels can be aligned to each other

## 12.1.11 Voltage Regulator (VREG)

The VREG module is a dual-output voltage regulator that provides two separate 2.5 V (typical) supplies differing in the amount of current that can be sourced. The regulator input voltage range is from 3.3 V up to 5 V (typical). The functional block diagram is shown in Figure 12.27.

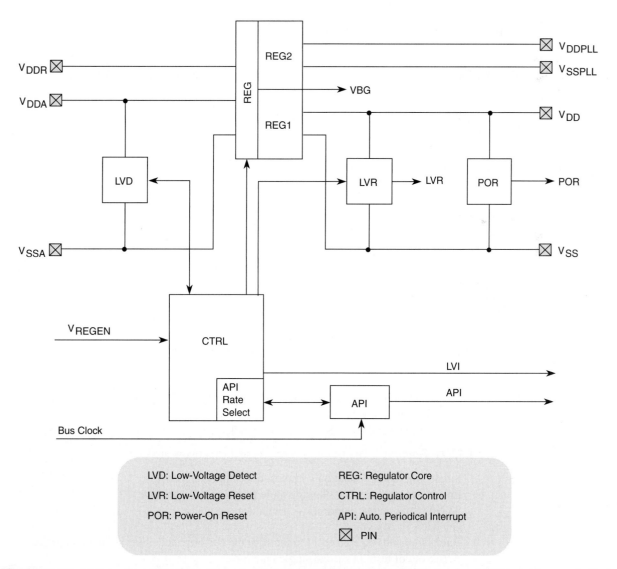

**FIGURE 12.27** Voltage Regulator Block Diagram (Copyright of Freescale Semiconductor, Inc. 2007, used by permission.)

***12.1.11.1  Features***　The VREG module includes these distinctive features:

Two parallel, linear voltage regulators
Low-voltage detect (LVD) with low-voltage interrupt (LVI)
Power-on reset (POR)
Low-voltage reset (LVR)
Autonomous periodical interrupt (API)

## 12.1.12  Background Debug Module (BDM)

The background debug module (BDM) subblock is a single-wire, background debug system implemented in on-chip hardware for minimal CPU intervention (Figure 12.28). All interfacing with the BDM is done via the BKGD pin.

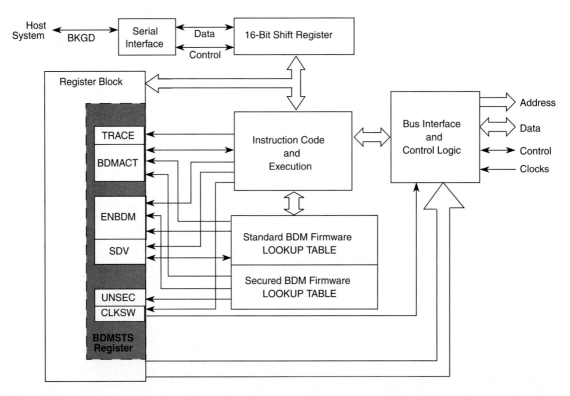

**FIGURE 12.28**　BDM Block Diagram (Copyright of Freescale Semiconductor, Inc. 2007, used by permission.)

The BDM has enhanced capability for maintaining synchronization between the target and host while allowing more flexibility in clock rates. This includes a sync signal to determine the communication rate and a handshake signal to indicate when an operation is complete.

***12.1.12.1  Features***　The BDM includes these distinctive features:

Single-wire communication with host development system
Enhanced capability for allowing more flexibility in clock rates
SYNC command to determine communication rate
GO_UNTIL command
Hardware handshake protocol to increase the performance of the serial communication
Active out of reset in special single-chip mode

Nine hardware commands using free cycles, if available, for minimal CPU intervention

Hardware commands not requiring active BDM

14 firmware commands execute from the standard BDM firmware lookup table

Software control of BDM operation during wait mode

Software selectable clocks

Global page access functionality

Enabled but not active out of reset in emulation modes

CLKSW bit set out of reset in emulation mode

When secured, hardware commands are allowed to access the register space in special single-chip mode, if the flash and EEPROM erase tests fail

Family ID readable from firmware ROM at global address 0x7FFF0F (value for HCS12X devices is 0xC1)

BDM hardware commands are operational until system stop mode is entered (all bus masters are in stop mode)

## 12.1.13 Interrupt Module (XINT)

The XINT module decodes the priority of all system exception requests and provides the applicable vector for processing the exception to either the CPU or the XGATE module (Figure 12.29). The XINT module supports

I bit and X bit maskable interrupt requests

A nonmaskable unimplemented opcode trap

A nonmaskable software interrupt (SWI) or background debug mode request

A spurious interrupt vector request

Three system reset vector requests

Each of the I-bit maskable interrupt requests can be assigned to one of seven priority levels supporting a flexible priority scheme. For interrupt requests that are configured to be handled by the CPU, the priority scheme can be used to implement nested interrupt capability where interrupts from a lower level are automatically blocked if a higher-level interrupt is being processed. Interrupt requests configured to be handled by the XGATE module cannot be nested because the XGATE module cannot be interrupted while processing.

### 12.1.13.1 Features

Interrupt vector base register (IVBR)

One spurious interrupt vector (at address vector base1 + 0x0010)

2–113 I-bit maskable interrupt vector requests (at addresses vector base + 0x0012–0x00F2)

Each I-bit maskable interrupt request has a configurable priority level and can be configured to be handled by either the CPU or the XGATE module2

I-bit maskable interrupts can be nested, depending on their priority levels

One X-bit maskable interrupt vector request (at address vector base + 0x00F4)

One nonmaskable software interrupt request (SWI) or background debug mode vector request (at address vector base + 0x00F6)

One nonmaskable unimplemented opcode trap (TRAP) vector (at address vector base + 0x00F8)

Three system reset vectors (at addresses 0xFFFA–0xFFFE)

Determines the highest priority DMA and interrupt vector requests, drives the vector to the XGATE module or to the bus on CPU request, respectively

Wakes up the system from stop or wait mode when an appropriate interrupt request occurs or whenever XIRQ is asserted, even if X interrupt is masked

XGATE can wake up and execute code, even with the CPU remaining in stop or wait mode

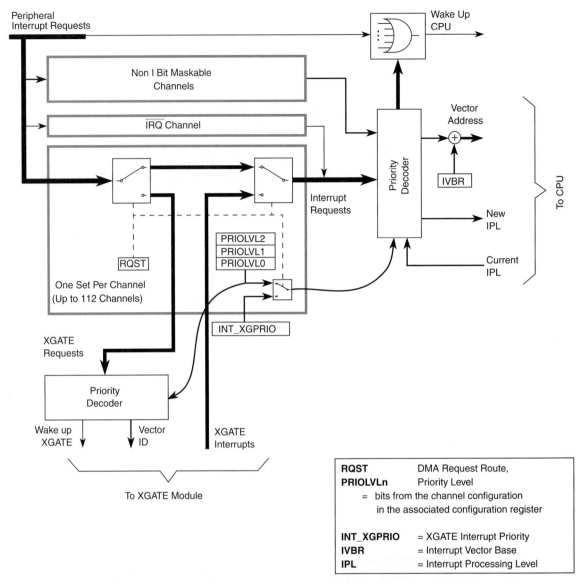

**FIGURE 12.29**   Interrupt Control Block Diagram (Copyright of Freescale Semiconductor, Inc. 2007, used by permission.)

***12.1.13.2 Interrupt Nesting***   The interrupt request priority-level scheme makes it possible to implement priority-based interrupt request nesting for the I-bit maskable interrupt requests handled by the CPU.

I-bit maskable interrupt requests can be interrupted by an interrupt request with a higher priority, so that there can be up to seven nested I-bit maskable interrupt requests at a time (refer to Figure 12.30 for an example using up to three nested interrupt requests).

I-bit maskable interrupt requests cannot be interrupted by other I-bit maskable interrupt requests per default. To make an interrupt service routine (ISR) interruptible, the ISR must explicitly clear the I bit in the CCR (CLI). After clearing the I bit, I-bit maskable interrupt requests with higher priority can interrupt the current ISR.

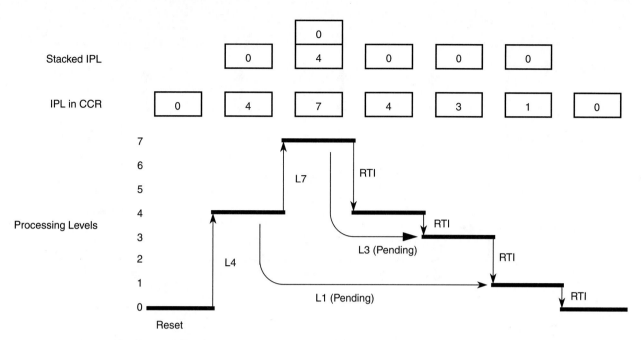

**FIGURE 12.30**   Nested Interrupt Example (Copyright of Freescale Semiconductor, Inc. 2007, used by permission.)

An ISR of an interruptible I-bit maskable interrupt request could basically look like this:

Service interrupt (e.g., clear interrupt flags, copy data)
Clear I bit in the CCR by executing the instruction CLI (thus allowing interrupt requests with higher priority)
Process data
Return from interrupt by executing the instruction RTI

## 12.1.14 Mapping Memory Control (MMC)

The block diagram of the MMC (S12XMMCV3) is shown in Figure 12.31. The MMC module controls the multimaster priority accesses, the selection of internal resources, and external space. Internal buses, including internal memories and peripherals, are controlled in this module. The local address space for each master is translated to a global memory space.

### 12.1.14.1  *Features*   The main features of the MMC are

Paging capability to support a global 8 Mbytes memory address space
Bus arbitration between the masters CPU, BDM, and XGATE
Simultaneous accesses to different resources1 (internal, external, and peripherals)
Resolution of target bus access collision
Access restriction control from masters to some targets (e.g., RAM write access protection for user-specified areas)
MCU operation mode control
MCU security control
Separate memory map schemes for each master CPU, BDM, and XGATE
ROM control bits to enable the on-chip FLASH or ROM selection
Port replacement registers access control
Generation of system reset when CPU accesses an unimplemented address (i.e., an address that does not belong to any of the on-chip modules) in single-chip modes

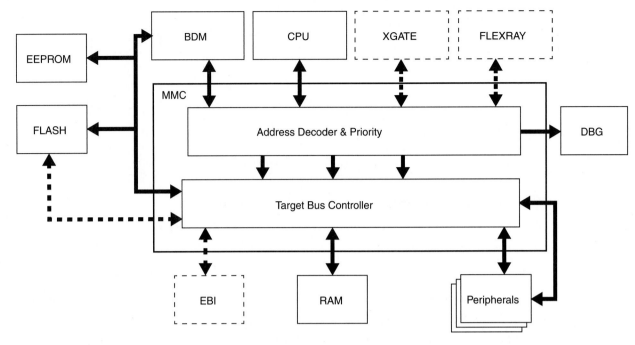

**FIGURE 12.31**   MMC Block Diagram (Copyright of Freescale Semiconductor, Inc. 2007, used by permission.)

## 12.1.15  Debug (DBG)

The DBG module provides an on-chip trace buffer with flexible triggering capability to allow nonintrusive debug of application software. The DBG module is optimized for the HCS12X 16-bit architecture and allows debugging of both CPU and XGATE module operations. The functional block diagram is shown in Figure 12.32.

Typically the DBG module is used in conjunction with the BDM module. It can be configured as the DBG module for a debugging session over the BDM interface. Once configured, the DBG module is armed, and the device leaves BDM mode returning control to the user program, which is then monitored by the DBG module.

Alternatively, the DBG module can be configured over a serial interface using SWI routines. Comparators monitor the bus activity of the CPU and XGATE modules. When a match occurs, the control logic can trigger the state sequencer to a new state or tag an op code. A tag hit, which occurs when the tagged op code reaches the execution stage of the instruction queue, can also cause a state transition.

On a transition to the final state, bus tracing is triggered and/or a breakpoint can be generated. Independent of comparator matches, a transition to final state with associated tracing and breakpoint can be triggered by the external TAGHI and TAGLO signals. This is done by an XGATE module S/W breakpoint request or by writing to the TRIG control bit. The trace buffer is visible through a 2-byte window in the register address map and can be read out using standard 16-bit word reads. Tracing is disabled when the MCU system is secured.

***12.1.15.1  Features***   The main features of the BDM are

Four comparators (A, B, C, and D):
Comparators A and C compare the full address and the full 16-bit data bus
Comparators A and C feature a data bus mask register
Comparators B and D compare the full address bus only
Each comparator can be configured to monitor either CPU or XGATE busses

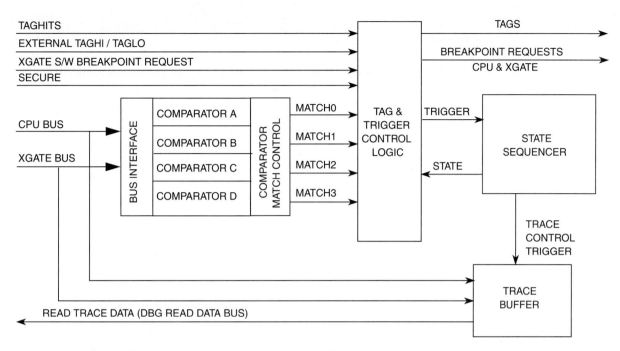

**FIGURE 12.32** Debug Block Diagram (Copyright of Freescale Semiconductor, Inc. 2007, used by permission.)

Each comparator features control of R/W and byte/word access cycles
Comparisons can be used as triggers for the state sequencer
Three comparator modes:
    Simple address/data comparator match mode
    Inside address range mode, Addmin ≤ Address ≤ Addmax
    Outside address range match mode, Address < Addmin or Address > Addmax
Two types of triggers:
    Tagged: triggers just before a specific instruction begins execution
    Force: triggers on the first instruction boundary after a match occurs.
Three types of breakpoints:
    CPU breakpoint entering BDM on breakpoint (BDM)
    CPU breakpoint executing SWI on breakpoint (SWI)
    XGATE breakpoint
Three trigger modes independent of comparators:
    External instruction tagging (associated with CPU instructions only)
    XGATE S/W breakpoint request
    TRIG bit immediate software trigger
Three trace modes:
    Normal: change of flow (COF) bus information is stored for change of flow definition
    Loop1: same as normal but inhibits consecutive duplicate source address entries
    Detail: address and data for all cycles except free cycles and opcode fetches are stored
Four-stage state sequencer for trace buffer control:
    Tracing session trigger linked to final state of state sequencer
    Begin, end, and mid alignment of tracing to trigger

The DBG module can be used in all MCU functional modes. During BDM hardware accesses and when the BDM module is active, CPU monitoring is disabled. Thus breakpoints, comparators, and bus tracing mapped to the CPU are disabled, but accessing the DBG registers, including comparator registers, is still possible. While in active BDM or during hardware BDM accesses, XGATE activity can still be compared, traced, and used to generate a breakpoint to the XGATE module. When the CPU enters active BDM mode through a BACKGROUND command, with the DBG module armed, the DBG remains armed.

The DBG module tracing is disabled if the MCU is in secure mode. Breakpoints can, however, still be generated if the MCU is secure.

## 12.1.16 External Bus Interface (XEBI)

The XEBI controls the functionality of a nonmultiplexed external bus (a.k.a. "expansion bus") in relationship with the chip operation modes. Dependent on the mode, the external bus can be used for data exchange with external memory, peripherals, or PRU and provide visibility to the internal bus externally in combination with an emulator (Figure 12.33).

**FIGURE 12.33** XEBI Block Diagram (Copyright of Freescale Semiconductor, Inc. 2007, used by permission.)

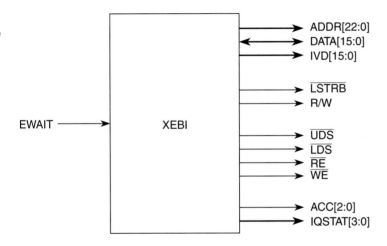

**12.1.16.1 Features** The XEBI includes the following features:

Output of up to 23-bit address bus and control signals to be used with a nonmultiplexed external bus
Bidirectional 16-bit external data bus with option to disable upper half
Visibility of internal bus activity

## 12.1.17 Port Integration Module (PIM)

The S12XD family port integration module establishes the interface between the peripheral modules including the nonmultiplexed external bus interface module (S12X_EBI) and the I/O pins for all ports (Figure 12.34). It controls the electrical pin properties as well as the signal prioritization and multiplexing on shared pins.

The PIM includes the following ports:

Port A, B used as address output of the S12X_EBI
Port C, D used as data I/O of the S12X_EBI
Port E associated with the S12X_EBI control signals and the IRQ, XIRQ interrupt inputs
Port K associated with address output and control signals of the S12X_EBI
Port T connected to the enhanced capture timer (ECT) module
Port S associated with 2 SCI and 1 SPI modules
Port M associated with 4 MSCAN modules and 1 SCI module
Port P connected to the PWM and 2 SPI modules—inputs can be used as an external interrupt source

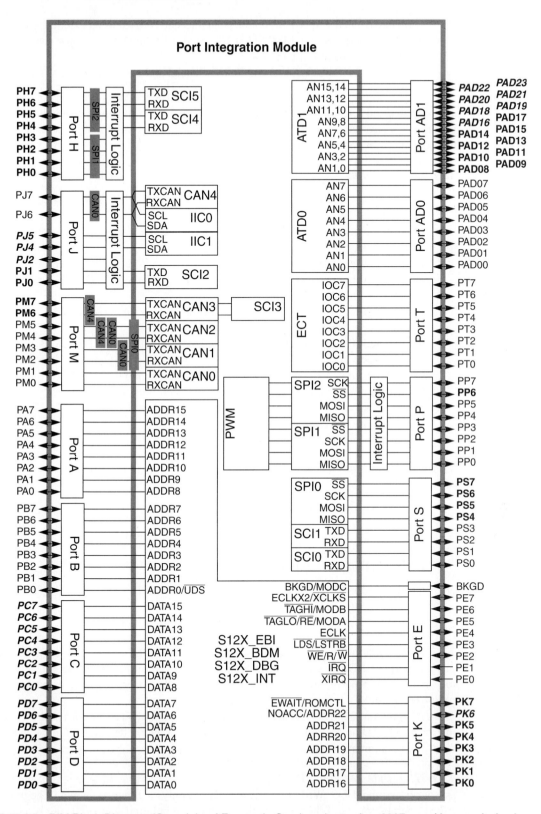

**FIGURE 12.34**  PIM Block Diagram (Copyright of Freescale Semiconductor, Inc. 2007, used by permission.)

Port H associated with 2 SCI modules—inputs can be used as an external interrupt source

Port J associated with 1 MSCAN, 1 SCI, and 2 IIC modules—inputs can be used as an external interrupt sourcert54ff

Port AD0 and AD1 associated with one 8-channel and one 16-channel ATD module

Most I/O pins can be configured by register bits to select data direction and drive strength and to enable and select pull-up or pull-down devices. Interrupts can be enabled on specific pins resulting in status flags. The I/O's of two MSCAN and all three SPI modules can be routed from their default location to alternative port pins.

### 12.1.17.1 Features    A full-featured PIM module includes these distinctive registers:

Data and data direction registers for Ports A, B, C, D, E, K, T, S, M, P, H, J, AD0, and AD1 when used as general-purpose I/O

Control registers to enable/disable pull-device and select pull-ups/pull-downs on Ports T, S, M, P, H, and J on per-pin basis

Control registers to enable/disable pull-up devices on Ports AD0, and AD1 on per-pin basis

Single-control register to enable/disable pull-ups on Ports A, B, C, D, E, and K on per-port basis and on BKGD pin

Control registers to enable/disable reduced output drive on Ports T, S, M, P, H, J, AD0, and AD1 on per-pin basis

Single-control register to enable/disable reduced output drive on Ports A, B, C, D, E, and K on per-port basis

Control registers to enable/disable open-drain (wired-OR) mode on Ports S and M

Control registers to enable/disable pin interrupts on Ports P, H, and J

Interrupt flag register for pin interrupts on Ports P, H, and J

Control register to configure IRQ pin operation

Free-running clock outputs

### 12.1.17.2 Port Pin    A standard port pin has the following minimum features:

Input/output selection

5 V output drive with two selectable drive strengths

5 V digital and analog input

Input with selectable pull-up or pull-down device

Port pin optional features include

Open drain for wired-OR connections

Interrupt inputs with glitch filtering

Reduced input threshold to support low-voltage applications

### 12.1.17.3 Functional Description    Each pin except PE0, PE1, and BKGD can act as general-purpose I/O. In addition, each pin can act as an output from the external bus interface module or a peripheral module or an input to the external bus interface module or a peripheral module.

A set of configuration registers is common to all ports with exceptions in the expanded bus interface and ATD ports (Figure 12.35). All registers can be written at any time; however, a specific configuration might not become active.

### 12.1.17.4 Data Register    This register holds the value driven out to the pin if the pin is used as a general-purpose I/O. Writing to this register has only an effect on the pin if the pin is used as general-purpose output. When reading this address, the buffered state of the pin is returned if the associated data direction register bit is set to "0."

| Port | Data | Data Direction | Input | Reduced Drive | Pull Enable | Polarity Select | Wired-OR Mode | Interrupt Enable | Interrupt Flag |
|------|------|----------------|-------|---------------|-------------|-----------------|---------------|------------------|----------------|
| A | yes | yes | — | yes | yes | — | — | — | — |
| B | yes | yes | — | | | — | — | — | — |
| C | yes | yes | — | | | — | — | — | — |
| D | yes | yes | — | | | — | — | — | — |
| E | yes | yes | — | | | — | — | — | — |
| K | yes | yes | — | | | — | — | — | — |
| T | yes | yes | yes | yes | yes | — | — | — | — |
| S | yes | yes | yes | yes | yes | yes | yes | — | — |
| M | yes | yes | yes | yes | yes | yes | yes | — | — |
| P | yes | yes | yes | yes | yes | yes | — | yes | yes |
| H | yes | yes | yes | yes | yes | yes | — | yes | yes |
| J | yes | yes | yes | yes | yes | yes | — | yes | yes |
| AD0 | yes | yes | — | yes | yes | — | — | — | — |
| AD1 | yes | yes | — | yes | yes | — | — | — | — |

[1] Each cell represents one register with individual configuration bits

**FIGURE 12.35**   Register Availability per Port (Copyright of Freescale Semiconductor, Inc. 2007, used by permission.)

If the data direction register bits are set to logic level "1," the contents of the data register is returned. This is independent of any other configuration as shown in Figure 12.36.

***12.1.17.5   Input Register***   This is a read-only register and always returns the buffered state of the pin as shown in Figures 12.36.

***12.1.17.6   Data Direction Register***   This register defines whether the pin is used as an input or an output. If a peripheral module controls the pin, the contents of the data direction register is ignored (Figures 12.36).

**FIGURE 12.36**   Illustration of I/O Pin Functionality (Copyright of Freescale Semiconductor, Inc. 2007, used by permission.)

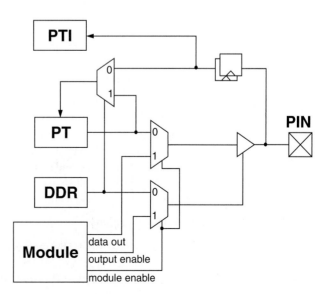

### 12.1.18  2 Kbyte EEPROM (EETX2K)

The EETX2K module is 2 Kbyte EEPROM (nonvolatile) memory block (Figure 12.37). A 4 Kbyte EEPROM block is also available. The EEPROM memory may be read as either bytes, aligned words, or misaligned words. Read access time is one bus cycle for bytes and aligned words and two bus cycles for misaligned words.

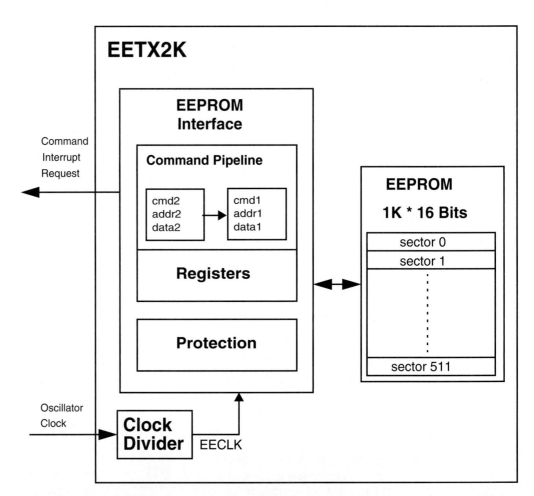

**FIGURE 12.37**    EEPROM Block Diagram (Copyright of Freescale Semiconductor, Inc. 2007, used by permission.)

The EEPROM memory is ideal for data storage for single-supply applications allowing for field reprogramming without requiring external voltage sources for program or erase. Program and erase functions are controlled by a command driven interface. The EEPROM module supports both block erase (all memory bytes) and sector erase (4 memory bytes). An erased bit reads 1 and a programmed bit reads 0. The high voltage required to program and erase the EEPROM memory is generated internally. It is not possible to read from the EEPROM block while it is being erased or programmed.

#### 12.1.18.1  Features

2 Kbytes of EEPROM memory divided into 512 sectors of 4 bytes
Automated program and erase algorithm
Interrupts on EEPROM command completion and command buffer empty

Fast sector erase and word program operation

Two-stage command pipeline

Sector erase abort feature for critical interrupt response

Flexible protection scheme to prevent accidental program or erase

Single power supply for all EEPROM operations, including program and erase

The EEPROM supports program, erase, and erase verify operations.

### 12.1.18.2 *Functional Description*

Write operations are used to execute program, erase, erase verify, sector erase abort, and sector modify algorithms described in this section. The program, erase, and sector modify algorithms are controlled by a state machine whose time base, EECLK, is derived from the oscillator clock via a programmable divider.

The command register as well as the associated address and data registers (Figure 12.38) operate as a buffer and a register (two-stage FIFO) so that a second command along with the

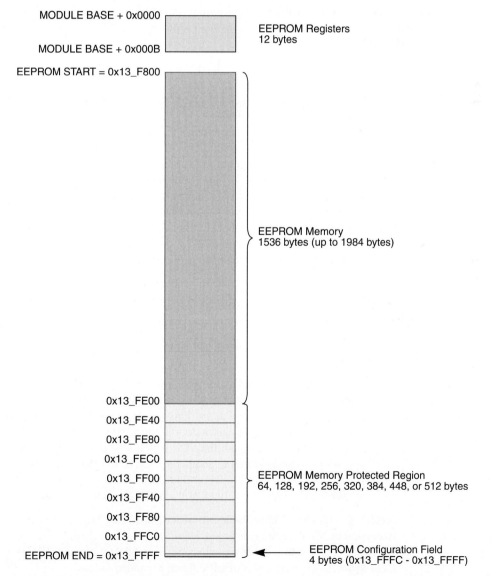

MODULE BASE + 0x0000

MODULE BASE + 0x000B

EEPROM Registers
12 bytes

EEPROM START = 0x13_F800

EEPROM Memory
1536 bytes (up to 1984 bytes)

0x13_FE00

0x13_FE40

0x13_FE80

0x13_FEC0

0x13_FF00

0x13_FF40

0x13_FF80

0x13_FFC0

EEPROM END = 0x13_FFFF

EEPROM Memory Protected Region
64, 128, 192, 256, 320, 384, 448, or 512 bytes

EEPROM Configuration Field
4 bytes (0x13_FFFC - 0x13_FFFF)

**FIGURE 12.38**    2 Kbyte EEPROM Memory Map (Copyright of Freescale Semiconductor, Inc. 2007, used by permission.)

necessary data and address can be stored to the buffer while the first command is still in progress. Buffer empty as well as command completion are signaled by flags in the EEPROM status register with interrupts generated, if enabled.

***12.1.18.3  EEPROM Module Security***    The EEPROM module does not provide any security information to the MCU. After each reset, the security state of the MCU is a function of information provided by the flash module.

## 12.1.19  512 Kbyte Flash Module (FTX512K4)

The flash memory is ideal for program and data storage for single-supply applications allowing for field reprogramming without requiring external voltage sources for program or erase (Figure 12.39). Program and erase functions are controlled by a command-driven interface. The flash module supports both block erase and sector erase. An erased bit reads 1, and a programmed bit reads 0. The high voltage required to program and erase the flash memory is generated internally. It is not possible to read from a flash block while it is being erased or programmed.

### 12.1.19.1   Features

    512 Kbytes of flash memory comprised of four 128 Kbyte blocks with each block divided
        into 128 sectors of 1024 bytes
    Automated program and erase algorithm
    Interrupts on flash command completion, command buffer empty
    Fast sector erase and word program operation
    Two-stage command pipeline for faster multiword program times
    Sector erase abort feature for critical interrupt response
    Flexible protection scheme to prevent accidental program or erase
    Single power supply for all flash operations, including program and erase
    Security feature to prevent unauthorized access to the flash memory
    Code integrity check using built-in data compression

The 512 Kbyte flash memory may be read as either bytes, aligned words, or misaligned words. Read access time is one bus cycle for bytes and aligned words and two bus cycles for misaligned words.

## 12.1.20  Security (SEC)

Hardware-based security is provided by this module. It is important to remember that part of the security must lie with the application code. An extreme example would be application code that dumps the contents of the internal memory. This would defeat the purpose of security. At the same time, it may be desirable to put a back door in the application program. An example would be downloading a security key through the SCI, which would allow access to a programming routine that updates parameters stored in another section of the flash memory.

***12.1.20.1  Features***    The security features of the S12X chip family (in secure mode) are

    Protect the contents of nonvolatile memories (flash, EEPROM)
    Execution of NVM commands is restricted
    Disable access to internal memory via background debug module (BDM)
    Disable access to internal flash/EEPROM in expanded modes
    Disable debugging features for CPU and XGATE

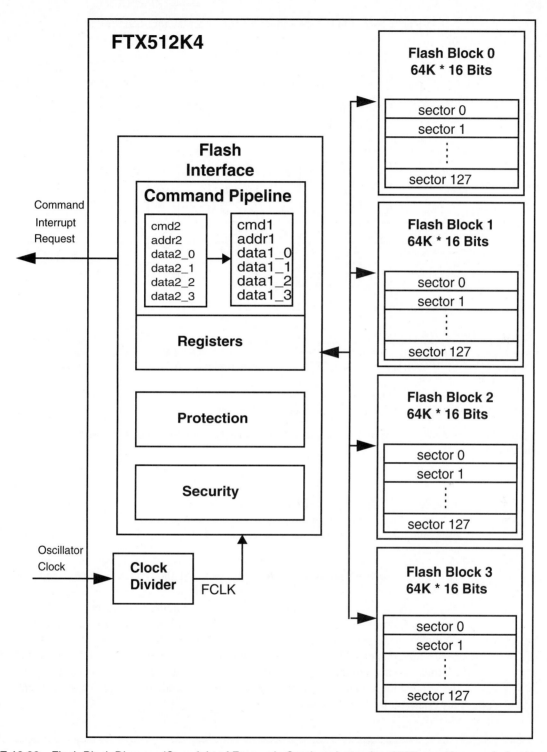

**FIGURE 12.39** Flash Block Diagram (Copyright of Freescale Semiconductor, Inc. 2007, used by permission.)

Secured operation has the following effects on the microcontroller:

| | Unsecure Mode | | | | | | Secure Mode | | | | | |
|---|---|---|---|---|---|---|---|---|---|---|---|---|
| | NS | SS | NX | ES | EX | ST | NS | SS | NX | ES | EX | ST |
| Flash Array Access | ✔ | ✔ | ✔[1] | ✔[1] | ✔[1] | ✔[1] | ✔ | ✔ | — | — | — | — |
| EEPROM Array Access | ✔ | ✔ | ✔ | ✔ | ✔ | ✔ | ✔ | ✔ | — | — | — | — |
| NVM Commands | ✔[2] | ✔ | ✔[2] | ✔[2] | ✔[2] | ✔ | ✔[2] | ✔[2] | ✔[2] | ✔[2] | ✔[2] | ✔[2] |
| BDM | ✔ | ✔ | ✔ | ✔ | ✔ | ✔ | — | ✔[3] | — | — | — | — |
| DBG Module Trace | ✔ | ✔ | ✔ | ✔ | ✔ | ✔ | — | — | — | — | — | — |
| XGATE Debugging | ✔ | ✔ | ✔ | ✔ | ✔ | ✔ | — | — | — | — | — | — |
| External Bus Interface | — | — | ✔ | ✔ | ✔ | ✔ | — | — | ✔ | ✔ | ✔ | ✔ |
| Internal status visible multiplexed on external bus | — | — | — | ✔ | ✔ | — | — | — | — | ✔ | ✔ | — |
| Internal accesses visible on external bus | — | — | — | — | — | ✔ | — | — | — | — | — | ✔ |

[1] Availability of Flash arrays in the memory map depends on ROMCTL/EROMCTL pins and/or the state of the ROMON/EROMON bits in the MMCCTL1 register. Please refer to the S12X_MMC block guide for detailed information.

[2] Restricted NVM command set only. Please refer to the FTX/EETX block guides for detailed information.

[3] BDM hardware commands restricted to peripheral registers only.

**FIGURE 12.40**   Features Availability in Unsecure and Secure Modes (Copyright of Freescale Semiconductor, Inc. 2007, used by permission.)

***12.1.20.2  Modes of Operation***   Once the flash and EEPROM have been programmed, the chip can be secured by programming the security bits located in the options/security byte in the flash memory array. These nonvolatile bits will keep the device secured through reset and power-down.

***12.1.20.3  Secured Microcontroller***   By securing the device, unauthorized access to the EEPROM and flash memory contents can be prevented. However, it must be understood that the security of the EEPROM and flash memory contents also depends on the design of the application program. For example, if the application has the capability of downloading code through a serial port and then executing that code (e.g., an application containing bootloader code), then this capability could potentially be used to read the EEPROM and flash memory contents, even when the microcontroller is in the secure state. In this example, the security of the application could be enhanced by requiring a challenge/response authentication before any code can be downloaded.

## 12.2    TEXAS INSTRUMENTS MSP430™ FAMILY

The MSP430™ 16-bit RISC CPU, peripherals and flexible clock system are combined by using a von Neumann common memory address bus (MAB) and memory data bus (MDB), as shown in Figure 12.41. The RISC-like instruction set executes the majority of instructions in a single

**FIGURE 12.41** MSP430 Architecture (Courtesy of Texas Instruments.)

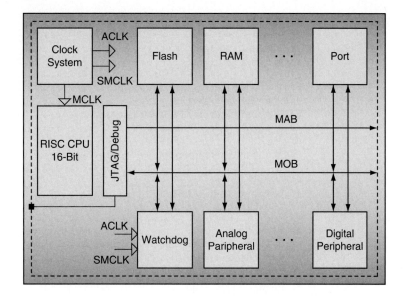

cycle. Incorporating a modern CPU with modular memory-mapped analog and digital peripherals, the MSP430 devices offer a comprehensive solution for mixed-signal applications.

The Texas Instruments MSP430 family of ultralow-power microcontrollers consists of several devices featuring different sets of peripherals targeted for various applications. The architecture, combined with five low-power modes, is optimized to achieve extended battery life in portable measurement applications. The device features a powerful 16-bit RISC CPU, 16-bit registers, and constant generators that contribute to maximum code efficiency.

The MSP430 MCU's orthogonal architecture provides the flexibility of 16 fully addressable, single-cycle 16-bit CPU registers (20-bit in the MSP430 arichitecture) and the power of a RISC (Figure 12.42). The modern design of the CPU offers versatility using only 27 easy-to-understand instructions and seven consistent-addressing modes. This results in a 16-bit low-power CPU that has more effective processing, is smaller-sized, and is more code efficient.

The basic device configurations include

1-KB to 120-KB ISP flash
RAM up to 10 KB
14- to 100-pin options

Ultralow-power features include

Multiple operating modes
0.1-μA power down

**FIGURE 12.42** MSP430 Register Set (Courtesy of Texas Instruments.)

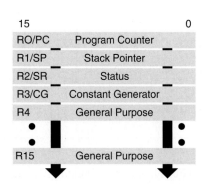

0.8-µA standby
250-µA/MIPS @ 3 V
Instant-on stable high-speed clock
1.8-V to 3.6-V operation
Zero-power BOR
<50-nA pin leakage
CPU that minimizes CPU cycles per task
Low-power peripheral options

Available integrated peripherals, depending on model, include

10-/12-bit SAR ADC
16-bit Sigma Delta ADC
12-bit DAC
Comparator
LCD driver
Supply voltage supervisor (SVS)
Operational amplifiers
16-bit and 8-bit timers
Ultralow power
Zero-power brown-out reset (BOR)
1-µs clock startup
<50-nA pin leakage
Watchdog timer
UART/LIN
I$^2$C
SPI
IrDA
Hardware multiplier
DMA controller
Temperature sensor

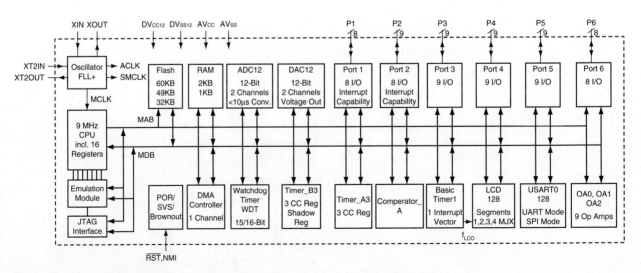

**FIGURE 12.43**   MSP430 Functional Block Diagram (Courtesy Texas Instruments.)

## 12.2.1 Low Power Design

The MSP430 is designed specifically for ultralow-power applications. A flexible clocking system, multiple operating modes, and zero-power always on brown-out reset (BOR) are implemented to reduce power consumption. Accompanied with appropriate programming methods, this can dramatically extend battery life. The MSP430 BOR function is always active, even in all low-power modes, to ensure the most reliable performance possible.

The MSP430 CPU architecture with 16 registers and 16-bit data and address buses minimize power-consuming fetches to memory, and a fast vectored-interrupt structure reduces the need for wasteful CPU software flag polling. Intelligent hardware peripheral feature design allows tasks to be completed more efficiently and independent of the CPU.

**FIGURE 12.44** MSP430 Power Consumption (Courtesy of Texas Instruments.)

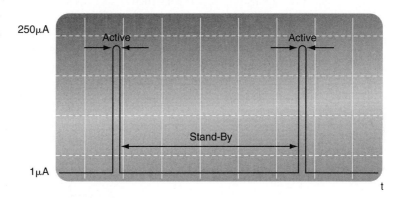

## 12.2.2 Flexible Clock System

The MSP430 MCU clock system is designed specifically for battery-powered applications. Multiple oscillators are utilized to support event-driven burst activity. A low-frequency Auxiliary Clock (ACLK) is driven directly from a common 32-kHz watch crystal or the internal very low power oscillator (VLO) with no additional external components required. The ACLK can be used for a background real-time clock self wake-up function.

An integrated high-speed digitally controlled oscillator (DCO) can source the master clock (MCLK) used by the CPU and submain clock (SMCLK) used by the high-speed peripherals. By design, the DCO is active and stable in 1 μs (F2xx) or 6 μs (x1xx, x4xx). This results in very high performance and ultralow power consumption.

**FIGURE 12.45** Multiple Oscillator Clock System (Courtesy of Texas Instruments.)

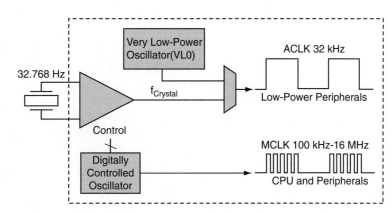

### 12.2.3 MSP430 CPU

The MSP430 CPU has a 16-bit RISC architecture that is highly transparent to the application. All operations, other than program-flow instructions, are performed as register operations in conjunction with seven addressing modes for source operand and four addressing modes for destination operand.

The MSP430 CPU is integrated with 16 16-bit registers that provide reduced instruction execution time. The MSP430X CPU has 20-bit registers as shown in Figure 12.46. The register-to-register operation execution time is one cycle of the CPU clock. Four of the registers, R0 to R3, are dedicated as program counter, stack pointer, status register, and constant generator, respectively. The remaining registers are general-purpose registers.

Peripherals are connected to the CPU using data, address, and control buses and can be handled with all instructions. The instruction set consists of 51 instructions (27 core and 24 emulated)

**FIGURE 12.46** MSP430X CPU Block Diagram (Courtesy of Texas Instruments.)

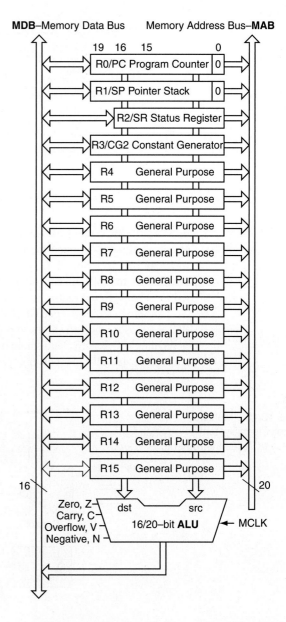

with three formats and seven address modes. Each instruction can operate on word and byte data. The core instructions are instructions that have unique op codes decoded by the CPU.

The emulated instructions are instructions that make code easier to write and read, but do not have op codes themselves; instead, they are replaced automatically by the assembler with an equivalent core instruction. There is no code or performance penalty for using emulated instruction.

There are three core-instruction formats:

Dual-operand
Single-operand
Jump

All single-operand and dual-operand instructions can be byte or word instructions. Byte instructions are used to access byte data or byte peripherals. Word instructions are used to access word data or word peripherals.

## 12.2.4 Operating Modes

The MSP430 has one active mode and five software selectable low-power modes of operation. An interrupt event can wake up the device from any of the five low-power modes, service the request, and restore back to the low-power mode on return from the interrupt program.

The following six operating modes can be configured by software:

Active mode (AM)
    All clocks are active
Low-power mode 0 (LPM0)
    CPU is disabled
        ACLK and SMCLK remain active, MCLK is disabled
        FLL + loop control remains active
Low-power mode 1 (LPM1)
    CPU is disabled
        FLL + loop control is disabled
        ACLK and SMCLK remain active, MCLK is disabled
Low-power mode 2 (LPM2)
    CPU is disabled
        MCLK, FLL + loop control, and DCOCLK are disabled
        DCO's dc-generator remains enabled
        ACLK remains active
Low-power mode 3 (LPM3)
    CPU is disabled
        MCLK, FLL + loop control, and DCOCLK are disabled
        DCO's dc-generator is disabled
        ACLK remains active
Low-power mode 4 (LPM4)
    CPU is disabled
        ACLK is disabled
        MCLK, FLL + loop control, and DCOCLK are disabled
        DCO's dc-generator is disabled
        Crystal oscillator is stopped

## 12.2.5 FLL$^+$ Clock Module

The frequency-locked loop (FLL$^+$) clock module supports low system cost and ultralow-power consumption. The functional block diagram is shown in Figure 12.47. Using three internal clock signals, the best balance of performance and low power consumption can be selected.

**FIGURE 12.47**  MSP430
Frequency-Locked Loop
(Courtesy of Texas Instruments.)

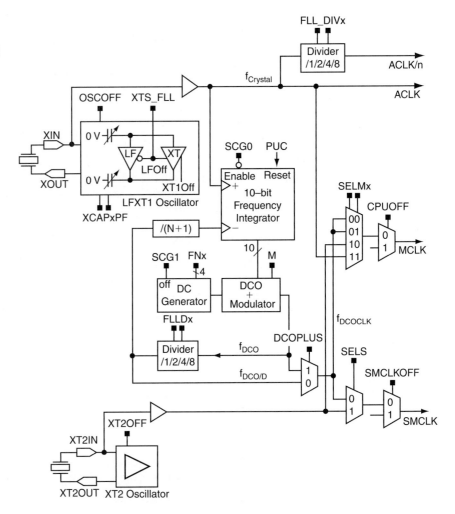

The FLL$^+$ features digital frequency-locked loop (FLL) hardware. The FLL operates together with a digital modulator and stabilizes the internal digitally controlled oscillator (DCO) frequency to a programmable multiple of the LFXT1 watch crystal frequency. The FLL$^+$ clock module can be configured to operate without any external components, with one or two external crystals, or with resonators under full software control.

Conflicting requirements typically exist in battery-powered MSP430 applications:

Low clock frequency for energy conservation and timekeeping
High clock frequency for fast reaction to events and fast burst processing capability
Clock stability over operating temperature and supply voltage

The FLL$^+$ clock module addresses the preceding conflicting requirements by allowing three available clock signals to be selected: ACLK, MCLK, and SMCLK. For optimal low-power performance, the ACLK can be configured to oscillate with a low-power 32,786-Hz watch crystal, providing a stable time base for the system and low power stand-by operation. The MCLK can be configured to operate from the on-chip DCO, stabilized by the FLL, and can activate when requested by interrupt events.

The digital frequency-locked loop provides decreased start-time and stabilization delay over an analog phase-locked loop. A phase-locked loop takes hundreds or thousands of clock cycles to start and stabilize. The FLL starts immediately at its previous setting.

## 12.2.6  Flash Memory Controller

The MSP430 flash memory is bit-, byte-, and word-addressable and programmable. The flash memory module has an integrated controller that controls programming and erase operations. The controller has three or four registers (see the device-specific data sheet), a timing generator, and a voltage generator to supply program and erase voltages.

MSP430 flash memory features include

Internal programming voltage generation
Bit, byte, or word programmable
Ultralow-power operation
Segment erase and mass erase

The block diagram of the flash memory and controller is shown in Figure 12.48.

**FIGURE 12.48**  Flash Memory Controller (Courtesy of Texas Instruments.)

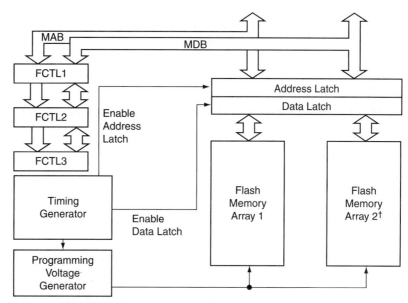

†MSP430FG461x devices only

MSP430FG461X devices have two flash memory arrays. Other MSP430 devices have one flash array. All flash memory is partitioned into segments. Single bits, bytes, or words can be written to flash memory, but the segment is the smallest size of flash memory that can be erased.

The flash memory is partitioned into main and information memory sections. There is no difference in the operation of the main and information memory sections. Code or data can be located in either section. The differences between the two sections are the segment size and the physical addresses. The information memory has four 64-byte segments on F47x devices or two 128-byte segments on all other 4xx devices. The main memory has two or more 512-byte segments. The segments are further divided into blocks. Figure 12.49 shows the flash segmentation using an example of 4-KB flash that has eight main segments and two information segments.

## 12.2.7  Hardware Multiplier

The hardware multiplier is a peripheral and is not part of the MSP430 CPU. This means that its activities do not interfere with the CPU activities. The multiplier registers are peripheral registers that are loaded and read with CPU instructions.

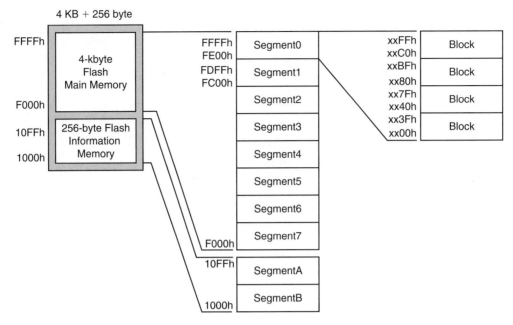

**FIGURE 12.49** Flash Memory Segments (4-KB) (Courtesy of Texas Instruments.)

The hardware multiplier supports

Unsigned multiply
Signed multiply
Unsigned multiply accumulate
Signed multiply accumulate
16×16 bits, 16×8 bits, 8×16 bits, 8×8 bits

The hardware multiplier supports unsigned multiply, signed multiply, unsigned multiply accumulate, and signed multiply accumulate operations. The type of operation is selected by the address the first operand is written to. The hardware multiplier block diagram is shown in Figure 12.50.

## 12.2.8 DMA Controller

The direct memory access (DMA) controller transfers data from one address to another, without CPU intervention, across the entire address range. For example, the DMA controller can move data from the ADC12 conversion memory to RAM. Devices that contain a DMA controller may have one, two, or three DMA channels available. Using the DMA controller can increase the throughput of peripheral modules. It can also reduce system power consumption by allowing the CPU to remain in a low-power mode without having to awaken to move data to or from a peripheral.

The DMA controller features include

Up to three independent transfer channels
Configurable DMA channel priorities
Requires only two MCLK clock cycles per transfer
Byte or word and mixed byte/word transfer capability
Block sizes up to 65535 bytes or words
Configurable transfer trigger selections

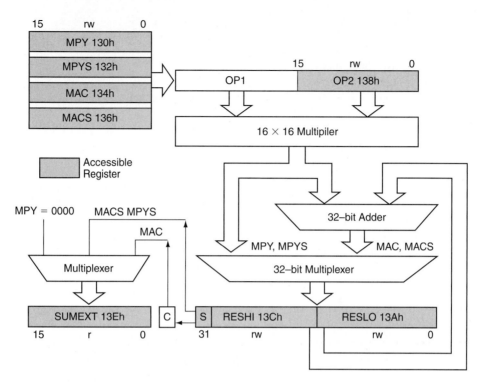

**FIGURE 12.50** Hardware Multiplier (Courtesy of Texas Instruments.)

Selectable edge or level-triggered transfer
Four addressing modes
Single, block, or burst-block transfer modes

## 12.2.9 Digital I/O

MSP430 devices have up to 10 digital I/O ports implemented, P1 to P10. Each port has eight I/O pins. Every I/O pin is individually configurable for input or output direction, and each I/O line can be individually read or written to.

Ports P1 and P2 have interrupt capability. Each interrupt for the P1 and P2 I/O lines can be individually enabled and configured to provide an interrupt on a rising edge or falling edge of an input signal. All P1 I/O lines source a single-interrupt vector, and all P2 I/O lines source a different single-interrupt vector.

The digital I/O features include

Independently programmable individual I/Os
Any combination of input or output
Individually configurable P1 and P2 interrupts
Independent input and output data registers

## 12.2.10 Watchdog Timer

The primary function of the watchdog timer (WDT) module is to perform a controlled system restart after a software problem occurs. If the selected time interval expires, a system reset is generated. If the watchdog function is not needed in an application, the module can be configured as an interval timer and can generate interrupts at selected time intervals.

Features of the watchdog timer module include

Four software-selectable time intervals
Watchdog mode
Interval mode
Access to WDT control register password protected
Control of RST/NMI pin function
Selectable clock source
Can be stopped to conserve power
Clock fail-safe feature in WDT+

After a PUC (Power Up Clear) condition, the WDT module is configured in the watchdog mode with an initial 32,768-cycle reset interval using the DCOCLK. The user must set up, halt, or clear the WDT prior to the expiration of the initial reset interval or another PUC will be generated. When the WDT is configured to operate in watchdog mode, either writing to WDTCTL with an incorrect password or expiration of the selected time interval triggers a PUC. A PUC resets the WDT to its default condition and configures the RST/NMI pin to reset mode.

## 12.2.11  Timers A and B

Timer_A is a 16-bit timer/counter with three or five capture/compare registers. Timer_A can support multiple capture/compares, PWM outputs, and interval timing. Timer_A also has extensive interrupt capabilities. Interrupts may be generated from the counter on overflow conditions and from each of the capture/compare registers.

Timer_A features include

Asynchronous 16-bit timer/counter with four operating modes
Selectable and configurable clock source
Three or five configurable capture/compare registers
Configurable outputs with PWM capability
Asynchronous input and output latching
Interrupt vector register for fast decoding of all Timer_A interrupts

A functional block diagram of Timer_A is shown in Figure 12.51. Timer_B is a 16-bit timer/counter with three or seven capture/compare registers. Timer_B can support multiple capture/compares, PWM outputs, and interval timing. Timer_B also has extensive interrupt capabilities. Interrupts may be generated from the counter on overflow conditions and from each of the capture/compare registers.

Timer_B features include

Asynchronous 16-bit timer/counter with four operating modes and four selectable lengths
Selectable and configurable clock source
Three or seven configurable capture/compare registers
Configurable outputs with PWM capability
Double-buffered compare latches with synchronized loading
Interrupt vector register for fast decoding of all Timer_B interrupts

Timer_B is identical to Timer_A with the following exceptions:

The length of Timer_B is programmable to be 8, 10, 12, or 16 bits.
Timer_B TBCCRx registers are double-buffered and can be grouped.
All Timer_B outputs can be put into a high-impedance state.
The SCCI bit function is not implemented in Timer_B.

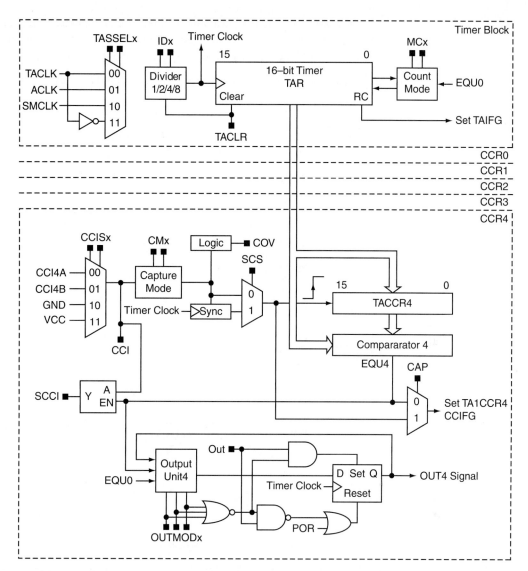

**FIGURE 12.51**    Timer_A Block Diagram (Courtesy of Texas Instruments.)

## 12.2.12  USART

The MSP430 has a Universal Synchronous/Asynchronous Receiver Transmitter that can operate in either UART or SPI mode. In synchronous mode, the USART connects to an external system via two external pins, URXD and UTXD. UART mode is selected when the SYNC bit is cleared.

UART mode features include

7- or 8-bit data with odd, even, or nonparity
Independent transmit and receive shift registers
Separate transmit and receive buffer registers
LSB-first data transmit and receive
Built-in idle-line and address-bit communication protocols for multiprocessor systems
Receiver start-edge detection for auto-wake up from LPMx modes

Programmable baud rate with modulation for fractional baud rate support
Status flags for error detection and suppression and address Detection
Independent interrupt capability for receive and transmit

Figure 12.52 shows the USART when configured for UART mode.

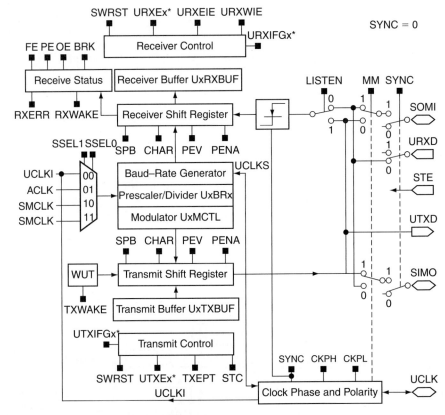

FIGURE 12.52   USART in UART Mode (Courtesy of Texas Instruments.)

In synchronous mode, the USART connects the MSP430 to an external system via three or four pins: SIMO, SOMI, UCLK, and STE. SPI mode is selected when the SYNC bit is set and the I²C bit is cleared.

SPI mode features include

7- or 8-bit data length
three-pin and four-pin SPI operation
Master or slave modes
Independent transmit and receive shift registers
Separate transmit and receive buffer registers
Selectable UCLK polarity and phase control
Programmable UCLK frequency in master mode
Independent interrupt capability for receive and transmit

## 12.2.13 USCI

The Universal Serial Communication Interface (USCI) modules support multiple serial communication modes. Different USCI modules support different modes. Each different USCI module is named with a different letter. For example, USCI_A is different from USCI_B, and so on. If more than one identical USCI module is implemented on one device, those modules are named with incrementing numbers. For example, if one device has two USCI_A modules, they are named USCI_A0 and USCI_A1.

The USCI_Ax modules support

UART mode
Pulse shaping for IrDA communications
Automatic baud rate detection for LIN communications
SPI mode

The USCI_Bx modules support

$I^2C$ mode
SPI mode

### 12.2.13.1  *UART Mode*

In asynchronous mode, the USCI_Ax modules connect the MSP430 to an external system via two external pins, UCAxRXD and UCAxTXD. UART mode is selected when the UCSYNC bit is cleared.

UART mode features include

7- or 8-bit data with odd, even, or nonparity
Independent transmit and receive shift registers
Separate transmit and receive buffer registers
LSB-first or MSB-first data transmit and receive
Built-in idle-line and address-bit communication protocols for multiprocessor systems
Receiver start-edge detection for auto wake-up from LPMx modes
Programmable baud rate with modulation for fractional baud rate support
Status flags for error detection and suppression
Status flags for address detection
Independent interrupt capability for receive and transmit

Figure 12.53 shows the USCI_Ax when configured for UART mode.

### 12.2.13.2  *SPI Mode*

In synchronous mode, the USART connects the MSP430 to an external system via three or four pins: SIMO, SOMI, UCLK, and STE. SPI mode is selected when the SYNC bit is set and the $I^2C$ bit is cleared.

SPI mode features include

7- or 8-bit data length
Three-pin and four-pin SPI operation
Master or slave modes
Independent transmit and receive shift registers
Separate transmit and receive buffer registers
Selectable UCLK polarity and phase control
Programmable UCLK frequency in master mode
Independent interrupt capability for receive and transmit

Figure 12.54 shows the USCI when configured for SPI mode.

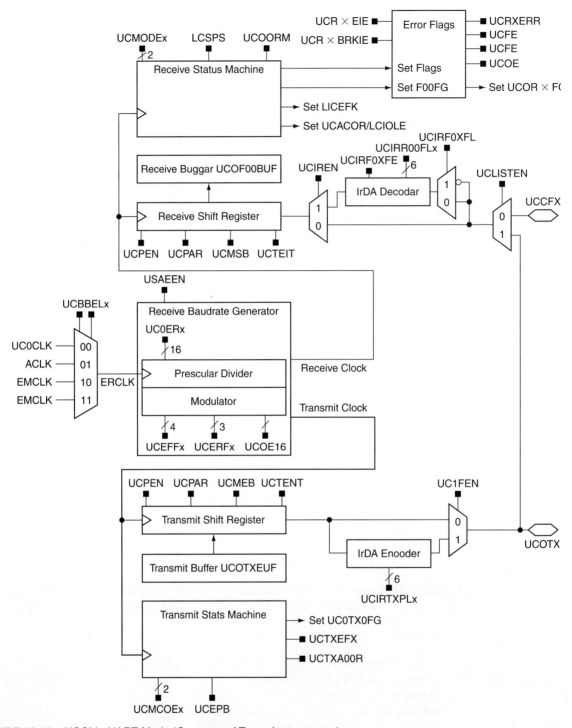

**FIGURE 12.53**   USCI in UART Mode (Courtesy of Texas Instruments.)

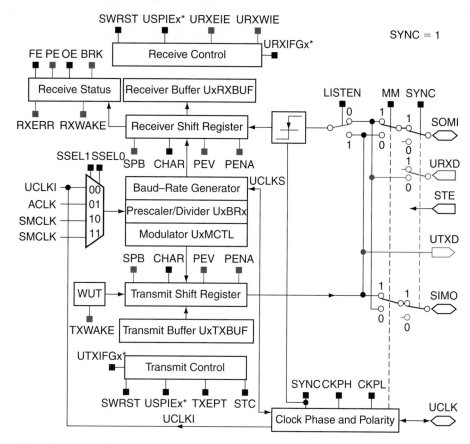

**FIGURE 12.54** USCI in SPI Mode (Courtesy of Texas Instruments.)

**12.2.13.3 I²C Mode** In I²C mode, the USCI module provides an interface between the MSP430 and I²C-compatible devices connected by way of the two-wire I²C serial bus. External components attached to the I²C bus serially transmit and/or receive serial data to/from the USCI module through the 2-wire I²C interface.

The I²C mode features include

Compliance to the Philips Semiconductor I²C specification v2.1
7-bit and 10-bit device addressing modes
General call
START/RESTART/STOP
Multimaster transmitter/receiver mode
Slave receiver/transmitter mode
Standard mode up to 100 kbps and fast mode up to 400 kbps support
Programmable UCxCLK frequency in master mode
Designed for low power
Slave receiver START detection for auto wake-up from LPMx Modes
Slave operation in LPM4

Figure 12.55 shows the USCI when configured in I²C mode.

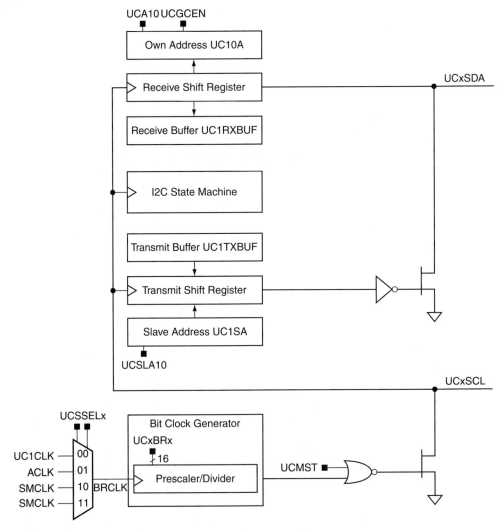

**FIGURE 12.55**   USCI in I²C Mode (Courtesy of Texas Instruments.)

## 12.2.14 ADC12 Function

The ADC12 module supports fast, 12-bit analog-to-digital conversions. The module implements a 12-bit SAR core, sample select control, reference generator, and a 16-word conversion-and-control buffer. The conversion-and-control buffer allows up to 16 independent ADC samples to be converted and stored without any CPU intervention.

ADC12 features include

Greater than 200-ksps maximum conversion rate
Monotonic 12-bit converter with no missing codes
Sample-and-hold with programmable sampling periods controlled by software or timers
Conversion initiation by software, Timer_A, or Timer_B
Software selectable on-chip reference voltage generation (1.5 V or 2.5 V)
Software selectable internal or external reference

Eight individually configurable external input channels (12 on MSP430FG43x and MSP430FG461x devices)

Conversion channels for internal temperature sensor, AVcc, and external references

Independent channel-selectable reference sources for both positive and negative references

Selectable conversion clock source

Single-channel, repeat-single-channel, sequence, and repeat-sequence conversion modes

ADC core and reference voltage can be powered down separately

Interrupt vector register for fast decoding of 18 ADC interrupts

16 conversion-result storage registers

The block diagram of ADC12 is shown in Figure 12.56.

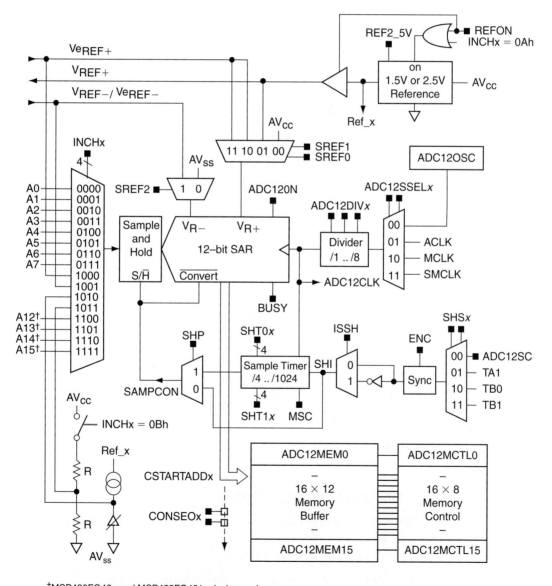

†MSP430FG43x and MSP430FG461x devices only

**FIGURE 12.56** ADC12 Functional Block Diagram (Courtesy of Texas Instruments.)

## 12.2.15 DAC12

The DAC12 module is a 12-bit, voltage output DAC. The DAC12 can be configured in 8-bit or 12-bit mode and may be used in conjunction with the DMA controller. When multiple DAC12 modules are present, they may be grouped together for synchronous update operation.

Features of the DAC12 include

12-bit monotonic output
8-bit or 12-bit voltage output resolution
Programmable settling time versus power consumption
Internal or external reference selection
Straight binary or 2s compliment data format
Self-calibration option for offset correction
Synchronized update capability for multiple DAC12s

Figure 12.57 shows the block diagram of the DAC12.

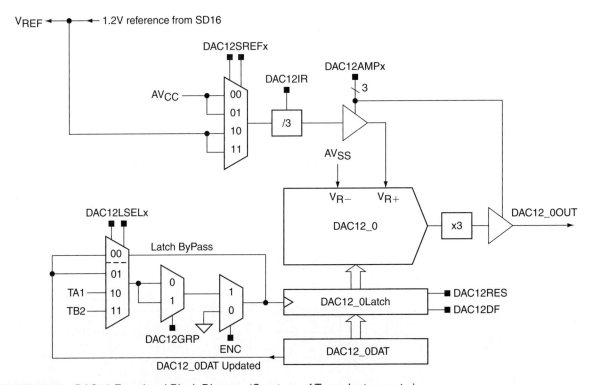

**FIGURE 12.57**    DAC12 Functional Block Diagram (Courtesy of Texas Instruments.)

## 12.2.16 Embedded Emulation Module

Every MSP430 flash-based microcontroller implements an embedded emulation module (EEM). It is accessed and controlled through JTAG. Each EEM is device specific.

In general, the following features are available:

Nonintrusive code execution with real-time breakpoint control
Single step, step into, and step over functionality
Full support of all low-power modes

Support for all system frequencies, for all clock sources

Up to eight (device dependent) hardware triggers/breakpoints on memory address bus (MAB) or memory data bus (MDB)

Up to two (device dependent) hardware triggers/breakpoints on CPU register write accesses

MAB, MDB, and CPU register access triggers can be combined to form up to eight (device dependent) complex triggers/breakpoints

Trigger sequencing (device dependent)

Storage of internal bus and control signals using an integrated trace buffer (device dependent)

Clock control for timers, communication peripherals, and other modules on a global device level or on a per-module basis during an emulation stop

Figure 12.58 shows a simplified block diagram of the largest 4xx EEM implementation.

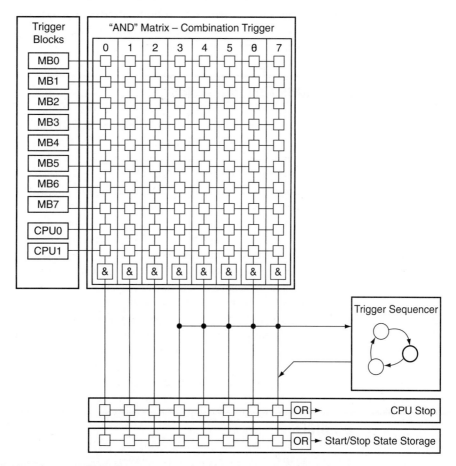

**FIGURE 12.58** EEM Block Diagram (Courtesy of Texas Instruments.)

***12.2.16.1 Triggers*** The event control in the EEM of the MSP430 system consists of triggers, which are internal signals indicating that a certain event has happened. These triggers may be used as simple breakpoints, but it is also possible to combine two or more triggers to allow detection of complex events and trigger various reactions besides stopping the CPU.

In general, the triggers can be used to control the following functional blocks of the EEM:

Breakpoints (CPU stop)
State storage
Sequencer

There are two different types of triggers, the memory trigger and the CPU register write trigger. Each memory trigger block can be independently selected to compare either the MAB or the MDB with a given value. Depending on the implemented EEM, the comparison can be $=$, $\neq$, $\geq$, or $\leq$. The comparison can also be limited to certain bits with the use of a mask. The mask is either bit-wise or byte-wise, depending on the device. In addition to selecting the bus and the comparison, the condition under which the trigger is active can be selected. The conditions include read access, write access, DMA access, and instruction fetch.

Each CPU register write trigger block can be independently selected to compare what is written into a selected register with a given value. The observed register can be selected for each trigger independently. The comparison can be $=$, $\neq$, $\geq$, or $\leq$. The comparison can also be limited to certain bits with the use of a bit mask. Both types of triggers can be combined to form more complex triggers. For example, a complex trigger can signal when a particular value is written into a user-specified address.

## QUESTIONS AND PROBLEMS

1. What is the primary reason to implement the XGATE function?
2. Why have the XGATE's R0 register value tied to zero?
3. How many external pins can support general-purpose I/O in the S12XD?
4. What does the MMC do in the S12XD?
5. Describe the advantage of the breakpoint function in the debug module.
6. Why have multiple channels on the PWM block?
7. Why use nonmaskable interrupts in a microcontroller based design?
8. Which architecture does the A/D converter in the MSP430 utilize?
9. Why use the I$^2$C bus in a multichip design?
10. List three reasons to use the SPI function instead of I$^2$C.
11. Describe an application using the S12XD MSCAN module.
12. Describe five low-power MSP430 applications.
13. What is the advantage to an orthogonal register file?
14. What is the purpose of the watchdog timer function in interrupt-driven design?
15. Why use DMA in an interrupt-driven design with the MSP430?
16. Describe an application where you would use the capture/compare function.
17. How does controlling clock frequency affect low power operation?

## SOURCES

*MC9S12XDP512 data sheet covers S12XD, S12XB & S12XA families, HCS12X microcontrollers, MC9S12XPD512, Rev. 2.17*. July 2007. Freescale Semiconductor.
*MSP430FG43X mixed signal microcontroller, SLAS380B*. June 2007. Texas Instruments.
*MSP430 ultra-low-power microcontrollers*. product brochure. SLAB043M. 2007. Texas Instruments.
*MSP430x4xx family user's guide, mixed signal products, SLAU056G*. 2007. Texas Instruments.
*S12XCPUV1 reference manual, HCS12X microcontrollers, S12XCPUV1, v01.01*. March 2005. Freescale Semiconductor.
*XGATE coprocessor fact sheet, XGATE COPROCFS REV 0*. Freescale Semiconductor.

# CHAPTER 13

## Intellectual Property SoC Cores

**OBJECTIVE: AN INTRODUCTION TO IP-BASED EMBEDDED CORE ARCHITECTURES**

The reader will learn about configurable core design and the features of two IP cores:

1. Overview of Configurable Core SoC Design.
2. MIPS32 4K Family Embedded Cores.
3. ARM10 Embedded Core.

## 13.0  SoC OVERVIEW

To remain competitive, system-on-a-chip (SoC) designers must keep pace with silicon technology's rapid evolution. New communication, consumer, and computer product designs must exhibit rapid increases in functionality, reliability, and bandwidth. Rapid declines in cost and power consumption must also be considered.

All these improvements dictate increasing use of high-integration silicon, in which designers traditionally use register-transfer-level (RTL) hardware to realize data-intensive capabilities, as shown in Figure 13.1. Three forces, the design productivity gap, the growing cost of nanometer-level semiconductor manufacturing, and the global time-to-market imperative, put intense pressure on chip designers to develop more complex systems more quickly and with lower cost.

One approach to speeding development of mega-gate SoCs uses multiple microprocessor cores to perform much of the processing currently relegated to RTL techniques, as shown in Figure 13.2. Although general-purpose embedded processors can handle many tasks, they often lack the bandwidth needed to perform particularly complex jobs, such as audio and video processing. Hence the historic rise of RTL use in SoC design.

Developers can configure a new class of processor, automatically generated extensible microprocessor cores such as Tensilica's Xtensa LX2, or user modifiable cores such as MIPS Technologies' M4K, to bring the required amount and type of processing bandwidth to bear on many embedded tasks. Because these configurable processors employ firmware instead of RTL-defined hardware for their control algorithm, designers can develop and verify processor-based task engines for many embedded SoC tasks more quickly and easily than they could develop and verify RTL-based hardware blocks that perform the same tasks.

**FIGURE 13.1** Fixed RTL SoC Implementation (Reprinted by permission of Tensilica Incorporated.)

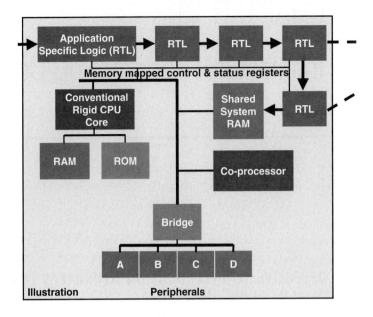

**FIGURE 13.2** Multi-Processor Xtensa SoC Implementation (Reprinted by permission of Tensilica Incorporated.)

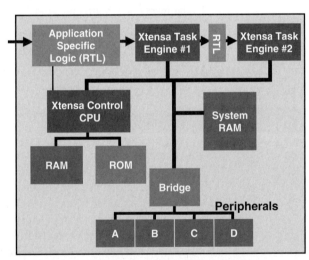

## 13.1          SoC DESIGN CHALLENGES

A few characteristics of typical deep-submicron integrated circuit (IC) design illustrate the challenge facing SoC design teams:

- In a generic 0.13 μ standard-cell foundry process, silicon density routinely exceeds 100,000 usable gates per square millimeter. Consequently, a low-cost chip with a core area of 50 square millimeters can carry 5 million logic gates. Simply because it's possible, a system designer somewhere will find a way to exploit this immense computational potential in any given market.
- In the past, silicon capacity and design-automation tools limited the practical size of an RTL block to fewer than 100,000 gates. Improved synthesis, place-and-route, and verification tools have raised that ceiling. Blocks of 500,000 gates are now within the capacity of these tools, but existing design and verification methods are not keeping pace with silicon fabrication capacity, which can now put millions of gates on an SoC.

- The design complexity of a typical logic block grows more rapidly than its gate count, and system complexity increases more rapidly than the number of constituent blocks. Verification complexity has also increased disproportionately with gate count. Consequently, many teams that have recently developed real-world designs report that they now spend as much as 90 percent of their development effort on block- or system-level verification.
- The cost of a design bug is increasing. Industry analysts make much of the rising cost of deep sub micron IC masks: The cost of a full mask set approaches $1 million. However, mask charges represent just the tip of the iceberg with respect to design-bug costs. The risk of bugs compounds the costs. The combination of larger teams required to create complex SoC designs, higher staff costs, bigger nonrecurring engineering fees, and lost profitability and market share make show-stopper design bugs intolerable. SoC design bugs can literally kill a company. As a result, design methods that reduce the occurrence of such showstoppers, or permit painless workarounds for them, pay for themselves rapidly.
- All embedded systems now contain significant amounts of software. Software integration is typically the last step in the system development process, and this step is routinely blamed for overall program delays. Analysts widely view earlier and faster hardware and software validation as a critical risk-reducer for new product development projects.
- Standard communication protocols are rapidly increasing in complexity. The need to conserve scarce communications spectrum, plus the inventiveness of modern protocol designers, has resulted in the creation of complex new standards such as IPv6 for packet forwarding, G.729 voice coding, JPEG2000 image compression, MPEG-4 video, and Rijndael AES encryption.

These new protocols, coupled with rising communication bit rates, demand much greater computational throughput than their predecessors.

Competitive pressures have pushed development of the next generation SoC, characterized by dozens of functions working together. Such designs illustrate the trend toward using many RTL-based logic blocks and mixing control and digital signal processors together on the same chip, as shown in Figure 13.3.

**FIGURE 13.3** RTL Proliferation (Reprinted by permission of Tensilica Incorporated.)

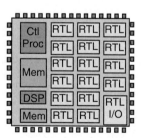

This ceaseless growth in integrated circuit complexity poses a central dilemma for SoC design. If developers could implement all these logic functions with multiple cheap, fast, and efficient heterogeneous processor blocks, a processor-based design approach would be ideal because using predesigned and preverified processor cores for an SoC's individual functional blocks moves the design effort largely to the coding of several relatively small software blocks.

This approach to SoC design permits bug fixes in minutes instead of months because changing and verifying software is much easier than altering RTL hardware, especially if the code is stored in on-chip RAM. Unfortunately, for the most computationally demanding problems, general-purpose processor cores fall far short with respect to application throughput, cost, and power efficiency.

At the same time, designing the custom RTL logic for complex functions and emerging standards takes too long, and once designed, the logic is too rigid to change easily. A closer look at the makeup of the typical RTL block, shown in Figure 13.4(a), gives insight into this paradox, while Figures 13.4(b) shows a more flexible alternative.

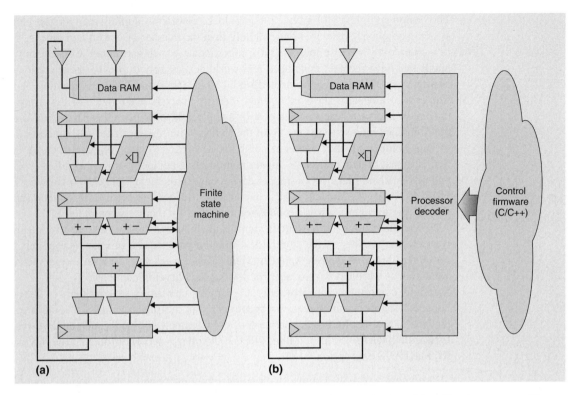

**FIGURE 13.4**    (a) Traditional RTL; (b) Configurable SoC (Reprinted by permission of Tensilica Incorporated.)

In most RTL designs, the data path consumes the vast majority of the logic block's gates. A typical data path can be as narrow as 16 or 32 bits or it can be hundreds of bits wide. The data path's width is generally sized to the task at hand. A data path typically contains many data registers, representing intermediate computational states, and often has significant blocks of RAM, or interfaces to RAM, that it shares with other RTL blocks. These basic data path structures reflect the data's nature and are largely independent of the finer details of the specific algorithm that operates on that data.

By contrast, the RTL logic block's finite state machine contains nothing but control details. This RTL block subsystem captures all the nuances of sequencing data through the data path, all exception and error conditions, and all handshakes with other blocks. The state machine may consume only a few percent of the block's gate count, but it embodies most of the design and verification risk due to its complexity. If developers make a late design change in an RTL block, the change is more likely to affect the state machine than the data path's structure, thus heightening the design risk.

Configurable, extensible processors are a fundamentally new form of microprocessor that provide a way of reducing the risk of state-machine design. This is accomplished by replacing state-machine logic blocks that are hard to design and verify with predesigned, preverified processor cores and application firmware.

## 13.1.1 Configurable Processors

Rapidly increasing logic complexity and technology scaling, as characterized by Moore's law, make multimillion-gate designs feasible. Fierce product competition in system features and capabilities generates demand for these advanced silicon designs. A well-recognized SoC design gap, which lies between the growth in chip complexity and the productivity growth in logic design tools, widens every year, as Figure 13.5 shows.

**FIGURE 13.5** Design Complexity and Designer Productivity (Reprinted by permission of Tensilica Incorporated.)

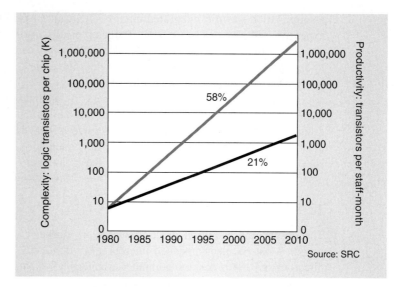

Moreover, market trends favoring high-performance, low-power systems also increase the number of SoC designs. Applications such as long-battery-life cell phones, 8-megapixel digital cameras, fast and inexpensive color printers, high-definition digital televisions, and 3-D video games push available design resources. Unless something closes this design gap, it will soon become impossible to bring enhanced versions of these system designs to market.

As Figure 13.6 shows, the conventional SoC design model closely follows that of its predecessor, the board-level combination of a standard microprocessor, memory, and logic built

**FIGURE 13.6** Board to SoC Transition (Reprinted by permission of Tensilica Incorporated.)

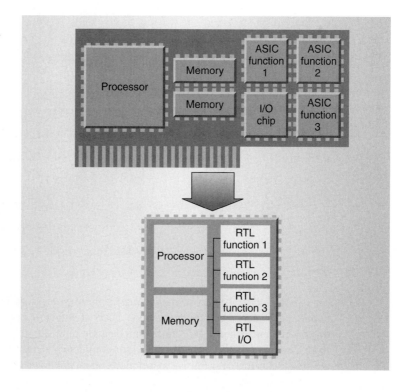

as application-specific integrated circuits (ASICs). Board-level, chip-to-chip interconnect is expensive and slow, so board-level designs typically use shared buses and narrow data paths. These are often only 32 bits wide. Designers frequently carry these relatively limited buses over to SoC designs because this approach provides the easiest solution to crafting an SoC architecture: just reuse what's been done before.

Combining all these system components on a single piece of silicon increases maximum achievable clock frequency and decreases power dissipation relative to the equivalent board-level design. System reliability and cost often improve as well. These benefits alone can justify investment in SoC design. However, the shift to SoC integration does not automatically change a design's organization or architecture. Thus, the architecture of these chips typically inherits the assumptions, limitations, and trade-offs of board-level design.

## 13.1.2  SoC Integration

The origins and evolution of microprocessors further constrain their use in traditional SoC design. Most popular embedded microprocessors, especially the 32-bit architectures, descend directly from 1980s desktop computer architectures such as ARM, MIPS, Freescale ColdFire, and Intel Viiv. Designed to serve general-purpose applications, these processors typically support only the most generic data types, such as 8-, 16-, 32-, and 64-bit integers. A typical processor layout is shown in Figure 13.7. They support only the most common operations, such as integer load, store, add, shift, compare, and bitwise logical operations.

**FIGURE 13.7** General-Purpose Processor (Reprinted by permission of Tensilica Incorporated.)

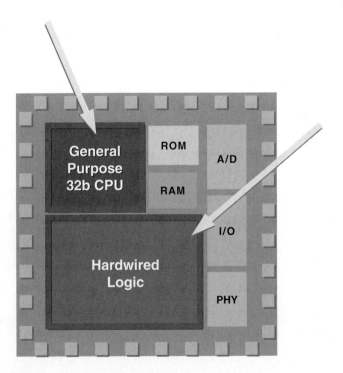

Their general-purpose nature makes these processors well suited to the diverse mix of applications run on computer systems. Their architectures perform equally well when running databases, spreadsheets, PC games, and desktop publishing. However, all these processors suffer from a common bottleneck: their need for complete generality dictates their ability to execute an arbitrary sequence of primitive instructions on an unknown range of data types. Put another way, general-purpose processors are not optimized to deal with the specific data types of any given embedded task, which results in inefficiencies.

Compared to general-purpose computer systems, embedded systems comprise a more diverse group and individually show more specialization. A digital camera must perform a variety of complex image processing tasks, but it never executes SQL database queries. A network switch must handle complex communications protocols at optical interconnect speeds, but it doesn't need to process 3-D graphics.

The specialized nature of individual embedded applications creates two issues for general-purpose processors in data-intensive embedded applications. First, the critical functions of many embedded applications and a processor's basic integer instruction set and register file are a poor match. Because of this mismatch, critical embedded applications require more computation cycles when they run on general-purpose processors.

Second, more focused embedded devices cannot take full advantage of a general-purpose processor's broad capabilities. Expensive silicon resources built into the processor go to waste because the specific embedded task that's assigned to the processor doesn't need them.

Many embedded systems interact closely with the real world or communicate complex data at high rates. A hypothetical general-purpose microprocessor running at tremendous speed could perform these data-intensive tasks. This is the basic assumption behind the use of multi-GHz processors in today's PCs. That is to say, throw a fast enough processor at a problem, regardless of the cost in dollars or power dissipation, and you can solve any problem. This is pictured in Figure 13.8.

**FIGURE 13.8** GP Processor Comparison to Embedded Processor (Reprinted by permission of Tensilica Incorporated.)

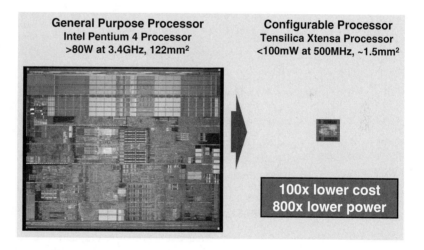

For many embedded tasks, however, no such processor exists today as a practical alternative because the fastest available processors typically cost orders of magnitude too much and dissipate orders of magnitude too much power to meet embedded-system design goals. Instead, embedded system hardware designers have traditionally turned to hardwired circuits to perform these data-intensive functions.

In the past 10 years, the wide availability of logic synthesis and ASIC design tools has made RTL design the standard for hardware developers. Reasonably efficient compared to custom transistor-level circuit design, RTL-based design can effectively exploit the intrinsic parallelism of many data-intensive problems. RTL design methods can often achieve tens or hundreds of times the performance a general-purpose processor achieves.

### 13.1.3  Extensible Processors

Like RTL-based design using logic synthesis, extensible-processor technology enables the design of high-speed logic blocks tailored to a specific task. The two technologies differ in that RTL designers realize both specialized data paths and the control state machines in hardware, but when building logic blocks with extensible processors, designers can create optimized data paths in hardware while implementing the control functions entirely in firmware (refer to Figures 13.4(a) and 13.4(b)).

A fully featured, configurable, and extensible processor consists of a processor design and the design tool environment for configuring that processor. The basic elements of the Xtensa design methodology are shown in Figure 13.9. This environment permits significant adaptation of the base processor design by letting a system designer change major processor functions, thus tuning the processor to specific application requirements. Typical configurability forms include additions, deletions, and modifications to memories, to external bus widths and handshake protocols, and to commonly used processor peripherals. An important superset of configurable processors, the extensible processor, lets the application developer extend the processor's instruction set and include features that the processor's original designers never considered or imagined.

**FIGURE 13.9**  Tensilica Xtensa Design Environment (Reprinted by permission of Tensilica Incorporated.)

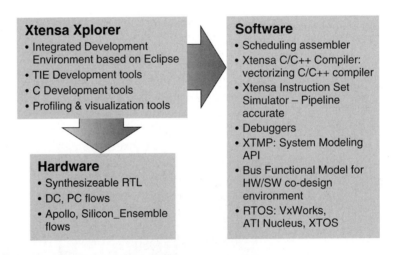

### 13.1.4  Extensible Processors as RTL Alternatives

Hardwired RTL design has many attractive characteristics, including small area, low power, and high throughput. However, RTL technology's liabilities, difficult design, slow verification, and poor scalability to complex problems have begun to overshadow its benefits now that designers work with millions of gates. A design methodology that retains most RTL efficiency benefits but reduces design time and risk has a natural appeal. Replacing complex RTL designs with application-specific processors can achieve this goal.

An application-specific processor can implement data path operations that closely match those of RTL functions. A chip architect can implement the processor's integer pipeline, plus additional execution units, registers, and other functions tailored to a specific application. In one example of defining such a processor using TIE, the Tensilica Instruction Extension language, a variant of Verilog, designers optimize a processor for high-level specification of data path functions in the form of instruction semantics and encoding. An example of TIE code is shown in Figure 13.10.

**FIGURE 13.10** Example TIE Code (Reprinted by permission of Tensilica Incorporated.)

```
regfile vec 160 16 v

operation MULA18.0 {inout vec acc, in vec m0, in vec m1} {} {
    wire [39:0] sum0 = m0[ 17:   0] * m1[ 17:   0] + acc[ 40:   0];
    wire [39:0] sum1 = m0[ 57:  40] * m1[ 57:  40] + acc[ 79:  40];
    wire [39:0] sum2 = m0[ 97:  80] * m1[ 97:  80] + acc[119:  80];
    wire [39:0] sum3 = m0[137:120] * m1[137:120] + acc[159:120];
    assign accum = {sum3, sum2, sum1, sum0}; }

schedule mula {MULA18.0} {
    use m0  4;  use m1  4;  use acc 5;  def acc 5; }
```

More concise than an RTL description, a TIE description omits all sequential logic, including state-machine descriptions, pipeline registers, and initialization sequences. The firmware programmer has access to the new processor instructions and registers described in TIE via the same compiler and assembler that employ the processor's base instructions and register set. Firmware uses the processor's normal instruction fetch, decode, and execution mechanisms to control all operation sequencing within the processor's data paths. Developers use a high-level language such as C or C++ to write this firmware.

Extended processors used as RTL-block replacements routinely have the same structures as traditional data-path-intensive RTL blocks: deep pipelines, parallel execution units, problem-specific state registers, and wide data paths to local and global memories. These extended processors can sustain the same high computation throughput and support the same low-level data interfaces as typical RTL designs. The control of extended-processor data paths works very differently, however. Instead of hardwired state machines, processor-based task engines use firmware for data path control.

## 13.1.5 Explicit Control Scheme

With firmware-controlled state transitions, designers do not fix the cycle-by-cycle control of the processor's data paths. Instead, they make the sequence of operations explicit in the firmware executed by the processor, as Figure 13.4b shows. The processor makes control-flow decisions explicitly in branches, makes memory references explicit in load and store operations, and makes sequences of computations explicit in sequences of general-purpose and application-specific computational operations. This design migration from hardwired state machine to firmware program control has the following implications:

*Flexibility.* Chip developers, system builders, and (when appropriate) end users can change the block's function just by changing the firmware.

*Software-based development.* Developers can use sophisticated, low-cost software development methods to implement most chip features.

*Faster, more complete system modeling.* RTL simulation is slow. For a 10-million-gate design, even the fastest software-based logic simulator may not exceed a few cycles per second. By contrast, firmware simulations for extended processors run at hundreds of thousands of cycles per second.

*Unification of control and data.* No modern system consists solely of hardwired logic: Such systems always incorporate a processor and some software. Moving functions previously handled by RTL functions into a processor removes the artificial distinction between control and data processing.

*Improved time to market.* Moving critical functions from RTL methods to application-specific processor engines simplifies SoC design, accelerates system modeling, and pulls in hardware finalization. Firmware-based engines easily accommodate changes to standards because hardware schedules are decoupled from finalization of product requirements details.

*Increased designer productivity.* Most important, migration from RTL-based design to using preverified, correct-by-construction, application-specific processors boosts the engineering team's productivity by reducing the resources needed for RTL development and verification. A processor-based SoC design approach sharply cuts risks of fatal logic bugs and permits graceful recovery when testers discover a bug.

Despite their attractions, application-specific processors may not be the best choice for all block designs. Consider these three exceptions:

*Small, fixed-state machines.* Some logic tasks are too trivial to warrant a processor. Bit-serial engines such as simple universal asynchronous receiver-transmitters fall into this category.

*Simple data buffering.* Similarly, some logic tasks amount to no more than storage control. Memory operations within a processor can emulate a first-in, first-out controller built with random-access memory and some wrapper logic, but a basic FIFO is faster and simpler.

*Very deep pipelines.* Some computation problems have so much regularity and so little state machine control that a single very deep pipeline provides the ideal implementation. The common examples (3-D graphics and magnetic-disk read-channel chips) sometimes have pipelines hundreds of clock stages deep. An application-specific processor could be used to control such deep pipelines, but the benefits of instruction-by-instruction control would be of less help in these applications.

Aside from these few caveats, firmware program control's advantages make it a wise design choice. The migration of functions from software to hardwired logic, over time, presents a well-known phenomenon. During early design exploration of prerelease protocol standards, processor-based implementations are common even for simple standards that clearly allow efficient logic-only implementations. Some common standards that have followed this path include popular video codecs such as MPEG-2, 3G wireless protocols such as W-CDMA, and encryption and security algorithms such as SSL and triple-DES.

However, the large gap in performance and design ease between software-based and RTL-based development has limited this migration. The emergence of configurable and extensible application-specific processors creates a new design path that's quick and easy enough for the development and refinement of new protocols and standards, yet efficient enough in silicon area and power to permit very-high-volume deployment.

## 13.2     THE MIPS32 4 K PROCESSOR CORE FAMILY

The MIPS32 4 K processor cores from MIPS Technologies are high-performance, low-power, 32-bit MIPS RISC cores intended for custom system-on-silicon applications. The cores (refer to Figure 13.11) are designed for semiconductor manufacturing companies, ASIC developers, and system OEMs who want to rapidly integrate their own custom logic and peripherals with a high-performance RISC processor. The cores are fully synthesizable to allow maximum flexibility; they are highly portable across processes and can easily be integrated into full system-on-silicon designs. This allows developers to focus their attention on end-user-specific characteristics of their product.

The 4KE family has three members: the 4KEc, 4KEm, and 4KEp cores. The cores incorporate aspects of both the MIPS Technologies R3000 and R4000 processors. The three devices differ mainly in the type of multiply-divide unit (MDU) and the memory management unit (MMU).

The 4KEc core contains a fully-associative translation lookaside buffer (TLB)-based MMU and pipelined MDU.

**FIGURE 13.11** MIPS 4KE Family Block Diagram (Reprinted by permission of MIPS.)

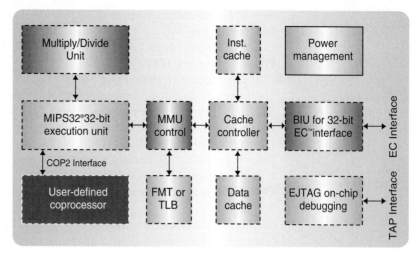

The 4KEm core contains a fixed mapping (FM) mechanism in the MMU that is smaller and simpler than the TLB-based implementation used in the 4KEc core. A pipelined MDU (like the 4KEc core) is used.

The 4KEp core contains the same FM-based MMU (like the 4KEm core), but a smaller non-pipelined MDU. Optional instruction and data caches are fully programmable from 0 to 64 Kbytes in size. In addition, each cache can be organized as direct-mapped, two-way, three-way, or four-way set associative. On a cache miss, loads are blocked only until the first critical word becomes available. The pipeline resumes execution while the remaining words are being written to the cache. Both caches are virtually indexed and physically tagged. Virtual indexing allows the cache to be indexed in the same clock in which the address is generated, rather than waiting for the virtual-to-physical address translation in the TLB.

All cores execute the MIPS32 Instruction Set Architecture (ISA). The MIPS32 ISA contains all MIPS II instructions as well as special multiply-accumulate, conditional move, prefetch, wait, and zero/one detect instructions. The R4000-style MMU of the 4KEc core contains a four-entry instruction TLB (ITLB), a four-entry data TLB(DTLB), and a 16 dual-entry joint TLB (JTLB) with variable page sizes.

The 4KEm and 4KEp processors cores contain a simplified fixed mapping (FM) mechanism where the mapping of address spaces is determined through bits in the CP0 configuration (select 0) register. The 4KEc and 4KEm multiply-divide unit (MDU) supports a maximum issue rate of one 32x16 multiply (MUL/MULT/MULTU), multiply-add (MADD/MADDU), or multiply-subtract (MSUB/MSUBU) operation per clock, or one 32x32 MUL, MADD, or MSUB every other clock.

The basic enhanced JTAG (EJTAG) features provide CPU run control with stop, single stepping and restart, and with software breakpoints through the SDBBP instruction. Additional EJTAG features—instruction and data virtual address hardware breakpoints, connection to an external EJTAG probe through the Test Access Port (TAP), and PC/Data tracing—may optionally be included.

## 13.2.1 Key Features of the 4KE Family

32-bit address and data paths
MIPS32-compatible instruction set
    All MIPSII instructions
    Multiply-add and multiply-subtract instructions (MADD, MADDU, MSUB, MSUBU)
    Targeted multiply instruction (MUL)

Zero and one detect instructions (CLZ, CLO)

Wait instruction (WAIT)

Conditional move instructions (MOVZ, MOVN)

Prefetch instruction (PREF)

MIPS16e Application Specific Extension

16-bit encodings of 32-bit instructions to improve code density

Special PC-relative instructions for efficient loading of addresses and constants

Data type conversion instructions (ZEB, SEB, ZEH, SEH)

Compact jumps (JRC, JALRC)

Stack frame set-up and tear down "macro" instructions (SAVE and RESTORE)

User-defined instructions (access to this feature requires a separate license)

Optional user-defined instructions added to the MIPS32 instructions set (as a build-time option)

Single or multicycle instructions

Source operations from register or immediate field

Destination to a register

Programmable cache sizes

Individually configurable instruction and data caches

Sizes from 0 up to 64 Kbytes

Direct-mapped, or two-, three-, four-way set associative

Loads that miss in the cache are blocked only until critical word is available

Supports write-back with write-allocation and write-through with or without write-allocation

128-bit (16-byte) cache line size, word sectored—suitable for standard 32-bit-wide single-port SRAM

Virtually indexed, physically tagged

Cache line locking support

Nonblocking prefetches

Scratchpad RAM support

Replace one way of instruction cache and/or data cache

Maximum 20-bit index (1M address)

Memory-mapped registers attached to scratchpad port can be used as a coprocessor interface

R4000 style privileged resource architecture

Count/compare registers for real-time timer interrupts

Instruction and data watch registers for software breakpoints

Separate interrupt exception vector

Programmable memory management unit (4KEc core only)

16 dual-entry MIPS32-style JTLB with variable page sizes

Four-entry instruction TLB

Four-entry data TLB

Programmable memory management unit (4KEm and 4KEp cores only)

Fixed mapping (no JTLB, ITLB, or DTLB)

Address spaces mapped using register bits

Simple bus interface unit (BIU)

All I/Os fully registered

Separate unidirectional 32-bit address and data buses

Two 16-byte collapsing write buffers

Full-featured coprocessor 2 interface

Almost all I/Os registered

Separate unidirectional 32-bit instruction and data buses
Support for branch on coprocessor condition
Processor to/from coprocessor register data transfers
Direct memory to/from coprocessor register data transfers
Multiply-divide unit (4KEc and 4KEm cores)
Maximum issue rate of one 32x16 multiply per clock
Maximum issue rate of one 32x32 multiply every other clock
Early-in divide control. Minimum 11, maximum 34 clock
Latency on divide
Multiply-divide unit (4KEp cores)
Iterative multiply and divide; 32 or more cycles for each instruction
Power control
No minimum frequency
Power-down mode (triggered by WAIT instruction)
Support for software-controlled clock divider
EJTAG debug support
CPU control with start, stop, and single stepping
Software breakpoints via the SDBBP instruction
Optional hardware breakpoints on virtual addresses
Four instruction and two data breakpoints, two instruction and one data breakpoint, or no breakpoints
Optional test access port (TAP) facilitates high-speed download of application code
Optional EJTAG trace hardware to enable real-time tracing of executed code

All MIPS cores contain both required and optional blocks. Required blocks for the 4KE are the lightly shaded areas of the block diagram in Figure 13.12 and must be implemented to remain MIPS-compliant. Optional blocks can be added to the cores based on the needs of the implementation.

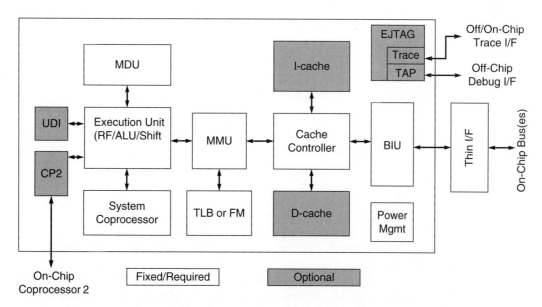

**FIGURE 13.12** MIPS 4KE Core Block Diagram (Reprinted by permission of MIPS.)

The required blocks are as follows:

Execution unit
Multiply-divide unit (MDU)
System control coprocessor (CP0)
Memory management unit (MMU)
Cache controller
Bus interface unit (BIU)
Power management

Optional blocks include

Instruction cache (I-cache)
Data cache (D-cache)
Enhanced JTAG (EJTAG) controller
Coprocessor 2 interface (CP2)
User-defined instructions (UDI)

The MMU can be implemented using either a translation lookaside buffer in the case of the 4KEc core, or a fixed mapping (FM) in the case of the 4KEm and 4KEp cores.

## 13.2.2  Execution Unit

The core execution unit implements a load-store architecture with single-cycle arithmetic logic unit (ALU) operations (logical, shift, add, subtract) and an autonomous multiply-divide unit (Figure 13.13). The core contains thirty-two 32-bit general-purpose registers (GPRs) used for scalar integer operations and address calculation. The register file consists of two read ports and one write port and is fully bypassed to minimize operation latency in the pipeline.

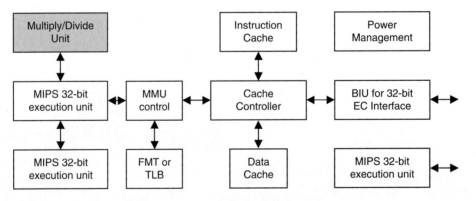

**FIGURE 13.13**  MIPS 4KE Execution Unit (Reprinted by permission of MIPS.)

The execution unit includes

32-bit adder used for calculating the data address
Address unit for calculating the next instruction address
Logic for branch determination and branch target address Calculation
Load aligner
Bypass multiplexers used to avoid stalls when executing instruction streams where data-producing instructions are followed closely by consumers of their results
Zero/one detect unit for implementing the CLZ and CLO Instructions
ALU for performing bitwise logical operations
Shifter and store aligner

## 13.2.3 Multiply/Divide Unit (MDU)

The multiply/divide unit performs multiply and divide operations (Figure 13.14). In the 4KEc and 4KEm processors, the MDU consists of a 32×16 booth-encoded multiplier, result-accumulation registers (HI and LO), multiply and divide state machines, and all multiplexers and control logic required to perform these functions. This pipelined MDU supports execution of a 16×16 or 32×16 multiply operation every clock cycle; 32×32 multiply operations can be issued every other clock cycle.

**FIGURE 13.14** MIPS 4KE Multiply/Divide Unit (Reprinted by permission of MIPS.)

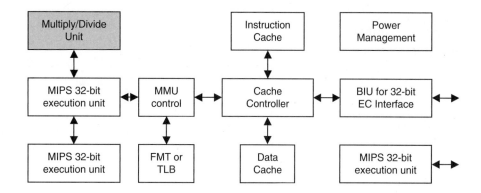

Appropriate interlocks are implemented to stall the issue of back-to-back 32×32 multiply operations. Divide operations are implemented with a simple 1-bit per clock iterative algorithm and require 35 clock cycles in worst case to complete. Early into the algorithm, it detects sign extension of the dividend; if its actual size is 24, 16, or 8 bit, the divider will skip 7, 15, or 23 of the 32 iterations. An attempt to issue a subsequent MDU instruction while a divide is still active causes a pipeline stall until the divide operation is completed.

In the 4KEp processor, the nonpipelined MDU consists of a 32-bit full-adder, result-accumulation registers (HI and LO), a combined multiply/divide state machine, and all multiplexers and control logic required to perform these functions. It performs any multiply using 32 cycles in an iterative 1 bit per clock algorithm. Divide operations are also implemented with a simple 1 bit per clock iterative algorithm (no early-in) and require 35 clock cycles to complete. An attempt to issue a subsequent MDU instruction while a multiply/divide is still active causes a pipeline stall until the operation is completed.

All cores implement an additional multiply instruction, MUL, which specifies that the lower 32 bits of the multiply result be placed in the register file instead of the HI/LO register pair. By avoiding the explicit move from LO (MFLO) instruction, required when using the LO register and by supporting multiple destination registers, the throughput of multiply-intensive operations is increased.

Two instructions, multiply-add (MADD/MADDU) and multiply-subtract (MSUB/MSUBU), are used to perform the multiply-add and multiply-subtract operations. The MADD instruction multiplies two numbers and then adds the product to the current contents of the HI and LO registers. Similarly, the MSUB instruction multiplies two operands and then subtracts the product from the HI and LO registers. The MADD/MADDU and MSUB/MSUBU operations are commonly used in digital signal processor (DSP) algorithms.

In the MIPS architecture, CP0 is responsible for the virtual-to-physical address translation, cache protocols, the exception control system, the processor's diagnostics capability, operating mode selection (kernel vs. user mode), and the enabling/disabling of interrupts. Configuration information such as cache size, set associativity, and EJTAG debug features are available by accessing the CP0 registers.

### 13.2.4  Memory Management Unit (MMU)

Each core contains an MMU that interfaces between the execution unit and the cache controller (Figure 13.15). Although the 4KEc core implements a 32-bit architecture, the memory management unit (MMU) is modeled after the MMU found in the 64-bit R4000 family, as defined by the MIPS32 architecture.

**FIGURE 13.15**  MIPS 4KE Memory Management Unit (Reprinted by permission of MIPS.)

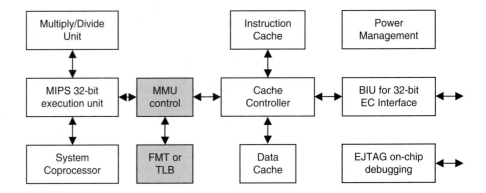

The 4KEc core implements its MMU based on a translation lookaside buffer (TLB), as shown in Figure 13.16. The TLB consists of three translation buffers: a 16-dual-entry fully associative Joint TLB (JTLB), a 4-entry fully associative Instruction TLB (ITLB), and a 4-entry fully associative data TLB (DTLB). The ITLB and DTLB, also referred to as the micro TLBs, are managed by the hardware and are not software visible. The micro TLBs contain subsets of the JTLB.

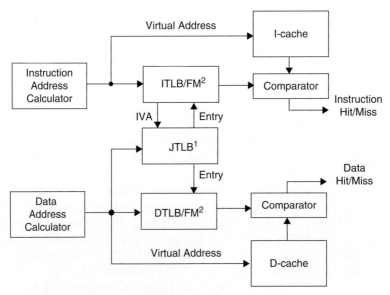

1. JTLB only exists in the 4KEc core
2. ITLB/DTLB implemented in the 4KEc core only. FM implemented inthe 4KEm and 4KEp cores.

**FIGURE 13.16**  MIPS 4KE Processor Memory Management Unit (Reprinted by permission of MIPS.)

When translating addresses, the corresponding micro TLB for instruction (I) or data (D) is accessed first. If there is not a matching entry, the JTLB is used to translate the address and refill the micro TLB. If the entry is not found in the JTLB, then an exception is taken. To minimize the micro TLB miss penalty, the JTLB is looked up in parallel with the DTLB for data references. This results in a one-cycle stall for a DTLB miss and a two-cycle stall for an ITLB miss.

## 13.2.5 Cache Controller

The data and instruction cache controllers support caches of various sizes, organizations, and set associativities (Figure 13.17). For example, the data cache can be 2 Kbytes in size and two-way set associative, whereas the instruction cache can be 8 Kbytes in size and four-way set associative.

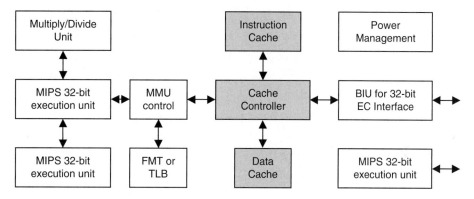

**FIGURE 13.17**   MIPS 4KE Cache Controller (Reprinted by permission of MIPS.)

There is a separate cache controller for the instruction cache and the data cache. Each cache controller contains and manages a one-line fill buffer. Besides accumulating data to be written to the cache, the fill buffer is accessed in parallel with the cache, and data can be bypassed back to the core. A more detailed discussion of cache memory design can be found at http://lwn.net/Articles/252125/.

## 13.2.6 Bus Interface Unit (BIU)

The bus interface unit (BIU) shown in Figure 3.18, controls the external interface signals. Additionally, it contains the implementation of a 32-byte collapsing write buffer. The purpose of this buffer is to hold and combine write transactions before issuing them to the external interface.

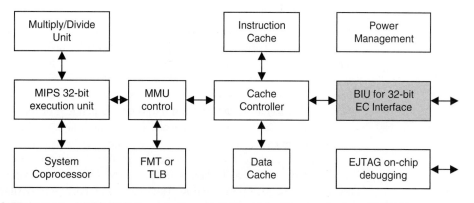

**FIGURE 13.18**   MIPS 4KE Bus Interface Unit (Reprinted by permission of MIPS.)

Because the data caches for all cores follow a write-through cache policy, the write buffer significantly reduces the number of write transactions on the external interface as well as reducing the amount of stalling in the core due to issuance of multiple writes in a short period of time.

The write buffer is organized as two 16-byte buffers. Each buffer contains data from a single 16-byte aligned block of memory. One buffer contains the data currently being transferred on the external interface, whereas the other buffer contains accumulating data from the core.

## 13.2.7 Power Management

The core offers a number of power management features, including low-power design, active power management, and power-down modes of operation (Figure 3.19). The core is a static design that supports a WAIT instruction designed to signal the rest of the device that execution and clocking should be halted, hence reducing system power consumption during idle periods.

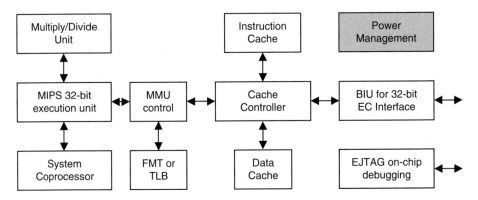

**FIGURE 13.19**    MIPS 4KE Power Management (Reprinted by permission of MIPS.)

The core provides two mechanisms for system-level, low-power support:

Register-controlled power management
Instruction-controlled power management

In register-controlled power management mode, the core provides 3 bits in the CP0 status register for software control of the power management function and allows interrupts to be serviced, even when the core is in power-down mode. In instruction-controlled, power-down mode execution of the WAIT instruction is used to invoke low-power mode.

## 13.2.8 Instruction Cache

The instruction cache is an optional on-chip memory array of up to 64 Kbytes (Figure 13.20). The cache is virtually indexed and physically tagged, allowing the virtual-to-physical address translation to occur in parallel with the cache access rather than having to wait first for the physical address translation. The tag holds 22 bits of the physical address, a valid bit, and a lock bit. There is a separate tag array that holds data used in the least recently used (LRU) replacement scheme. The LRU array ranges from 0 to 6 bits, depending on associativity.

All cores support instruction cache locking. Cache locking allows critical code to be locked into the cache on a "per-line" basis, enabling the system designer to maximize the efficiency of the system cache. Cache locking is always available on all instruction cache entries. Entries can be marked as locked or unlocked (by setting or clearing the lock bit) on a per-entry basis using the CACHE instruction. The LRU array must be bit-writable. The tag and data arrays only need to be word-writable.

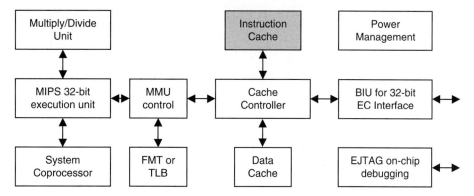

**FIGURE 13.20**   MIPS 4KE Instruction Cache (Reprinted by permission of MIPS.)

## 13.2.9 Data Cache

The data cache is an optional on-chip memory array of up to 64 Kbytes (Figure 13.21). The cache is virtually indexed and physically tagged, allowing the virtual-to-physical address translation to occur in parallel with the cache access. The tag holds 22 bits of the physical address, a valid bit, and a lock bit. A separate array holds the dirty and LRU bits; this array ranges from 0 to 10 bits, depending on the associativity.

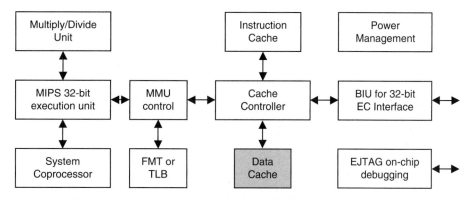

**FIGURE 13.21**   MIPS 4KE Data Cache (Reprinted by permission of MIPS.)

In addition to instruction cache locking, all cores also support a data cache locking mechanism identical to the instruction cache, with critical data segments to be locked into the cache on a "per-line" basis. The locked contents cannot be selected for replacement on a cache miss, but can be updated on a store hit.

Cache locking is always available on all data cache entries. Entries can be marked as locked or unlocked on a per-entry basis using the CACHE instruction. The physical data cache memory must be byte writable to support subword store operations. The LRU/dirty bit array must be bit-writable.

## 13.2.10 EJTAG Controller

All cores provide basic EJTAG support with debug mode, run control, single step, and software breakpoint instruction as part of the core (Figure 13.22). These features allow for the basic software debug of user and kernel code. Optional EJTAG features include hardware

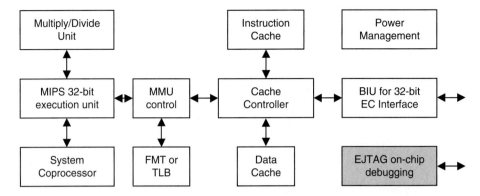

**FIGURE 13.22**   MIPS 4KE EJTAG Controller (Reprinted by permission of MIPS.)

breakpoints. A 4KE core may have four instruction breakpoints and two data breakpoints, two instruction breakpoints and one data breakpoint, or no breakpoints. The hardware instruction breakpoints can be configured to generate a debug exception when an instruction is executed anywhere in the virtual address space.

These breakpoints are not limited to code in RAM like the software instruction breakpoint. The data breakpoints can be configured to generate a debug exception on a data transaction. The data transaction may be qualified with both virtual address, data value, size, and load/store transaction type.

An optional TAP, enabling communication between an EJTAG probe and the CPU through a dedicated port, may also be applied to the core. This provides the possibility for debugging without debug code in the application and for download of application code to the system. Another optional block is EJTAG trace, which enables real-time tracing capability. The trace information can be stored to either an on-chip trace memory or to an off-chip trace probe. The trace of program flow is highly flexible and can include instruction program counter as well as data addresses and data values. The trace feature provides a powerful software debugging mechanism.

## 13.2.11  System Coprocessor

The optional system coprocessor (CP2) interface provides a full-featured interface for a coprocessor (Figure 13.23). It provides full support for all the MIPS32 COP2 instructions, with the exception of the 64-bit load/store instructions (LDC2/SDC2).

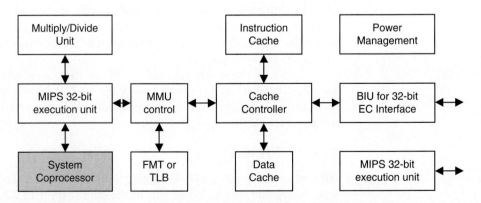

**FIGURE 13.23**   MIPS 4KE System Coprocessor (Reprinted by permission of MIPS.)

## 13.2.12  User-Defined Instructions (UDI)

This optional module contains (if implemented) support for user-defined instructions. These instructions must be defined at build time for the MIPS 4KE core. There are 16 instructions in the opcode map that are available for UDI, and each instruction can have single or multicycle latency. A UDI instruction can operate on any one or two general-purpose registers or immediate data contained within the instruction and must always write the result of each instruction back to a general-purpose register.

## 13.2.13  Instruction Pipeline

The MIPS32 4KE processor cores implement a five-stage pipeline. This pipeline allows the processor to achieve high frequency while minimizing device complexity, reducing both cost and power consumption.

The pipeline consists of five stages:

Instruction (I stage)
Execution (E stage)
Memory (M stage)
Align (A stage)
Writeback (W stage)

All three cores implement a "bypass" mechanism that allows the result of an operation to be sent directly to the instruction that needs it without having to write the result to the register and then read it back. Figure 13.24 shows the operations performed in each pipeline stage of the 4KEc processor.

**FIGURE 13.24**  MIPS 4KEc Instruction Pipeline Stages (Reprinted by permission of MIPS.)

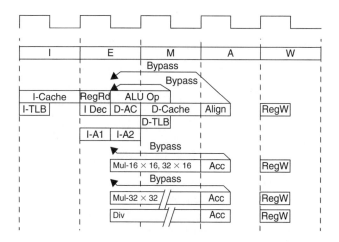

### 13.2.13.1  *Instruction Fetch*   During the instruction fetch (I) stage:

An instruction is fetched from the instruction cache.
The I-TLB performs a virtual-to-physical address translation (4Kec core only).

### 13.2.13.2  *Execution*   During the execution (E) stage:

Operands are fetched from the register file.
Operands from the M and A stage are bypassed to this stage.

The arithmetic logic unit (ALU) begins the arithmetic or logical operation for register-to-register instructions.

The ALU calculates the data virtual address for load and store instructions.

The ALU determines whether the branch condition is true and calculates the virtual branch target address for branch instructions.

Instruction logic selects an instruction address

All multiply and divide operations begin in this stage

**13.2.13.3  *Memory Fetch***   During the memory fetch (M) stage:

The arithmetic or logic ALU operation completes.

The data cache fetch and the data virtual-to-physical address translation are performed for load and store instructions.

Data TLB (4KEc core only) and data cache lookup are performed, and a hit/miss determination is made.

A 16×16 or 32×16 MUL operation completes in the array and stalls for one clock in the this stage to complete the carry-propagate-add in the stage (4KEc and 4KEm cores).

A 32×32 MUL operation stalls for two clocks in the this stage to complete the second cycle of the array and the carry-propagate-add in the stage (4KEc and 4KEm cores).

Multiply and divide calculations proceed in the MDU. If the calculation completes before the IU moves the instruction past the this stage, then the MDU holds the result in a temporary register until the instruction unit moves the instructions to the align stage (and it is consequently known that it won't be killed).

**13.2.13.4  *Align***   During the align (A) stage:

A separate aligner aligns loaded data with its word boundary.

An MUL operation makes the result available for writeback. The actual register writeback is performed in the Writeback stage (all 4KE cores).

From this stage load data or a result from the MDU are available in the execution stage for bypassing.

**13.2.13.5  *Writeback***   During the writeback (W) stage:

For register-to-register or load instructions, the result is written back to the register file.

## 13.2.14 Instruction Cache Miss

When the instruction cache is indexed, the instruction address is translated to determine if the required instruction resides in the cache. An instruction cache miss occurs when the requested instruction address does not reside in the instruction cache. When a cache miss is detected in the instruction fetch stage, the core transitions to the execution stage.

The pipeline stalls in the execution stage until the miss is resolved. The bus interface unit must select the address from multiple sources. If the address bus is busy, the request will remain in this arbitration stage until the bus is available. The core drives the selected address onto the bus, and the number of clocks before data is returned is then determined by the array containing the data.

Once the data is returned to the core, the critical word is written to the instruction register for immediate use. The bypass mechanism allows the core to use the data as soon as it arrives, as opposed to having the entire cache line written to the instruction cache, then reading out the required word.

Figure 13.25 shows a timing diagram of an instruction cache miss.

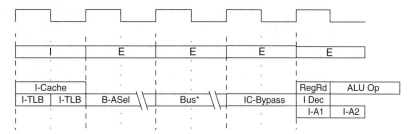

* Contains all of the cycles that address and data are utilizing the bus.

**FIGURE 13.25**   MIPS 4KE Instruction Cache Miss Timing (Reprinted by permission of MIPS.)

## 13.2.15  Data Cache Miss

When the data cache is indexed, the data address is translated to determine if the required data resides in the cache. A data cache miss occurs when the requested data address does not reside in the data cache. When a data cache miss is detected in the memory stage (D-TLB), the core transitions to the align stage. The pipeline stalls in the align stage until the miss is resolved (requested data is returned). Figure 13.26 shows a timing diagram of a data cache miss.

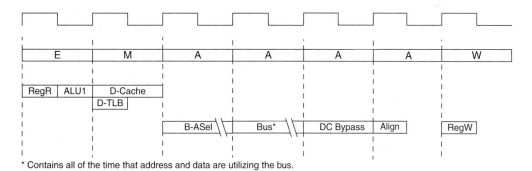

* Contains all of the time that address and data are utilizing the bus.

**FIGURE 13.26**   MIPS 4KE Load/Store Cache Miss Timing (Reprinted by permission of MIPS.)

The bus interface unit arbitrates between multiple requests and selects the correct address to be driven onto the bus. The core drives the selected address onto the bus. The number of clocks before data is returned is then determined by the array containing the data. Once the data is returned to the core, the critical word of data passes through the aligner before being forwarded to the execution unit. The bypass mechanism allows the core to use the data as soon as it arrives, as opposed to having the entire cache line written to the data cache, then reading out the required word.

## 13.2.16  Multiply/Divide Operations

All three cores implement the standard MIPS II multiply and divide instructions. Additionally, several new instructions are added for enhanced performance. The targeted multiply instruction, MUL, specifies that multiply results be placed in the general-purpose register file instead of the HI/LO register pair. By avoiding the explicit MFLO instruction, required when using the LO register, and by supporting multiple destination registers, the throughput of multiply-intensive operations is increased.

Four instructions, multiply-add (MADD), multiply-add-unsigned (MADDU) multiply-subtract (MSUB), and multiply-subtract-unsigned (MSUBU), are used to perform the multiply-accumulate and multiply-subtract operations. The MADD/MADDU instruction multiplies two numbers and then adds the product to the current contents of the HI and LO registers. Similarly, the MSUB/MSUBU instruction multiplies two operands and then subtracts the product from the HI and LO registers.

The MADD/MADDU and MSUB/MSUBU operations are commonly used in DSP algorithms. All multiply operations (except the MUL instruction) write to the HI/LO register pair. All integer operations write to the general-purpose registers (GPR). Because MDU operations write to different registers than integer operations, following integer instructions can execute before the MDU operation has completed. The MFLO and MFHI instructions are used to move data from the HI/LO register pair to the GPR file. If an MFLO or MFHI instruction is issued before the MDU operation completes, it will stall to wait for the data.

## 13.2.17 Branch Delay

The pipeline has a branch delay of one cycle. The one-cycle branch delay is a result of the branch decision logic operating during the execution pipeline stage. This allows the branch target address to be used in the instruction stage of the instruction following two cycles after the branch instruction. By executing the first instruction following the branch instruction sequentially before switching to the branch target, the intervening branch delay slot is utilized. This avoids bubbles being injected into the pipeline on branch instructions. Both the address calculation and the branch condition check are performed in the execution stage.

The pipeline begins the fetch of either the branch path or the fall-through path in the cycle following the delay slot. After the branch decision is made, the processor continues with the fetch of either the branch path (for a taken branch) or the fall-through path (for the nontaken branch). The branch delay means that the instruction immediately following a branch is always executed, regardless of the branch direction. If no useful instruction can be placed after the branch, then the compiler or assembler must insert an NOP instruction in the delay slot. Figure 13.27 illustrates the branch delay.

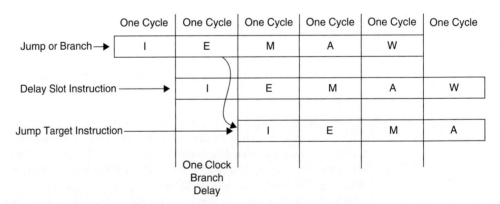

**FIGURE 13.27**    MIPS 4KE Pipeline Branch Delay (Reprinted by permission of MIPS.)

## 13.2.18 Memory Management

The MIPS32 4KE processor cores contain a memory management unit (MMU) that interfaces between the execution unit and the cache controller. The MIPS32 4KEc core contains a translation lookaside buffer (TLB), whereas the MIPS32 4KEm and MIPS32 4KEp cores implement a simpler fixed mapping (FM) style MMU.

***13.2.18.1  MMU Overview***    The MMU in a 4KE processor core will translate any virtual address to a physical address before a request is sent to the cache controllers for tag comparison or to the bus interface unit for an external memory reference. This is shown in Figure 13.28. This

**FIGURE 13.28** MIPS 4KE
Address Translation during a
Cache Access 4KEc Core
(Reprinted by permission of
MIPS.)

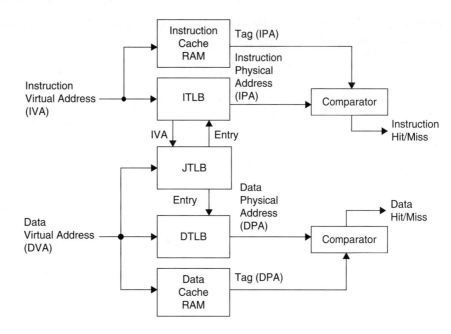

translation is a very useful feature for operating systems when trying to manage physical memory to accommodate multiple tasks active in the same memory, possibly on the same virtual address, but of course in different locations in physical memory (4KEc core only).

Other features handled by the MMU are protection of memory areas and defining the cache protocol. In the 4KEc processor core, the MMU is TLB based.

In the 4KEm and 4KEp processor cores, the MMU is based on a simple algorithm to translate virtual addresses into physical addresses via a fixed mapping (FM) mechanism. These translations are different for various regions of the virtual address space.

## 13.2.19 Modes of Operation

There are three modes of operation:

> User mode
> Kernel mode
> Debug mode

User mode is most often used for application programs. Kernel mode is typically used for handling exceptions and privileged operating system functions, including CP0 management and I/O device accesses. Debug mode is used for software debugging and most likely occurs within a software development tool. The address translation performed by the MMU depends on the mode in which the processor is operating.

*13.2.19.1  Virtual Memory Segments*   The virtual memory segments are different depending on the mode of operation. Figure 13.29 shows the segmentation for the 4 GByte virtual memory space addressed by a 32-bit virtual address for the three modes of operation. The core enters kernel mode both at reset and when an exception is recognized. While in kernel mode, software has access to the entire address space, as well as all CP0 registers.

User mode accesses are limited to a subset of the virtual address space (0x0000_0000 to 0x7FFF_FFFF) and can be inhibited from accessing CP0 functions. In user mode, virtual addresses 0x8000_0000 to 0xFFFF_FFFF are invalid and cause an exception if accessed. Debug mode is entered on a debug exception.

**FIGURE 13.29** MIPS 4KE
Virtual Memory Map (Reprinted
by permission of MIPS.)

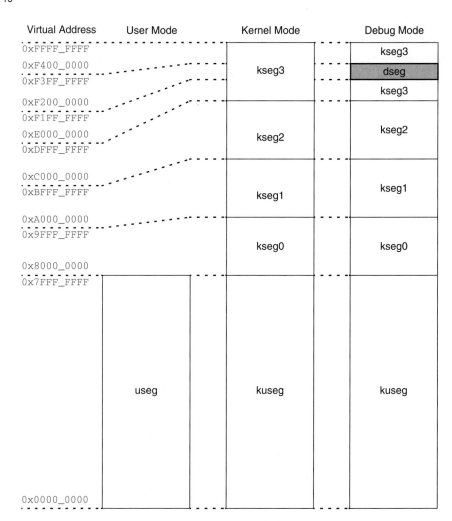

While in debug mode, the debug software has access to the same address space and CP0 registers as for kernel mode. In addition, while in debug mode, the core has access to the debug segment dseg. This area overlays part of the kernel segment kseg3. The dseg access in debug mode can be turned on or off, allowing full access to the entire kseg3 in debug mode, if so desired.

**13.2.19.2 *User Mode*** In user mode, a single 2 GByte uniform virtual address space called the user segment (useg) is available. Figure 13.30 shows the location of user mode virtual address space.

**FIGURE 13.30** MIPS 4KE User
Mode Virtual Address Space
(Reprinted by permission of
MIPS.)

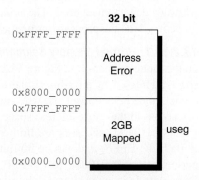

### 13.2.19.3  Kernel Mode

The processor operates in kernel mode when the DM bit in the debug register is 0 and the Status register contains one or more of the following values: UM = 0, ERL = 1, or EXL = 1

When a nondebug exception is detected, the processor will enter kernel mode. At the end of the exception handler routine, an Exception Return (ERET) instruction is generally executed. The ERET instruction jumps to the Exception PC, clears ERL, and clears EXL if ERL = 0. This may return the processor to user mode.

Kernel mode virtual address space is divided into regions differentiated by the high-order bits of the virtual address, as shown in Figure 13.31.

**FIGURE 13.31**  MIPS 4KE Kernel Mode Virtual Address Space (Reprinted by permission of MIPS.)

```
0xFFFF_FFFF  ┌──────────────────────────────┐
             │  Kernel virtual address space │
             │        Mapped, 512MB          │  kseg3
0xE000_0000  ├──────────────────────────────┤
0xDFFF_FFFF  │                               │
             │  Kernel virtual address space │
             │        Mapped, 512MB          │  kseg2
0xC000_0000  ├──────────────────────────────┤
0xBFFF_FFFF  │                               │
             │  Kernel virtual address space │
             │  Unmapped, Uncached, 512MB    │  kseg1
0xA000_0000  ├──────────────────────────────┤
0x9FFF_FFFF  │                               │
             │  Kernel virtual address space │
             │        Unmapped, 512MB        │  kseg0
0x8000_0000  ├──────────────────────────────┤
0x7FFF_FFFF  │                               │
             │                               │
             │                               │
             │                               │
             │                               │
             │       Mapped, 2048MB          │  kuseg
             │                               │
             │                               │
             │                               │
0x0000_0000  └──────────────────────────────┘
```

### 13.2.19.4  Debug Mode

Debug mode address space is identical to kernel mode address space with respect to mapped and unmapped areas, except for kseg3. In kseg3, a debug segment dseg coexists in the virtual address range 0xFF20_0000 to 0xFF3F_FFFF. The layout is shown in Figure 13.32.

**FIGURE 13.32**  MIPS 4KE Debug Mode Virtual Address Space (Reprinted by permission of MIPS.)

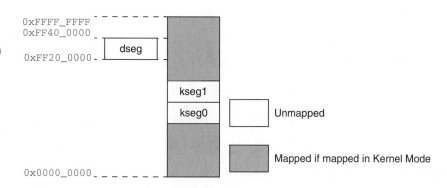

## 13.3        OVERVIEW OF THE ARM1022E PROCESSOR

The ARM1022E processor incorporates the ARM10E integer core, which implements the ARMv5TE architecture, as shown in Figure 13.33. It is a high-performance, low-power, cached processor that provides full virtual memory capabilities. It is designed to run high-end embedded applications and sophisticated operating systems such as JavaOS, Linux, and Microsoft WindowsCE. It supports the ARM and Thumb instruction sets and includes EmbeddedICE-RT logic and JTAG software debug features.

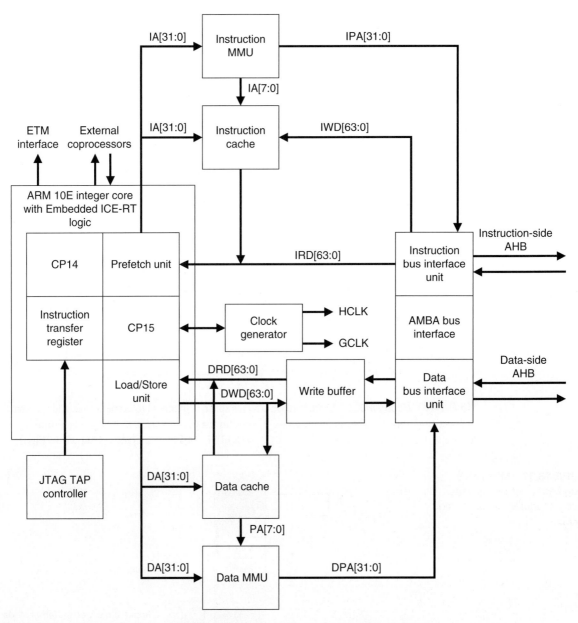

**FIGURE 13.33**   ARM10E Block Diagram (Reprinted by permission of ARM Limited.)

The ARM1022E processor consists of:

ARM10E integer core:
- Load/store unit
- Prefetch unit
- Integer unit
- Embedded ICE-RT logic for JTAG-based debug
- External coprocessor interface and coprocessors CP14 and CP15
- Memory management unit (MMU)
- Instruction and data caches
- Write-back physical address (PA) TAG RAM
- Write buffer and hit-under-miss (HUM) buffer
- Advanced micro bus architecture (AMBA) high-performance bus interface
- Embedded trace macrocell (ETM) interface

Features of the ARM1022E processor include

A six-stage pipeline
Branch prediction that supports branch folding with zero cycle branches
32-KB level 1 cache with a 16-KB instruction cache and a 16-KB data cache
Full-64-bit interfaces between the integer core and caches, write buffer, and bus interface units on both instruction and data sides and coprocessors
Multilayer AHB support through independent 64-bit AHB interfaces for instruction and data sides
Parallel execution of data processing instructions under load-and-store multiple instructions
A HUM buffer that supports execution of load hits underneath an outstanding load miss
Nonblocking caches that support execution of data processing instructions under load misses
Additional register read-and-write ports to support reading of up to four registers and writing of three registers in one cycle
Improved power management support
Enhanced debug support

## 13.3.1 Components of the Processor

The following sections introduce the main blocks of the ARM1022E processor. Their descriptions are given in the following sections. The main blocks of the ARM1022E processor include:

- Integer core
- Memory management unit
- Instruction and data caches
- Cache power-down capabilities
- Branch prediction and prefetch unit
- AMBA interface
- Coprocessor interface
- Debug
- Instruction cycle summary and interlocks
- Design-for-test features
- Power management
- Clocking and PLL

**13.3.1.1 *Integer Unit*** The ARM1022E processor is built around the ARM10E integer core in an ARMv5TE implementation that runs the 32-bit ARM and 16-bit compressed Thumb instruction sets (Figure 13.34). You can balance high performance against code size and extract maximum performance from 8-bit, 16-bit, and 32-bit memory. The processor includes Embedded ICE-RT logic for JTAG software debugging and is supported by the Multi-ICE JTAG debug interface.

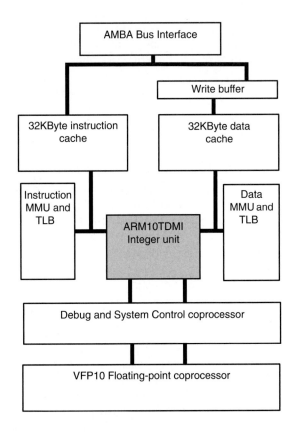

**FIGURE 13.34** ARM1022E Integer Unit Block Diagram (Reprinted by permission of ARM Limited.)

## 13.3.2 Registers

The ARM1022E processor has 37 32-bit registers (Figure 13.35):

    16 general-purpose registers
    1 status register
    15 banked, mode-specific, general-purpose registers
    5 banked, mode-specific status registers

These registers are not all accessible at the same time. The processor state and processor operating mode determine which registers are available to the programmer.

## 13.3.3 Integer Core

By overlapping the various stages of operation, the integer core (refer to Figure 13.36) maximizes the clock rate achievable to execute each instruction. Because it has multiple execution units, the integer core enables multiple instructions to exist in the same pipeline stage, enabling simultaneous execution of some instructions. As a result, it delivers a peak throughput approaching one instruction per cycle.

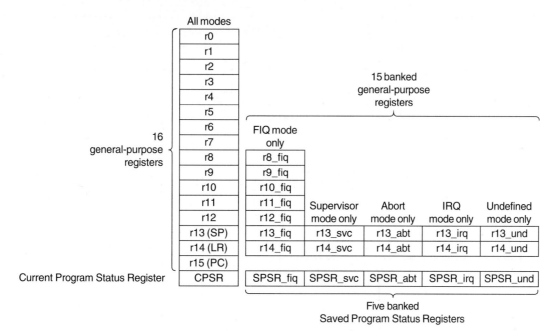

**FIGURE 13.35**   ARM1022E Register Set (Reprinted by permission of ARM Limited.)

## 13.3.4  Integer Core Pipeline

The ARM10 integer pipeline consists of six stages to maximize instruction throughput. As shown in Figure 13.37, the pipeline incorporates the following functions:

> Fetch instruction cache access. Branch prediction for instructions that have already been fetched.
>
> Issue initial instruction decode.
>
> Decode final instruction decode, register reads for arithmetic/logic unit (ALU) operation, forwarding, and initial interlock resolution.
>
> Execute data access address calculation, data processing shift, shift and saturate, ALU operation, first stage of multiplications, flag setting, condition code check, branch mispredict detection, and store data register read.
>
> Memory data cache access, second stage of multiplications, and saturations. Write register writes, instruction retirement.
>
> The fetch stage uses a first-in-first-out (FIFO) prefetch buffer that can hold up to three instructions. Here a path to fetch along is predicted ahead of execution of branch instructions.

The issue and decode stages can contain a predicted branch in parallel with one instruction. The execute, memory, and write stages can simultaneously contain all of the following:

> A predicted branch
>
> An ALU or multiply instruction
>
> Ongoing multicycle load or store multiple instructions
>
> Ongoing multicycle coprocessor instructions

***13.3.4.1   Prefetch Unit***   The prefetch unit operates in the fetch stage of the pipeline. It can fetch 64 bits every cycle from the instruction-side cache. It can only issue one 32-bit instruction per cycle to the integer unit. Because it can fetch more instructions than it can issue, the prefetch

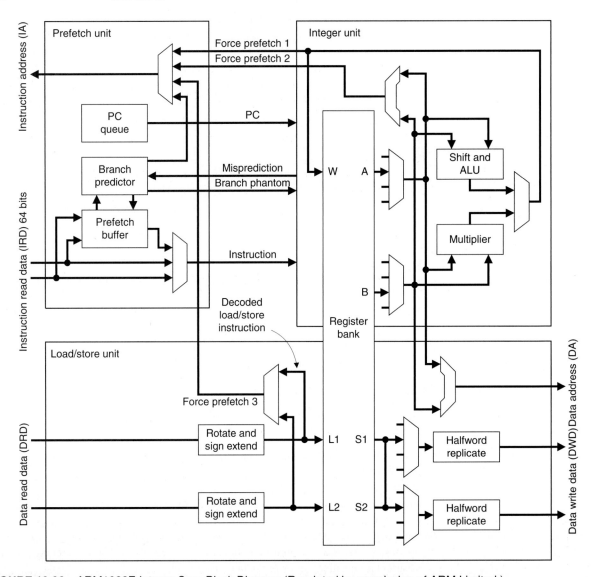

**FIGURE 13.36**   ARM1022E Integer Core Block Diagram (Reprinted by permission of ARM Limited.)

unit puts pending instructions in the prefetch buffer. While an instruction is in the prefetch buffer, the branch prediction logic can decode it to see if it is a predictable branch. Where possible, the branch prediction logic removes branches from the instruction stream. If the branch is predicted to be taken, then the instruction address is redirected to the branch target address. If the branch is predicted not to be taken, then the instruction address continues to progress through the instructions following the branch instruction.

Often in these cases, if the instruction following the branch is already in the prefetch buffer, it can be issued in place of the branch, and the branch effectively takes no cycles. When there is not enough time to completely remove the branch, the fetch address is redirected anyway, because this still helps to reduce the branch penalty.

The integer unit executes unpredicted or unpredictable branches. To get the address out quickly, it uses a dedicated fast branch adder whose inputs do not pass through the barrel shifter. A multiplexer in the LSU sends loaded data straight to the prefetch unit. This updates the fetch address after loads to the program counter (PC).

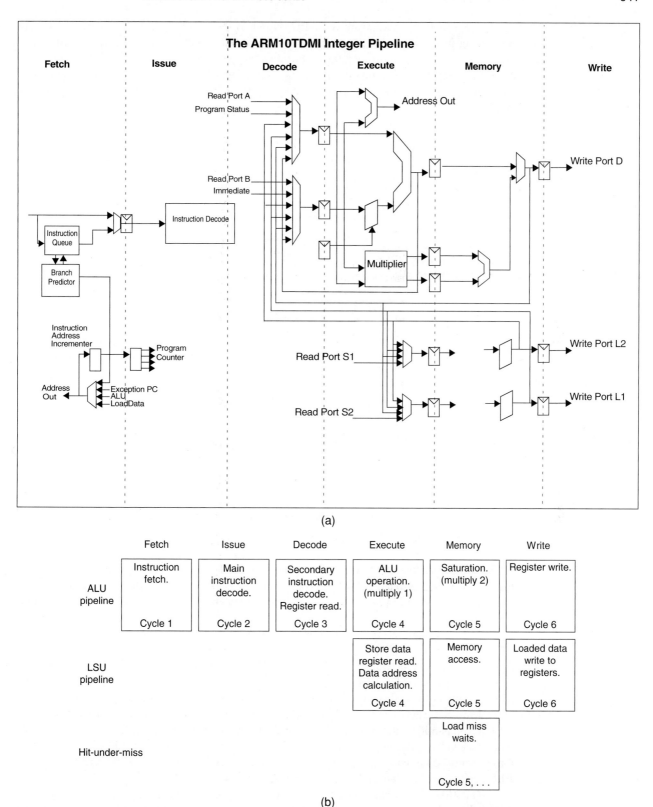

**FIGURE 13.37** (a) ARM1022E Integer Core Six-Stage Pipeline; (b) ARM1022E Six-Stage Pipeline Operation (Reprinted by permission of ARM Limited.)

There is also a path from the ALU output to the prefetch unit. This is used for data processing instructions that write to the PC. Because the path through the barrel shifter and ALU is slower than that through the dedicated adder, these instructions usually take one more cycle than branches. The one exception is a simple move that does not require a shift, for example, MOV PC R14. For optimum performance, this uses the fast branch adder rather than the ALU.

Figure 13.38 shows the stages of a typical multiply operation. The MUL loops in the execute stage until it passes through the first part of the multiplier array enough times. Then it progresses to the memory stage, where it passes once through the second half of the array to produce the final result.

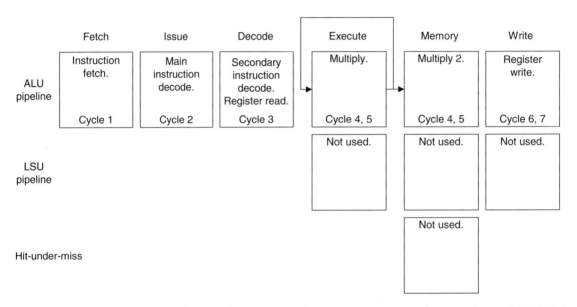

**FIGURE 13.38**    ARM1022E Pipeline Stages of a Typical Multiply Operation (Reprinted by permission of ARM Limited.)

### 13.3.4.2    Load/Store Unit
If the data address is 64-bit aligned, the LSU (loadstore unit) can load or store two 64-bit words per transfer. This does not speed up LDR (load word) or STR (store word) instructions, but it does considerably speed up LDM (load multiple) and STM (store multiple) instructions.

Accesses that are not 64-bit aligned have to take place over two cycles. If an LDM or STM address is not 64-bit aligned, then the first access transfers only one 32-bit word. After that, two words can be transferred in each cycle. Single loads and stores work in cooperation with the integer unit. The first cycle of multiple loads and stores works in cooperation with the integer unit, but the LSU can finish ongoing multiple loads and stores autonomously.

The LSU calculates the address for the data access using a dedicated adder. This adder evaluates in parallel with the adder in the ALU. The adder in the ALU calculates a base register write-back value if it is required. The A and B register ports of the integer unit read the operands for both adders. For complex scaled-register addressing modes that require the barrel shifter, the ALU has to calculate data addresses. This costs one extra cycle.

The LSU has two dedicated register bank read ports, S1 and S2, and two dedicated write ports, L1 and L2. These are used to read data to be stored and to write data that is loaded. The steps in a load/store operation are shown in Figure 13.39.

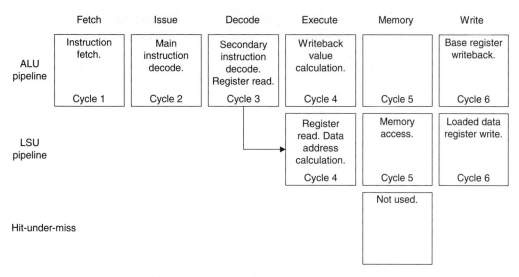

**FIGURE 13.39** ARM1022E Pipeline Stages of a Load or Store Operation (Reprinted by permission of ARM Limited.)

### 13.3.5  Memory Management Unit

The MMU has separate instruction and data Translation Lookaside Buffers (TLBs). It is backward-compatible with the ARM v4 architecture MMU of StrongARM and ARM920T. The MMU includes a 1-KB tiny page mapping size to enable a smaller RAM and ROM footprint for embedded systems and operating systems such as Windows CE that have many small mapped objects. The ARM1022E processor implements the fast context switching extension (FCSE) and high vectors extension that are required to run Microsoft Windows CE.

MMU features include the following key functions:

Standard MMU mapping sizes, domains, and access protection
1-KB, 4-KB, 64-KB, and 1-MB mapping sizes
Access permissions for 1-MB sections
Separate access permissions for one-quarter page subpages of 64-KB large pages and
    4-KB small pages
16 domains
Separate 64-entry instruction and data TLBs
Independent lockdown of instruction and data TLBs
Hardware translation table walks
Round-robin replacement algorithm
Support for soft page tables

### 13.3.6  Caches and Write Buffer

The ARM processor includes

An instruction cache
A data cache
A write buffer
A hit-under-miss (HUM) buffer

The 16-KB I-cache and 16-KB D-cache have the following features:

Eight segments, each containing 64 lines.
Virtually-addressed 64-way associativity.

Eight words per line with one valid bit, one dirty bit, and one write-back bit per line.

Write-through and write-back D-cache operation, selected per memory region by the C and B bits in the MMU translation tables.

Pseudorandom or round-robin replacement, selectable by the RR bit in CP15 c1 Control Register 1.

Low-power CAM-RAM implementation.

Independently lockable caches with granularity of 1/64th of the cache, that is 64 words (256 bytes) to a maximum of 63/64ths of the cache.

For compatibility with Microsoft Windows CE, and to reduce interrupt latency, the physical address corresponding to each D-cache entry is stored in the D-cache PA tag RAM for use during cache line write-backs, in addition to the VA tag stored in the cache CAMs. This means that the MMU is not involved in cache write-back operations, removing the possibility of MMU misses related to the write-back address.

Cache maintenance operations to provide efficient cleaning of the entire D-cache and to provide efficient cleaning and invalidation of small regions of virtual memory. The latter enables I-cache coherency to be efficiently maintained when small code changes occur, for example, self-modifying code and changes to exception vectors. The write buffer can hold eight 64-bit packets of data, each with an associated address element.

### 13.3.7 Bus Interface

The ARM10 processor is designed for the advanced microcontroller bus architecture (AMBA) and uses the AMBA high-performance bus (AHB) as its interface to memory and peripherals.

The ARM10 processor uses separate AHB bus interfaces for instructions and data:

Instruction bus interface unit (IBIU)
Data bus interface unit (DBIU).

Separate bus interfaces enhance the ability to fetch and execute instructions in parallel with a data cache miss. There is no sharing of any AHB signals between the two interfaces. The ARM10 AHB interface is always driven. When either bus master is not granted the bus, that master drives zeros onto the bus to prevent bus contention.

The ARM10 processor has unidirectional inputs, outputs, and control signals. For a complete description of AMBA, including the AHB bus and the AMBA test methodology, see the AMBA specification. The bus interface handles all data transfers and instruction transfers between the core clock domain and the AMBA bus clock domain (Figure 13.40). Any request from the prefetch unit or the LSU that has to go outside the ARM10 processor is handled by the bus interface in a way that is transparent to the prefetch unit and the LSU.

**FIGURE 13.40** ARM1022E Bus Interface Units (Reprinted by permission of ARM Limited.)

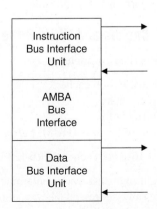

The following requests from the caches and MMU drive the bus interface:

Translation table walks generated by the MMU
Noncachable reads
Nonbuffered writes
Linefills
Buffered writes
CP15 empty write buffer and clean index operations

## 13.3.8  Topology

The bus interface consists of two completely separate blocks typical of a RISC architecture. The instruction bus interface unit (IBIU) handles all instruction fetches and linefills. The data bus interface unit (DBIU) performs all data loads and stores.

Both the IBIU and DBIU perform translation table walks when required. Figure 13.41 shows the structure of the bus interface. The DBIU is on the left with control, read, write, and address data-path. The IBIU on the right has a read and an address data-path only because no writes ever happen on the instruction side.

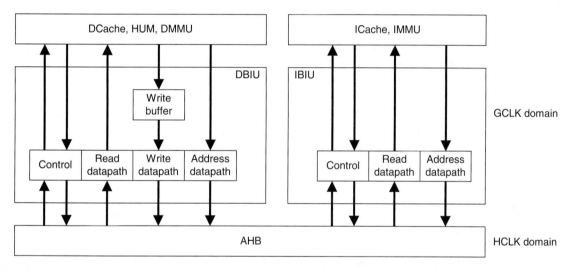

**FIGURE 13.41**   ARM1022E Bus Interface Block Diagram (Reprinted by permission of ARM Limited.)

Both the IBIU and the DBIU have a similar layer for transferring data or instructions to and from the HCLK domain and further on to the rest of the AMBA system. The arrows illustrate the flow of requests and data or instructions.

The DBIU and the IBIU are independent of each other. There is no efficient way of communicating between the data and the instruction side of the ARM processor, making self-modifying code difficult to accommodate.

## 13.3.9  Coprocessor Interface

The coprocessor interface enables you to attach multiple coprocessors (CPs) to the ARM10 processor. To limit the number of connections required by the interface, each CP tracks the progress of instructions in the ARM10 pipeline. To enable optimum performance from CPs, the ARM10 processor issues CP instructions as early as possible. This means that the instructions are issued speculatively, and they can be canceled later in the pipeline if, for example, an exception or branch misprediction occurs. As a result, CPs must be able to cancel instructions in late stages of the ARM10 pipeline.

Simple CPs track the ARM10 pipeline only until they are certain that a given instruction is not going to be canceled. At this point the CP starts to execute the instruction. More complex CPs make extensive use of the early issue of the instruction. At certain points in the pipeline, a CP sends back signals to the ARM10 processor. These can indicate that the CP requires more time to execute or to indicate that the Undefined instruction exception must be taken.

### 13.3.10  Coprocessor Pipeline

The CP pipeline runs one cycle behind the ARM10 pipeline. This enables pipeline holds from the ARM10 processor to be registered before they are sent to the CPs. Figures 13.42(a) and (b) show the ARM10 and CP pipeline stages.

### 13.3.11  Debug Unit

The ARM1022E processor is designed to be embedded into large SoC designs. The Embedded ICE-RT logic debug facilities, AMBA on-chip system bus, and test methodology are all designed for efficient use of the processor when integrated into a larger IC.

The debug unit assists in debugging software (Figure 13.43). The debug hardware, in combination with a software debugger program, can be used to debug:

Application software
Operating systems
ARM10-based hardware systems

The debug unit enables you to:

Stop program execution
Examine and alter processor and coprocessor state
Examine and alter memory and input/output peripheral state
Restart the processor core

The debug unit provides several ways to stop execution. The most common is for execution to halt at an instruction fetch breakpoint or at a data access watchpoint. When execution has stopped, one of two modes is entered: halt mode or monitor debug mode.

### 13.3.12  Halt Mode

All processor execution halts, and can only be restarted with hardware connected to the external JTAG interface. You can examine and alter all processor registers, coprocessor registers, memory, and input/output locations through the JTAG interface.

This mode is intentionally invasive to program execution. In halt mode you can debug the processor irrespective of its internal state. Halt mode requires external hardware to control the JTAG interface. A software debugger provides the user interface to the debug hardware.

### 13.3.13  Monitor Debug-Mode

In monitor debug-mode the processor stops execution of the current program and starts execution of a debug abort handler. The state of the processor is preserved in the same manner as all ARM exceptions.

The abort handler communicates with a debugger application to access processor and coprocessor state and to access memory contents and input/output peripherals. Monitor debug-mode requires a debug monitor program to interface between the debug hardware and the software debugger.

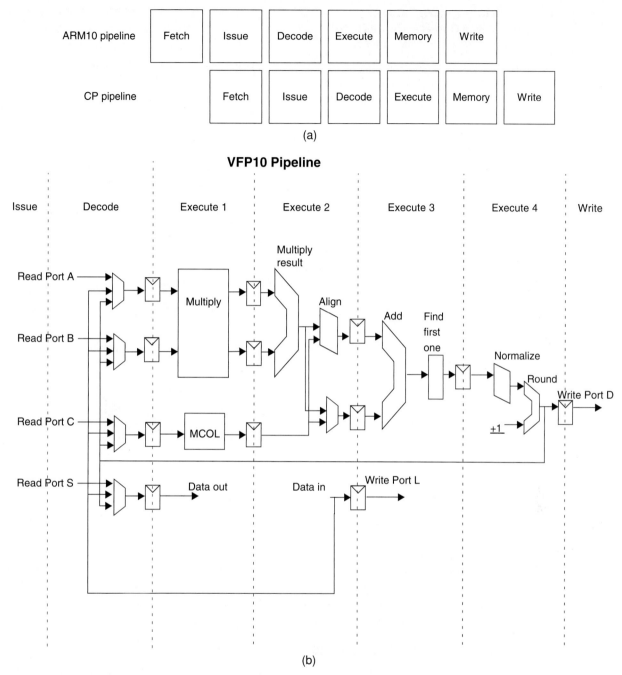

**FIGURE 13.42**   (a) ARM1022E and Coprocessor Pipeline Stages; (b) ARM1022E Floating Point Coprocessor Pipeline (Reprinted by permission of ARM Limited.)

## 13.3.14 Clocking and PLL

The ARM1022E processor has two clock inputs:

    GCLK
    HCLK

**FIGURE 13.43** ARM1022E
Debug Controller (Reprinted by
permission of ARM Limited.)

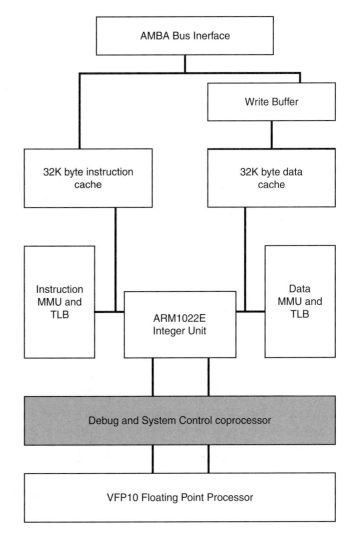

The design is fully static. When both these clocks (Figure 13.44) are stopped, the internal state of the processor is preserved indefinitely. GCLK drives the internal logic in the processor. HCLK drives the bus interface. Most input and output timings are specified with respect to HCLK.

Note that typically, GCLK frequency is higher than that of HCLK. The two clocks must have a fixed-phase relationship. HCLK is usually derived by dividing down the source of GCLK.

### 13.3.15  ETM Interface Logic

An optional external ETM (external trace module), shown in Figure 13.45, can be connected to the ARM1022E processor to provide real-time tracing of instructions and data in an embedded system. The processor includes the logic and interface to enable you to trace program execution and data transfers using the ETM10.

### 13.3.16  Operating States

The ARM1022E processor has two operating states. The register map within the ARMv5TE architecture depends on the operating mode.

**ARM state**—The processor executes 32-bit, word-aligned ARM instructions.

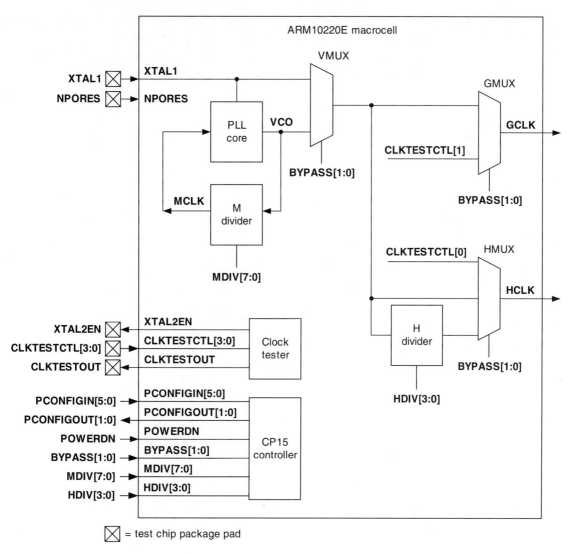

☒ = test chip package pad

**FIGURE 13.44**  ARM1022E Clock Generator (Reprinted by permission of ARM Limited.)

**FIGURE 13.45**  ARM1022E ETM Interface Logic (Reprinted by permission of ARM Limited.)

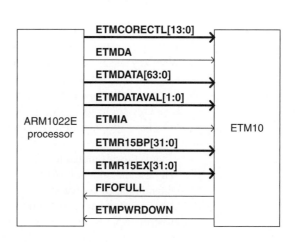

**Thumb state**—The processor executes 16-bit, halfword-aligned Thumb instructions. The program counter (PC) uses bit 1 to select between alternate halfwords.

Note that the transition between ARM and Thumb states does not affect the processor mode or the register contents.

### 13.3.17 Switching State

You can switch the operating state of the processor between ARM state and Thumb state by:

Using the BX and BLX instructions
Loading the PC with the L4 bit cleared in CP15 c1 Control Register 1

The processor begins all exception handling in ARM state. If an exception occurs in Thumb state, the processor changes to ARM state. The change back to Thumb state occurs automatically on return from exception handling.

### 13.3.18 Switching State During Exception Handling

An exception handler can put the processor in Thumb state, but it must return to ARM state to enable the exception handler to terminate correctly.

### 13.3.19 Operating Modes

There are seven processor modes of operation:

*User:* The nonprivileged mode for normal program execution
*Fast interrupt (FIQ):* The privileged exception mode for handling fast interrupts
*Interrupt (IRQ):* The privileged exception mode for handling regular interrupts
*Supervisor:* The privileged mode for operating system functions
*Abort:* The privileged exception mode for handling data aborts and prefetch aborts
*System:* The privileged user mode for operating system functions
*Undefined:* The privileged exception mode for handling undefined instructions

Modes other than user mode are collectively known as privileged modes. Privileged modes are used to service exceptions or to access protected resources.

---

## QUESTIONS AND PROBLEMS

1. List three key reasons to use an SoC instead of a GPM.
2. What are limitations of RTL block design?
3. What key issues need to be addressed when transferring a board-level design to an SoC?
4. List three challenges to using large RTL blocks.
5. List six challenges to migrating RTL blocks to SoC.
6. List three applications not well suited to an SoC.
7. Why have special multiply-accumulate instructions in the MIPS 4Ke?
8. Why have optional blocks in the MIPS 4Ke?
9. What is the advantage to having options for the MIPS 4Ke instruction and data caches?
10. Why use the optional MIPS 4Ke EJTAG feature?
11. Diagram the basic five-stage pipeline of the MIPS 4ke.
12. Describe what happens when the MIPS I-cache miss occurs.
13. Describe what happens when a MIPS D-cache miss occurs.

14. What are the three modes of operation of the MIPS 4Ke?
15. List the key features of the ARM10E processor.
16. What are the two instruction sets executed by the ARM10E core?
17. What does branch mean for the instruction flow in the ARM10E pipeline?
18. List the seven operating modes of the ARM10E.
19. Diagram the ARM10E six-stage pipeline.
20. Which bus does the ARM10E use for peripheral and memory interfacing?
21. List the four key features of the ARM10E debug unit.

## SOURCES

*AMBA bus architecture specification.*

*ARM1022E technical reference manual.* Revision: r0p2. ARM Limited.

*ARM10 Thumb family, product overview.* ARM DVI 0014A, ARM Limited.

Courtright, David. 2002. *MIPS32 M4K core for multi-CPU applications.* Embedded Processor Forum. MIPS Technologies.

Francis, Hedley. 2001. *ARM DSP-enhanced extensions.* ARM White Paper. ARM Ltd.

Kimelman, Paul. 2002. *Developing embedded software in multi-core SoCs.* ARM Ltd.

MIPS. 2002. *Microprocessor architecture, performance headroom to compete.* MIPS Technologies.

MIPS. *MIPS low-power synthesizable cores, a competitive comparison.* MIPS Technologies.

*MIPS32 4KE of synthesizable processor cores.* Product Brief, MIPS Technologies.

*MIPS32 4Kc processor core datasheet.* March 6, 2002. MIPS Technologies.

*MIPS32 4KE Processor core family software user's manual.* Document Number: MD00103, Revision 01.08, January 30, 2002. MIPS Technologies.

Rowen, Chris. 2002. *Reducing SoC simulation and development time.* Computer, December 2002, reprint.

Thekkath, Radhika. 2004. *Digital signal processing on the industry-standard MIPS architecture presentation.* Fall Processor Forum. MIPS Technologies.

# CHAPTER 14

## Tensilica Configurable IP Core

**OBJECTIVE: UNDERSTAND CONFIGURABLE PROCESSOR CORE TECHNOLOGY**

The reader will learn about the architecture and features of the

1. Tensilica Xtensa LX2 Configurable Processor.

## 14.0 INTRODUCTION: MOORE'S LAW REVISITED

In 1965, Gordon Moore prophesized that integrated circuit density would double roughly every one to two years. Figure 14.1 plots this as standard cell density and clock rate over time. The universal acceptance and relentless tracking of this trend set a grueling pace for all chip developers. This trend makes transistors ever cheaper and faster but also invites system buyers to expect constant improvements to functionality, battery life, throughput, and cost. The moment a new function is technically feasible, the race is on to deliver it.

**FIGURE 14.1** Moore's Law with Gate Density (Reprinted by permission of Tensilica Incorporated.)

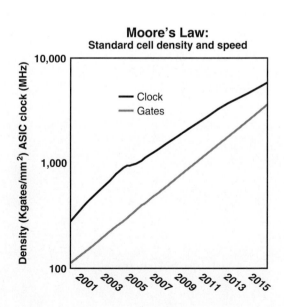

The remarkable characteristics of CMOS silicon scaling allow the cost, size, performance, and power for a given function to all improve simultaneously. This scaling allows continuous improvement in end-product benefits: longer battery life, smaller size, more functionality, and higher user productivity. This scaling has been a primary driver for the parallel revolutions in digital consumer electronics, personal computing, and the Internet.

The growth in available transistors creates a fundamental role for concurrency in system-on-a-chip (SoC) designs. Different tasks, such as audio and video processing and network-protocol stack management, can operate largely independent of one another. Complex tasks with inherent internal execution parallelism can be decomposed into a tightly coupled collection of subtasks operating in parallel to perform the same work as the original nonparallel task implementation. This kind of concurrency offers the potential for significant improvements in application latency, data bandwidth, and energy efficiency when compared to serial execution of the same collection of tasks with a single computational resource.

The design task must be recognized as correspondingly difficult. Three forces work together to make chip design tougher and tougher. First, the astonishing success of semiconductor manufacturers to track Moore's law gives designers twice as many gates to play with every two years. Second, the continuous improvement in process geometry and circuit characteristics motivates chip builders to design with new IC fabrication technologies as they come available. Third, and perhaps most important, the end markets for electronic products, consumer, computing, and communications systems, are in constant churn demanding a constant stream of new functions and performance to justify new purchases.

As a result, the design "hill" keeps getting steeper. Certainly, improved chip-design tools help faster RTL simulation; higher-capacity logic synthesis and better block placement and routing all mitigate some of the difficulties. Similarly, the movement toward systematic logic design reuse can reduce the amount of new design that must be done for each chip. But all these improvements fail to close the design gap. This well-recognized phenomenon is captured in the Semiconductor Research Corporation's simple comparison of the growth in logic complexity and designer productivity in Figure 14.2.

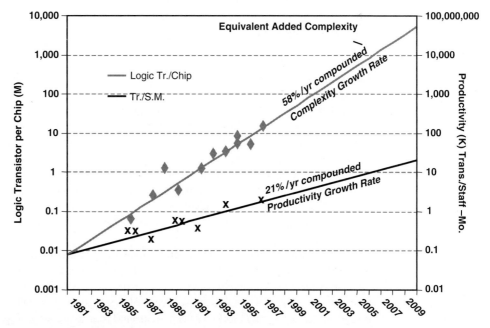

**FIGURE 14.2** Design Complexity and Designer Productivity (Reprinted by permission of Tensilica Incorporated.)

Even as designers wrestle with the growing resource demands of advanced chip design, they face two additional worries:

How do design teams ensure that the chip specification really satisfies customer needs? How do design teams ensure that the chip really meets those specifications?

Further, a good design team will also anticipate future needs of current customers and potential future customers as it has a built-in road map.

## 14.1        CHIP DESIGN PROCESS

If the design team fails on the first criterion listed earlier, the chip may work perfectly, but will have inadequate sales to justify the design expense and manufacturing effort. Changes in requirements may be driven by demands of specific key customers or may reflect overall market trends such as the emergence of new data-format standards or new feature expectations across an entire product category. Although most SoC designs include some form of embedded control processor, the limited performance of those processors often precludes them from being used for essential data-processing tasks, so software usually cannot be used to add or change fundamental new features.

### 14.1.1  Building the Wrong Chip

If the design team fails on the second criterion listed, additional time and resources must go toward changing or fixing the design. This resource diversion delays market entry and causes companies to miss key customer commitments. The failure is most often realized as a program delay. This delay may come in the form of missed integration or verification milestones, or it may come in the form of hardware bugs, explicit logic errors that are not caught in the limited verification coverage of typical hardware simulation.

The underlying cause might be a subtle error in a single design element, or it might be a miscommunication of requirements, subtle differences in assumptions between hardware and software teams, between design and verification teams, or between SoC designer and SoC library or foundry supplier. In any case, the design team may often be forced into an urgent cycle of redesign, reverification, and refabrication of the chip. These design "spins" rarely take less than six months, causing significant disruption to product and business plans. To improve the design process, we must consider simultaneous changes in all three interacting dimensions of the design environment: design elements, design tools, and design methodology.

Design elements are the basic building blocks, the silicon structures, and the logical elements that form the basic vocabulary of design expression. Historically, these blocks have been basic logic functions (NAND and NOR gates and flip-flops), plus algorithms written in C and assembly code running on reduced instruction set compating (RISC) microprocessors and digital signal processors.

Design tools are the application programs and techniques that designers use to capture, verify, refine, and translate design descriptions for particular tasks and subsystems. Historically, tools such as register transfer level (RTL) compilation and verification, code assemblers and compilers, and standard-cell placement and routing have comprised the essential toolbox for complex chip design.

Design methodology is the design team's strategy for combining the available elements and tools into a systematic process for implementing the target silicon and software. A methodology specifies what elements and tools are available, describes how the tools are used at each step of the design refinement, and outlines the sequence of design steps as shown in Figure 14.3. The current dominant SoC design methodology is built around four major steps, typically implemented in the following order: hardware-software partitioning; detailed RTL block design and verification; chip integration of RTL blocks, processors, and memories; and postsilicon software bring-up.

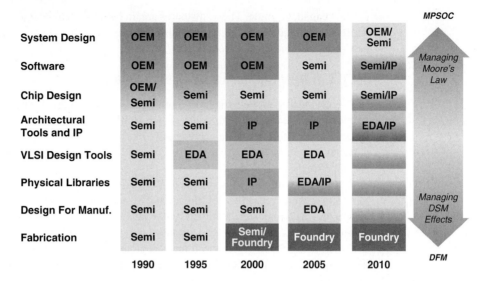

| | 1990 | 1995 | 2000 | 2005 | 2010 |
|---|---|---|---|---|---|
| System Design | OEM | OEM | OEM | OEM | OEM/Semi |
| Software | OEM | OEM | OEM | Semi | Semi/IP |
| Chip Design | OEM/Semi | Semi | Semi | Semi | Semi/IP |
| Architectural Tools and IP | Semi | Semi | IP | IP | EDA/IP |
| VLSI Design Tools | Semi | EDA | EDA | EDA | |
| Physical Libraries | Semi | Semi | IP | EDA/IP | |
| Design For Manuf. | Semi | Semi | Semi | EDA | |
| Fabrication | Semi | Semi | Semi/Foundry | Foundry | Foundry |

**FIGURE 14.3**   Merging EDA and IP (Reprinted by permission of Tensilica Incorporated.)

Changes in any one dimension are unlikely to prevent the pitfalls of SoC design of "building the wrong chip" or "building the chip wrong." Piecemeal improvements in RTL design or software development tools cannot solve the larger design problem. Instead, it is necessary change the design problem itself. The design elements, key tools, and the surrounding methodology must all change together.

## 14.1.2  Fundamental Trends of SoC Design

Several basic trends suggest that the engineering community needs a new approach for SoC design. The first trend is the seemingly inexorable growth in silicon density, which underlies the fundamental economics of building electronic products in the 21st century. At the center of this trend is the fact that the semiconductor industry seems willing and able to continue to push chip density by consistent, sustained innovation through smaller transistor sizes, smaller interconnect geometries, higher transistor speed, significantly lower cost, and lower power dissipation over a long period of time. Technical challenges for scaling abound. Issues of power dissipation, nanometer lithography, signal integrity, and interconnect delay all will require significant innovation. Past experience suggests that these challenges, at worst, will only marginally slow down the pace of scaling.

This silicon scaling trend stimulates the second trend, the drive to take this available density and actually integrate into one piece of silicon. This required by the enormous diversity and huge number of functions required by modern electronic products. The increasing integration level creates the possibility of taking all the key functions associated with a network switch, or a digital camera, or a personal information appliance, and putting these functions into one piece of silicon. That is, all the logic, all the memory, all the interfaces, in fact almost everything electronic in the end product.

The benefits of high silicon integration levels are clear. Tight integration drives the end product's form factor, making complex systems small enough to put into your pocket, inside your television, or in your car. High integration levels also drive down power dissipation, making more end products battery powered, fanless, or available for use in a much wider variety of environments. Ever-increasing integration levels drive the raw performance upward in terms of how quickly a product will accomplish tasks, or in terms of the number of different functions that a product can incorporate. These attributes are, in fact, likely to become even more important product features, ideally enough to make the average consumer rush to their favorite retailer to buy new products to replace the old ones.

### 14.1.3  A New SoC for Every System Is a Bad Idea

The resulting silicon specialization stemming from higher and higher integration creates an economic challenge for the product developer. If all of the electronics in a system are embodied in roughly one chip, that chip is increasingly likely to be a direct reflection of the end product the designer is trying to define. Such a chip design lacks flexibility. It cannot be used in a wide variety of products.

In the absence of some characteristic that make that highly integrated chip significantly more flexible and reusable, SoC design moves toward a direct 1:1 correspondence between chip design and system design. Ultimately, if SoC design were to really go down this road, the time to develop a new system and the amount of engineering resources required to build a new system will, unfortunately, become at least as great as the time and costs to build new chips. A better way is to create a flexible SoC that can then be targeted to multiple applications, as shown in Figure 14.4.

**FIGURE 14.4**  Single SoC Multiple Applications (Reprinted by permission of Tensilica Incorporated.)

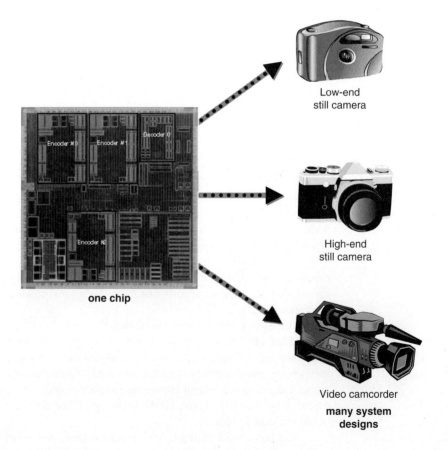

This can result in significant cost savings when multiple applications can be derived from a single flexible SoC design. This is shown dramatically in Figure 14.5. It can easily be seen that the up-front design cost is amortized over the cost of multiple applications with significant per chip cost reduction.

**FIGURE 14.5** SoC Flexibility
Versus Per-Unit Cost (Reprinted
by permission of Tensilica
Incorporated.)

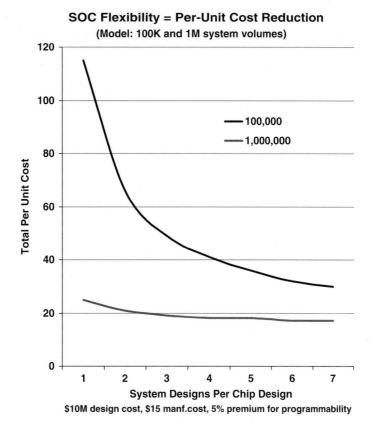

In the past, product designers built systems by combining chips onto large printed circuit boards (PCBs). Different systems used different combinations of (mostly) off-the-shelf chips soldered onto system-specific PCBs. The approach worked because a wide variety of silicon components were available and because PCB design and prototyping was easy. System reprogrammability was relatively unimportant because system redesign was relatively cheap and quick.

### 14.1.4 Nanometer Technology

In the world of nanometer silicon technology, the situation is dramatically different. Demands for smaller physical system size, greater energy efficiency, and lower manufacturing cost all make the large PCB obsolete. Volume-oriented end-product requirements can only be satisfied with SoC designs. Even when appropriate "virtual components" are available as SoC building blocks, SoC design integration and prototyping are more than two orders of magnitude more expensive than PCB design and prototyping. Moreover, SoC design changes take months, whereas PCB changes take just days. SoC design is mandatory to reap the benefits of nanometer silicon, but to make SoC design practical SoCs cannot be built like PCBs.

The problem of SoC inflexibility must be addressed. Chip-level inflexibility is really a crisis in reusability of the chip's hardware design. Despite substantial industry attention to the benefits of block-level hardware reuse (IP reuse), the growth in internal complexity of blocks coupled with the complex interactions among blocks has limited the systematic and economical reuse of IP blocks. Figure 14.6 shows the SoC hardware and software design process.

Too often customer requirements, implemented standards, and the necessary interfaces to other functions must evolve with each product variation. These boundaries constrain successful block reuse to two categories: simple blocks that implement stable interface functions and inherently

**FIGURE 14.6** SoC Design Process (Reprinted by permission of Tensilica Incorporated.)

| | Hardware | Software | |
|---|---|---|---|
| **System Planning** | System Specification and Architecture | Algorithm Invention | *Application Centric* |
| **Partitioning** | Communication and Computation | Application Implementation | |
| **Hardware/ Software Integration** | Processor to Bus & Memory Organization | SW Integration | *Processor Centric* |
| **Block Integration** | IP Reuse | System Testbench | |
| **Function Block** | RTL Synthesis | RTL Simulation | *Transistor Centric* |
| **Physical Implementation** | Place & Route | Timing, Signal, Yield Modelling | |
| | Creation | Verification | |

flexible functions that can be implemented in processors, whose great flexibility and adaptability are realized via software programmability.

On the other hand, a requirement to build new chips for every system would be an economic disaster for system developers because there's no question that building chips is hard. We can improve the situation somewhat with better chip-design tools, but in the absence of some significant innovation in chip-design methodology, the situation's not getting better very fast.

In fact, it would appear that in the absence of some major innovation, the efforts required to design a chip will increase more rapidly than the transistor complexity of the chip itself. We're losing ground in systems design because innovation in design methodology is lacking. We cannot afford to lose ground on this problem, as system and chip design grow closer together.

## 14.1.5  SoC Design Reform

System developers are trying to solve two closely related problems:

> To develop system designs with significantly fewer resources by making it much, much easier to design the chips in those systems
>
> To make SoCs more adaptable so not every new system design requires a new SoC design

The way to solve these two problems is to make the SoC sufficiently programmable so that one chip design will efficiently serve 10, or 100, or even 1000 different system designs while giving up none or perhaps just a few of the benefits of integration. Solving these problems means having chips available off the shelf to satisfy the requirements of the next system design and amortize the costs of chip development over a large number of system designs, as shown in Figure 14.7.

**FIGURE 14.7** Multiple System Designs

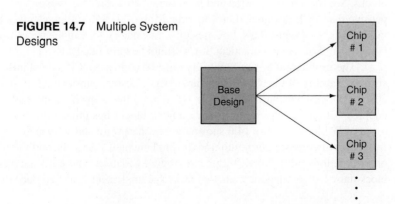

These trends constitute the force behind the need for a fundamental shift in IC design. That fundamental shift will ideally provide both a big improvement in the effort needed to design SoCs (not just in the silicon but also the required software), and it will increase the intrinsic flexibility of SoC designs so that the design effort can be shared across many system designs.

Economic success in the electronics industry hinges on the ability to make future SoCs more flexible and more highly optimized at the same time. The core dilemma for the SoC industry and for all the users of SoC devices is really simultaneous management of flexibility and optimality. SoC developers are trying to minimize chip design costs and trying to get closer and closer to the promised benefits of high-level silicon integration at the same time. Consequently, they need to take full advantage of what high-density silicon offers, and at the same time, they need to overcome or mitigate the issues created by the sheer complexity of those SoC designs and the high costs and risks associated with the long SoC development cycle. Programmability allows SoC designers to substantially mitigate the costs and risks of complex SoC designs, by accelerating the initial development effort and by easing the effort to accommodate subsequent revision of system requirements.

### 14.1.6 SoC Programmability

Rising system complexity also makes programmability essential to SoCs, and the more efficient programming becomes, the more pervasive it will be. The market already offers a wide range of possible ways to achieve system programmability, including field-programmable gate arrays (FPGAs), standard microprocessors, and reconfigurable logic.

Programmability's benefits come at two levels. First, programmability increases the likelihood that a preexisting design can meet the performance, efficiency, and functional requirements of the system. If there is a fit, no new SoC development is required; an existing platform will serve. Second, programmability means that even when a new SoC must be designed, more of the total functions are implemented in a programmable fashion, reducing the design risk and effort. The successes of both the FPGA and processor markets are traceable to these factors.

The programming models for different platforms differ widely. Traditional processors (including DSPs) can execute applications of unbounded complexity, though as complexity grows, performance typically suffers. Processors typically use sophisticated pipelining and circuit-design techniques to achieve high clock frequency, but achieve only modest parallelism, one (or a few) operations per clock cycle. FPGAs, by contrast, have finite capacity. Once the problem grows beyond some level of complexity, the problem will not fit in an FPGA at all (refer to Figure 14.8).

**FIGURE 14.8** FPGA Gate Count Limitations

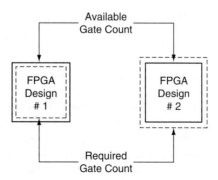

On the other hand, FPGAs can implement algorithms with very high levels of intrinsic parallelism, sometimes performing the equivalent of hundreds of operations per cycle. FPGAs typically operate at more modest clock rates than processors and tend to have larger die sizes and higher chip costs than processors used in the same applications.

### 14.1.7 Programmability Versus Efficiency

All these flavors of programmability allow the underlying silicon design to be somewhat generic while permitting configuration or personalization for a specific situation at the time that the system is booted or during system operation. The traditional problem with programmability is that there is a tremendous gap in efficiency and/or performance between a hardwired design and a design with the same function implemented with programmable technology. This gap could be called the "programmability overhead."

This overhead may be defined as the increased area for implementation of a function using programmable methods, compared to a hardwired implementation with the same performance (refer to Figure 14.9). Alternatively, the overhead may be defined as the increase in execution time of a programmable design solution, compared to a hardwired implementation of the same silicon area.

**FIGURE 14.9** Programmability Gap

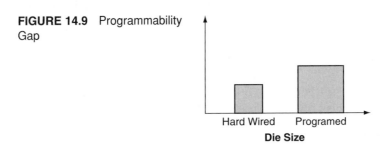

As a rule of thumb, the overhead for FPGA or generic processor programmability is more than a factor of ten and can reach a factor of one hundred. For example, hardwired logic solutions are typically about 100x faster for security applications such as DES and AES encryption than the same tasks implemented with a general-purpose RISC processor. An FPGA implementation of these encryption functions may run only at 3 to 4x lower clock frequency than hardwired logic, but may require 10 to 20x more silicon area.

These inefficiencies stem from the excessive generality of the universal digital substrates: FPGAs and general-purpose processors. The designers of these general-purpose substrates are working to construct platforms to cover all possible scenarios. Unfortunately, the creation of truly general-purpose substrates requires a super abundance of basic facilities from which to fabricate specific computational functions and connection paths to move data among computation functions. Silicon efficiency is constrained by the limited reuse or "time-multiplexing" of the transistors implementing an application's essential functions.

In fact, if you look at either an FPGA or a general-purpose processor performing an "add" computation, you will find a group of logic gates comprising an adder surrounded by a large number of multiplexers and wires to deliver the right data to the right adder at the right moment. The circuit overhead associated with storing and moving the data and selecting the correct sequence of functions leads to much higher circuit delays and a much larger number of required transistors and wires than a design where the sequence of operations to be performed is known.

General-purpose processors rely on time multiplexing of a small and basic set of function units for basic arithmetic and logical operations and memory references. Most of the processor logic serves as hardware to route different operands to the small set of shared hardwire function units. Communication among functions is implicit in the reuse of processor registers and memory locations by different operations. FPGA logic, by contrast, minimizes the implicit sharing of hardware among different functions. Instead, each function is statically mapped to a particular region of the FPGA silicon, so each transistor typically performs a single function repeatedly, as shown in Figure 14.10. Communication among functions is explicit in the static configuration of interconnect among functions.

**FIGURE 14.10** Fixed FPGA Functions (Reprinted by permission of Tensilica Incorporated.)

The more that is known about the required computation, the more the transistors involved in the computation can be interconnected with dedicated wires to enable high utilization of computational units. Both general-purpose processors and general-purpose FPGA technologies have overhead, but an exploration of software programmability highlights the hidden overhead of field hardware programmability.

Modern software programmability's power really stems from two complementary characteristics. One of these is abstraction. Software programs allow developers to deal with computation in a form that is more concise, more readily understood at a glance, and more easily enhanced independent of implementation details. Abstraction yields insight into overall solution structure by hiding the implementation details.

Modest-sized software teams routinely develop, reuse, and enhance applications with hundreds of thousands of lines of source code, including extensive reuse of operating systems, application libraries, and middleware software components. In addition, sophisticated application-analysis tools have evolved to help teams debug and maintain these complex applications. By comparison, similar-sized hardware teams consider logic functions with tens of thousands of lines of Verilog or VHDL code to be quite large and complex. Blocks are modified only with the greatest care. Coding abstraction is limited to simple registers, memories, and primitive arithmetic functions. They may constitute the most complex generic reusable hardware functions in a block.

The second characteristic is software's ease of modification. System functionality changes when you first boot the system, and it changes dynamically when you switch tasks. In a software-driven environment, if a task requires a complete change to a subsystem's functionality, the system can load a new software-based personality for that subsystem from memory in a few microseconds in response to changing system demands. This is a key point: the economic benefits of software flexibility as shown in Figure 14.11 appear both in the development cycle (what happens between product conception and system "power-on") and during the expected operational life of a design (what happens between freezing the product specification and the moment when the last variant of the product is shipped to the last customer).

**FIGURE 14.11** Product Lifetime Costs

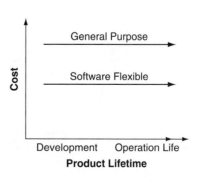

Continuous adaptability to new product requirements plays a central role in improving product profitability. If the system can be reprogrammed quickly and cheaply, the developers reduce the risk of failing to meet design specifications and have greater opportunity to quickly adapt the product to new customer needs. Field-upgrading software has become routine with PC systems, and the ability to upgrade software in the field is starting to find its way into embedded products. For example, products such as Cisco network switches get regular software upgrades. The greater the flexibility is (at a given cost) the greater the number of customers and the higher the SoC volume.

In contrast, hardwired design choices must be made very early in the system design cycle. If, at any point in the development cycle—design, prototyping, field trials, upgrades in the field, or second-generation product development—the system developers decide to change key computation or communication decisions, then it's back to square one for the system design. Hardwiring key design decisions also narrows the potential range of customers and systems into which the SoC might fit and limits the potential volume shipments.

Making systems more programmable has benefits and liabilities. The benefits are seen in development agility and efficiency. Designers don't need to decide exactly how the computational elements relate to each other until later in the design cycle, when the cost of change is low. Whether programmability comes through loading a net list into an FPGA or software into processors, the designer doesn't have to decide on a final configuration until system power up.

If programmability is realized through software running on processors, developers may defer many design decisions until system bring-up. Some decisions can be deferred until the eve of product shipment. The liabilities of programmability have historically been cost, power efficiency, and performance. Conventional processor and FPGA programmability carries an overhead of thousands of transistors and thousands of microns of wire between the computational functions. This overhead translates into large die size and low clock frequency for FPGAs and long execution time for processors, compared to hardwired logic implementing the same function.

Fixing the implementation hardware in silicon can dramatically improve unit cost and system performance, but dramatically raises design risk and design cost. The design team faces trying choices. Which functions should be implemented in hardware? Which in software? Which functions are most likely to change? How will communication among blocks evolve?

This trade-off between efficiency and performance, on one hand, and programmability and flexibility, on the other hand, is a recurring theme in this book. Figure 14.12 gives a conceptual view of this trade-off.

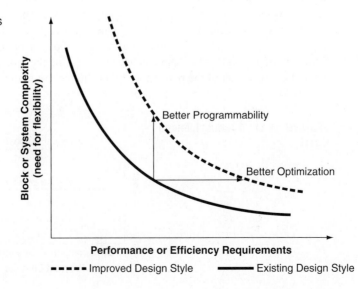

**FIGURE 14.12**  The Essentials of Design Trade-off (Reprinted by permission of Tensilica Incorporated.)

The vertical axis indicates the intrinsic complexity of a block or of the whole system. The horizontal axis indicates the performance or efficiency requirement of a block or the whole system. One recurring dilemma of SoC design is this: solutions that have the flexibility to support complex designs sacrifice performance, efficiency, and throughput; solutions with high efficiency or throughput often sacrifice the flexibility necessary to handle complex applications.

The curves in Figure 14.12 represent overall design styles or methodologies. Within a methodology, a variety of solutions are possible for each block or for the whole system, but the design team must trade off complexity and high efficiency. An improved design style or methodology may still exhibit trade-offs in solutions, but with an overall improvement in the "flexibility-efficiency product." In moving to an improved design methodology, the team may focus on improving programmability (to get better flexibility at a given level of performance), or it may focus on improving performance (to get better efficiency and throughput at a given level of programmability). In a sense, the key to efficient SoC design is managing uncertainty. If the design team can optimize all dimensions of the product for which requirements are stable and leave flexible all dimension of the product that are unstable, they will have a cheaper, more durable, and more efficient product than their competitors.

## 14.1.8 The Key to SoC Design Success

The goal of SoC development is to strike an optimal balance between getting just enough flexibility in the SoC to meet changing demands within a problem, and yet to still realize the efficiency and optimality associated with targeting an end application. Therefore, what's needed is an SoC design methodology that permits a high degree of system and subsystem parallelism, an appropriate degree of programmability, and rapid design. It is not necessary for SoC-based system developers to use a completely universal piece of silicon, as most SoCs ship in enough volume to justify specialization. For example, a designer of digital cameras doesn't need to use the same chip that's used in a high-end optical network switch. One camera chip, however, can support a range of related consumer imaging products, as shown in Figure 14.13.

**FIGURE 14.13** One Camera SoC into many Camera Systems (Reprinted by permission of Tensilica Incorporated.)

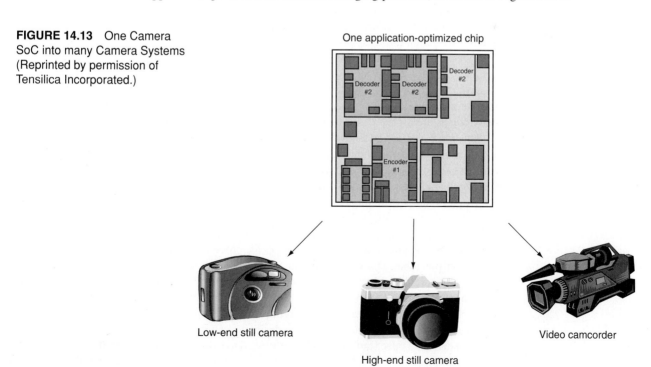

The difference in benefit derived from a chip shared by ten similar designs, versus one shared by 1,000 designs, is relatively modest. If each camera design's volume is 200,000 units, and the shared SoC design costs $10M, then the SoC design contributes $5 to final camera cost (−5%). Sharing the SoC design across 1000 designs could save $5 in amortized design cost, but would almost certainly require such generality in the SoC that SoC production costs would increase by far more than $5. SoC designs need not be completely universal, as high-volume products can easily afford to have a chip-level design platform that is appropriate to their application domain, yet flexible within it.

If designers have sufficient flexibility within an SoC to adapt to any tasks they are likely to encounter during that design's lifetime, then they essentially have all the relevant benefits of universal flexibility without much of the overhead of universal generality. If the platform design is done correctly, the cost for this application-specific flexibility is much lower than the flexibility derived from a truly universal device, such as an FPGA and a high-performance, general-purpose processor.

In addition, a good design methodology should enable as broad a population of hardware and software engineers as possible to design and program the SoCs. The larger the talent pool, the faster the development and the lower the project cost.

The key characteristics for such a SoC design methodology are

Support for concurrent processing.

Appropriate application efficiency.

Ease of development by people who are not necessarily SoC design specialists. (That's not to say that the people using this design methodology to develop SoCs may not be IC-design specialists, but that they are far more likely to be specialists in the specific application domain of interest to the design team.)

## 14.1.9  An Improved Design Methodology for SoC Design

A fundamentally new way to speed development of mega-gate SoCs is emerging. First, processors replace hardwired logic to accelerate hardware design and bring full chip-level programmability. Second, those processors are extended, often automatically, to run functions very efficiently with high throughput, low power dissipation, and modest silicon area. Blocks based on extended processors often have characteristics that rival those of the rigid RTL blocks they replace. Third, these processors become the basic building blocks for complete SoCs, where the rapid development, flexible interfacing, and easy programming and debugging of the processors accelerate the overall design process. Finally, and perhaps most important, the resulting SoC-based products are highly efficient and highly adaptable to changing requirements.

This improved SoC design flow allows full exploitation of the intrinsic technological potential of nanometer semiconductors (parallelism, pipelining, fast transistors, application-specific operations) and the benefits of modern software development methodology. A sketch of the new SoC design paradigm appears in Figure 14.14.

The flow starts from the high-level requirements, especially the external input and output requirements for the new SoC platform and the set of tasks that the system performs on the data flowing through the system. The computation within tasks and the communication among tasks and interfaces are optimized using application-specific processors and quick function-, performance-, and cost-analysis tools. The flow makes an accurate system model available early in the design schedule, so detailed VLSI and software implementations can proceed in parallel. Early and accurate modeling of both hardware and software reduces development time and minimizes expensive surprises late in the development and bring-up of the entire system.

Using this design approach means that designers can move through the design process with fewer dead ends and false starts and without the need to back up and start over. It means that SoC designers can make a much fuller and more detailed exploration of the design possibilities early

**FIGURE 14.14** New SoC Design Paradigm (Reprinted by permission of Tensilica Incorporated.)

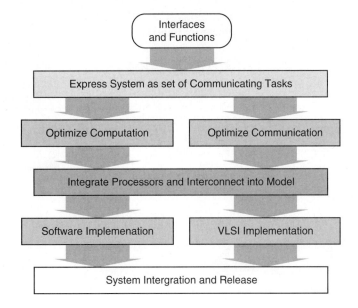

in the design cycle. Using this approach, they can better understand the design's hardware costs, application performance, interface, programming model, and all the other important characteristics of an SoC's design.

Taking this approach to designing SoCs means that the possible, efficient uses of the silicon platform will be as large as possible with the fewest compromises in the cost and the power efficiency of that platform. The more a design team uses the application-specific processor as the basic SoC building block, as opposed to hardwired logic written as RTL, the more the SoC will be able to exploit the flexibility inherent in a software-centric design approach.

### 14.1.10  The Configurable Processor as the Building Block

The basic building block of this methodology is a new type of microprocessor: the configurable, extensible microprocessor core. These processors are created by a generator that transforms high-level application domain requirements (in the form of instruction-set descriptions or even examples of the application code) into an efficient hardware design and software tools. The "sea of processors" approach to SoC design allows engineers without microprocessor design experience to specify, evaluate, configure, program, interconnect, and compose those basic building blocks into combinations of processors that together create the essential digital electronics for SoC devices.

To develop a processor configuration using one of these configurable microprocessor cores, the chip designer or application expert comes to the processor-generator interface (shown in Figure 14.15) and selects or describes the application source, instruction-set options, memory hierarchy, closely coupled peripherals, and interfaces required by the application. It takes about one hour to fully generate the hardware design in the form of standard RTL languages, EDA tool scripts and test benches, and the software-development environment (C and C++ compilers, debuggers, simulators, RTOS code, and other support software).

The generation process provides immediate availability of a fab-portable hardware implementation and software development environment. This timely delivery of the hardware and software infrastructure permits rapid tuning and testing of the software applications on that processor design. The completeness of software largely eliminates the issues of software porting and enables rapid design iteration. The tools can even be configured to automatically explore a wide range of possible processor configurations from a common application base to reveal more optimal hardware solutions, as measured by target application requirements.

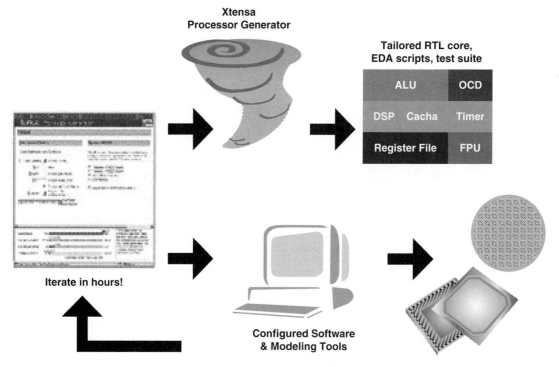

**Xtensa Processor Generator**

**Tailored RTL core, EDA scripts, test suite**

| ALU | OCD |
|-----|-----|
| DSP  Cacha | Timer |
| Register File | FPU |

**Iterate in hours!**

**Configured Software & Modeling Tools**

**FIGURE 14.15**    Basic Processor Design Flow (Reprinted by permission of Tensilica Incorporated.)

The application-specific processor performs all the same tasks that a microcontroller or a high-end RISC processor can perform: run applications developed in high-level languages, implement a wide variety of real-time features, and support complex protocol stacks, libraries, and application layers. Application-specific processors perform generic integer tasks very efficiently, even as measured by traditional microprocessor power, speed, area, and code-size criteria. But because these application-specific processors are able to incorporate the data paths, instructions, and register storage for the idiosyncratic data types and computation required by an embedded application, they can also support virtually all the functions that chip designers have historically implemented as hardwired logic.

## 14.1.11  Rapid SoC Development Using Automatically Generated Processors

The system architect faces a number of important decisions in creating the best SoC structure. Good choices early in the design process reduce silicon cost and power, increase system performance, and improve development and verification efficiency. This design flow encourages wide use of processors as the default for implementing tasks and focuses on how to balance cost, performance, and flexibility within an SoC design framework. The foundations for the design flow are these:

Work top-down from the system's essential I/O interfaces and computation requirements.

Use processors pervasively to implement tasks when tasks have specific computational patterns, optimize the processor to fit the tasks.

When a task exceeds the capacity of an optimized processor, parallelize the task across processors.

When a group of tasks fits together within a processor's capacity limit, map the tasks together onto one processor to minimize hardware cost, power, and communications overhead.

Measure the communications traffic patterns and optimize the software and hardware interconnects around those patterns.

Start with early, rough simulation of communication tasks and refine the system into detailed implementations of processors, software, and other blocks, all running in increasingly accurate simulations.

### 14.1.12 The Starting Point: Essential Interfaces and Computation

The first system design step is identification of the chip's essential input/output interfaces and computation. The target product's marketing requirements usually establish the mandatory physical interfaces and necessary functions. These mandatory elements form the starting point for all other decisions such as implementation decisions about these functions and decisions about inclusion of other supporting functions.

**FIGURE 14.16**  Design Process

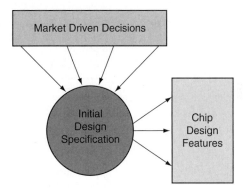

Not all interfaces and computations are equally fundamental to the design, of course. For example, in an integrated disk-drive controller, the external interface to the read-head, and the servomotor are essential to the function, but external interface to buffer memory is not. Buffer memory could be implemented either on-chip or off-chip, depending on more detailed analysis of cost and bandwidth trade-offs. Similarly, a secure socket layer security chip effectively requires implementation of the RSA (Rivest-Shamir-Adelman) algorithm for public/private key encryption, but may or may not include other TCP/IP protocol-processing functions.

### 14.1.13 Parallelizing a Task

When a single task shows particularly high computational demands, the designer must either use a faster general-purpose processor or pursue a more parallel hardware implementation. Fast general-purpose processors often require very high clock rates, for example, greater than 3 GHz. High clock rate processors are likely to be both unacceptably power hungry and difficult to design and integrate into an SoC without specialized skills. For embedded applications, an approach using parallel hardware resources is more likely to fit the embedded SoC's needs.

Historically, hardwired logic using RTL-based design was the only viable choice for parallel computation, but this design approach typically limited the complexity of the algorithms that could be implemented. Extensible processors open up a simple but highly efficient means to exploit parallelism, especially for fine-grained (instruction-level) parallelism.

The basic algorithm analysis and instruction-design process is systematic but complex. Compiler-like tools that design application-specific instruction sets can automate much of this process. In fact, the more advanced compiler algorithms for code selection, software pipelining, register allocation, and long-instruction-word operation scheduling can also be applied to discovery and implementation of new instruction definitions and code generation using those instructions. Automatic processor generation builds on the basic flow for application-specific processor generation but adds the creation of automatic processor architecture, as shown in Figure 14.17.

**FIGURE 14.17** Tensilica XPRES Processor Generation Flow (Reprinted by permission of Tensilica Incorporated.)

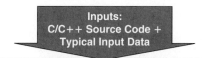

Inputs:
C/C++ Source Code +
Typical Input Data

### Prepare Application

- Compile C/C++ application for baseline architecture and profile on sample data
- Use the profile to rank code regions by number of execution cycles, determine vectorizable loops, generate dataflow graphs for every important region
- Generate hints for code manual improvements (e.g., vectorization)

### Evaluate Possible Architectures

- For each code region, automatically generate potential extensions across the space of configurations:
  - SIMD: Vector width
  - Combination: Number of slots in long instruction word
  - Fusion: Automatically generated sequences of dependent operations
- Estimate hardware cost and performance benefit of each set of extensions across all code regions
- Select best set of extensions within cost and performance goals.

### Manually Tune Extensions (Optional)

- Examine implementations for further hardware optimizations
- Add additional instructions to broaden applicability beyond input application set

### Use Generated Extension to Generate Code

- Generalized "graph matcher" finds code sequences that match generated instruction semantics
- Code Scheduler, vectorizer, and software pipeliner generate optimized code for long-instruction-word slots
- Simulator, assembler, debugger, and RTOS ports all created directly from generated instruction description

Output:
Optimized Application Binary + New
Instruction Set Description
New Compiler

However, automatic processor generation offers benefits beyond simple discovery of improved architectures. Compared to human-designed instruction-extension methods, automation tools eliminate any need to manually incorporate new data types and intrinsic functions into the application source code and provide effective acceleration of applications that may be too large or complex for a human programmer to assess.

As a result, this technology holds tremendous promise for transforming the development of processor architectures for embedded applications. The essential goals for automatic processor generation are these:

A software developer of average experience should be able to easily use the tool and achieve consistently good results.

No source-code modification should be required to take advantage of generated instruction sets. (Note that some ways of expressing algorithms are better than others in exposing the latent parallelism, especially for SIMD optimization, so source code tuning can help. The automatic processor generator should highlight opportunities to the developer to improve the source code.)

The generated instruction sets should be sufficiently general purpose and robust so that small changes to the application code do not degrade application performance. The architecture design-automation environment should provide guidance so that advanced developers can further enhance automatically generated instruction-set extensions to achieve better performance.

The development tool must be sufficiently fast so that a large range of potential instruction-set extensions can be assessed—on the order of thousands of architectures per minute.

The requirement for generality and reprogrammability mandates two related use models for the system:

Initial SoC development: C/C++ in, instruction-set description out

Software development for an existing SoC: C/C++ and generated instruction-set description in, binary code out Tensilica's XPRES (Xtensa Processor Extension System) compiler implements automated processor instruction-set generation.

A more detailed explanation of the XPRES flow will help explain the use and capability of this further level of processor automation. Figure 14.17 shows the four steps implemented by XPRES compiler. All these steps are machine automated, except for optional manual steps as noted.

The generation of a tailored C/C++ compiler adds significantly to the usefulness of the automatically generated processor. Even when the source application evolves, the generated compiler looks aggressively for opportunities to use the extended instruction set. In fact, this method can even be effective for generating fairly general-purpose architectures. So long as the basic set of operations is appropriate to another application, even if that application is unrelated to the first, the generated compiler will often use the extended architecture effectively.

The automatic processor generator internally enumerates the estimated hardware cost and application performance benefit of each of thousands of configurations, effectively building a pareto curve, such as that shown in Figure 14.18. Each point on the curve represents the best performance level achieved at each level of added gate count. This image is a screen capture from Tensilica's Xplorer development environment, for XPRES results on a simple video motion-estimation routine (sum-of-absolute-differences).

Automatic generation of instruction-set extensions applies to a very wide range of potential problems. It yields the most dramatic benefits for data-intensive tasks, where much of the processor-execution time is spent in a few hot spots and where SIMD, wide-instruction, and operation-fusion techniques can sharply reduce the number of instructions per loop iteration. Media- and signal-processing tasks often fall squarely in the sweet spot of automatic architecture generation.

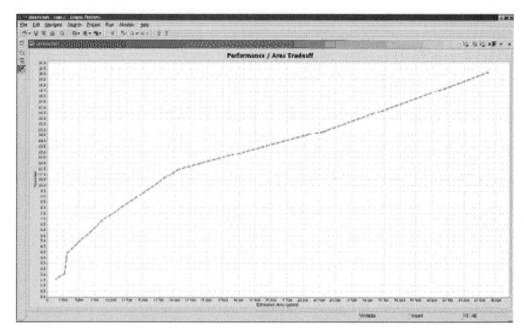

**FIGURE 14.18** Automatic Generation of Architectures (Reprinted by permission of Tensilica Incorporated.)

The automatic generator also handles applications where the developer has already identified key application-specific functions, implemented those functions in TIE, and used those functions in the C source code. Figure 4.19 shows the results of automatic processor generation for three applications using the XPRES compiler, including one fairly large application: an MPEG4 video encoder.

| Application | MPEG-4 Encoder | Radix-4 FFT | GSM Encoder |
|---|---|---|---|
| Speed-Up | 3.0x | 10.6x | 3.9x |
| Baseline Code Size | 111 KB | 1.5 KB | 17 KB |
| Code Size with Acceleration | 136 KB | 3.6 KB | 20 KB |
| Configurations Evaluated | 1,830,796 | 175,796 | 576,722 |
| Generator Run Time (minutes) | 30 | 3 | 3 |

**FIGURE 14.19** Application Comparisons

The figure includes code-size results for the baseline Xtensa processor architecture and the automatically optimized Xtensa processor architecture for each application. Using aggressively optimized instruction sets generally increases code size slightly, but in all cases, the optimized code remains significantly smaller than that for conventional 32-bit RISC architectures. The figure also shows the number of configurations evaluated, which increases with the size of the application. The automatic processor generator run time also increases along with the size of the application, but averages about 100,000 evaluated configurations per minute on a 4 GHz PC running Linux.

The figure also shows one example of generated-architecture generality. The GSM Encoder source code was compiled and run, not using an architecture optimized for the GSM Encoder, but for the architecture optimized for the FFT. Although both are DSP-style applications, they have no source code in common. Nevertheless, the compiler automatically generated for the FFT-optimized processor could recognize ways to use the processor's FFT-optimized instruction set to accelerate the GSM Encoder by 80% when compared to the performance of code compiled for the baseline Xtensa processor instruction set.

Completely automatic instruction-set extension carries two important caveats:

Programmers may know certain facts about the behavior of their application, which are not made explicit in the C or C++ code. For example, the programmer may know that a variable can only take on a certain range of values, or that two indirectly referenced data structures can never overlap.

The absence of that information from the source code may inhibit automatic optimizations in the machine code and instruction extensions. Guidelines for using the automatic instruction-set generator should give useful hints on how to better incorporate that application-specific information into the source code. The human creator of instruction extensions may know this information and be able to exploit this additional information to create instruction sets and corresponding code modifications.

Expert architects and programmers can sometimes develop dramatically different and novel alternative algorithms for a task. A different inner-loop algorithm may lend itself much better to accelerated instructions than the original algorithm captured in the C or C++ source code. Very probably, there will always be a class of problems where the expert human will outperform the automatic generator, though the human will take longer (sometimes much longer) to develop an optimized architecture.

## 14.1.14 Implications of Automatic Instruction-Set Generation

The implications of automatic instruction-set generation are wide ranging. First, this technology opens up the creation of application-specific processors to a broad range of designers. It is not even necessary to have a basic understanding of instruction-set architectures. The basic skill to run a compiler is sufficient to take advantage of the mechanisms of automatic instruction-set extension.

Second, automatic instruction-set generation deals effectively with complex problems where the application's performance bottleneck is spread across many loops or sections of code. An automated, compiler-based method is easily able to track the potential for sharing instructions among loops, the relative importance of different code sections based on dynamic execution profiles, and the cumulative hardware cost estimate. Global optimization is more difficult for the human designer to track.

Third, automatic instruction-set generation ensures that newly created instructions can be used by the application without source-code modification. The compiler-based tool knows exactly what combination of primitive C operations corresponds to each new instruction, so it is able to instantiate that new instruction wherever it benefits performance or code density. Moreover, once the instruction set is frozen and the SoC is built, the compiler retains knowledge of the correspondence between the C source code and the instructions. The compiler can utilize the same extended instructions even as the C source is changed.

Fourth, the automatic generator may make better instruction-set-extension decisions than human architects. The generator is not affected by the architect's prejudice against creating new instructions (design inertia) or influenced by architectural folklore on rumored benefits of certain instructions. It has complete and quite accurate estimates of gate count and execution cycles and can perform comprehensive and systematic cost/benefit analysis. This combination of benefits therefore fulfills both of the central promises of application-specific processors: cheaper and more rapid development of optimized chips and easier reprogramming of that chip once it's built to accommodate the evolving system requirements.

## 14.2        TENSILICA XTENSA ARCHITECTURE OVERVIEW

Processors have long played an important role in embedded systems design, but fundamental changes in the performance and complexity of those systems are changing that role. Across every embedded application (including data communications, telephony, storage, imaging, and consumer systems), algorithms and protocols are becoming more varied and complex, and the data types are growing in richness. In these new systems, a new class of processors is required. Paradoxically, each new processor must fully support the specific new algorithms, datatypes, and bandwidth of its target application, yet must be smaller and more power efficient than traditional processors (refer to Figure 14.20).

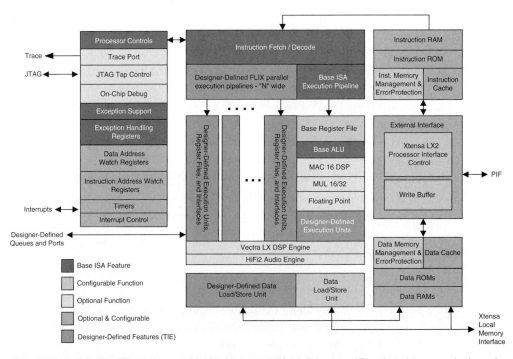

**FIGURE 14.20**   Tensilica Xtensa LX2 Architectural Block Diagram (Reprinted by permission of Tensilica Incorporated.)

Resolving this paradox requires a new approach to processors' automatic generation of application-specific processors. Tensilica has designed the Xtensa Xplorer design environment and the Xtensa Processor Generator for this purpose. Xtensa Processor Generator, shown in Figure 14.21, allows chip architects to rapidly explore alternative design approaches by quickly describing the key instruction, memory, peripheral and interface functions required by their processor. The Xtensa Processor Generator produces a complete processor hardware design, verification, and software development environment for that custom processor in minutes. The resulting

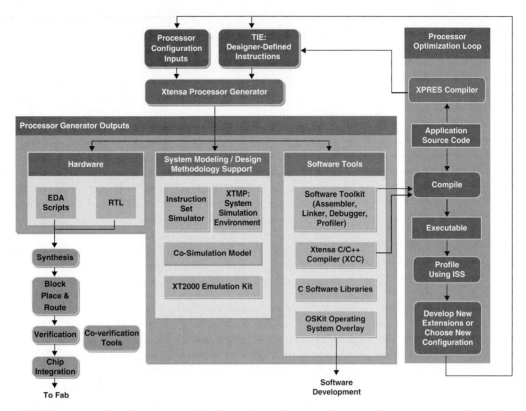

**FIGURE 14.21** Tensilica Xtensa LX2 Core Block Diagram (Reprinted by permission of Tensilica Incorporated.)

processor designs consistently execute applications faster, dissipate less power, and use less code than general-purpose embedded processors.

The essential foundation of all these processors is the Xtensa Instruction Set Architecture. The base architecture is common to all Xtensa processors. This instruction set was built from the ground up to satisfy the unique requirements of high-performance, high-volume embedded communication and consumer applications. The Xtensa architecture was designed with four key goals:

> Reduce code size
> Improve general-purpose embedded processor performance
> Reduce power dissipation
> Enable seamless extension of the architecture in support of demanding new applications

Furthermore, it is appropriate today to build a new architectural foundation because of the three dimensions of progress in underlying technology since the emergence of conventional RISC architecture more than 15 years ago:

> The shift in silicon technology from roughly 0.15 μm to 65 nm transistor geometry as the "sweet-spot" for implementation
> The shift from general-purpose computing on the desktop to data-centric computing in embedded systems as the primary domain of use
> Accumulated progress in architectural insights, compilation technology, and programming languages and practices
> The need to reduce power for mobile consumer applications

## 14.3      PRINCIPLES OF INSTRUCTION SET DESIGN

The design of processor instruction sets is a well-established art. Most instruction set features are not new in themselves, but features can be combined in new and unique ways that advance the state of the art. In particular, when instruction set design is optimized for a different use than prior instruction sets, significant improvements result.

Instruction set architecture (ISA) design needs to balance many competing goals, including the following:

The size of the machine code required to encode various algorithms
The extensibility and adaptability of the ISA for new algorithms and applications
The performance of processors that employ this ISA on such algorithms
The power consumption of processors that employ this ISA on such algorithms
The cost of processors that employ the ISA
The ISA's suitability for multiple future processor implementations
The design complexity of processors that employ the ISA
The ISA's suitability as a target for compilation from high-level programming languages

The instruction set architecture has one direct and two indirect influences on processor performance. The ISA directly determines the number of instructions required to implement a given algorithm. Other components of processor performance include the minimum possible clock period and the average number of clocks per instruction. These are primarily attributes of the implementation of the instruction set, but instruction set features may affect the ability of the implementer to simultaneously meet time per clock and clocks per instruction goals. For example, a certain encoding choice might mandate additional logic in series with the rest of instruction execution, which an implementer would address either by increasing the time per clock or by adding an additional pipeline stage, which will increase the number of clocks per instruction (instruction latency).

## 14.4      TENSILICA XTENSA PROCESSOR UNIQUENESS

The RISC (reduced instruction set computing) processor design philosophy emerged in the 1980s. RISC ISAs allow implementers to reduce a processor's cycles per instruction and clock period significantly without seriously increasing the number of instructions required to execute a program. RISC ISAs improve the performance of processors, lower design complexity, allow lower-cost processor implementations at a given performance level, and are well suited to compilation from high-level programming languages.

Curiously, there is no single, completely comprehensive or satisfactory definition of the term RISC, but RISC processors typically include

Fixed-size instruction words
Large uniform register files for computation operations
Simple and fixed instruction-field encoding
Memory access via loads and stores of registers
A small number (often one, usually less than four) of memory addressing modes
Avoidance of features that would make pipelined execution of instructions difficult (variable latency and microcoded instructions)

On the other hand, most RISC ISAs are designed for high-performance desktop computing environments where a large hard disk storage capacity is a given. They are not optimized for producing compact machine code. In particular, RISC instruction sets usually require more program bits to encode an application than pre-RISC ISAs. In many embedded applications today, the cost of code storage (on-chip RAM/ROM) is often greater than the cost of the processor (gate count), so the use of RISC processors is sometimes limited in the most cost-sensitive applications. An ISA that combines the advantages of RISC with reduced code size would be useful in many embedded applications. This combination is one of the underlying themes for the development of the Xtensa ISA.

The Xtensa instruction set borrows the best features of established RISC architectures and adds new instruction set architecture developments. The Xtensa ISA derives most of its features from RISC, but it has targeted areas in which older CISC architectures have been strongest, such as compact code.

The Xtensa core ISA is implemented as a set of 24-bit instructions that perform 32-bit operations. The instruction width was chosen primarily with code-size economy in mind. The instructions themselves were selected for the utility in a wide range of embedded applications. The core ISA has many powerful features, such as compound operation instructions that enhance its fit to embedded applications, but avoids features that would benefit some applications at the expense of cost or power on others (for example, features that require extra register-file ports). Such features can be implemented in the Xtensa architecture using options and coprocessors specifically targeted at a particular application area.

The Xtensa ISA is organized as a core set of instructions with various optional packages that extend the functionality for specific application areas. This allows the designer to include only the required functionality in the processor core, maximizing the efficiency of the solution. The core ISA provides the functionality required for general control applications and excels at decision making and bit and byte manipulation. The core also provides a target for third-party software, and for this reason deletions from the core are not supported. On the other hand, numeric computing applications such as Digital Signal Processing are best done with optional ISA packages appropriate for specific application areas, such as the MAC16 option for integer filters or the Floating-Point Coprocessor option for high-end audio processing.

The baseline Xtensa architecture builds on many of the principles of RISC, but introduces new techniques to improve both the number of instructions required to encode a program and the average number of bits per instruction. These techniques hold the promise to both improve performance and reduce cost relative to previous architectures. The Xtensa ISA starts with the premise that it must provide good code density in a fixed-length, high-performance encoding based on RISC principles, including a general register file and a load/store architecture.

To achieve exemplary code density, Xtensa processors add a simple variable-length encoding scheme that doesn't compromise performance. The Xtensa architecture further optimizes the cost of processor implementation, by balancing such features as register files, control-flow operations, arithmetic and logic instructions, and load/store capabilities in favor of operations that are frequent in modern embedded software and small and fast in modern deep-sub-micron implementation.

## 14.5    REGISTERS

To maintain performance, a RISC instruction set must support at least two source register fields and one distinct destination register field. General register instruction sets that optimize only for code density are sometimes designed around two register fields, with one used for source only and one used for both source and destination. This design approach sometimes reduces code size, but there is no way to compensate for the increase in the number of instructions required to execute a program. Instruction sets that specify fewer registers use narrower register fields and save bits per instruction. However, these instruction sets increase the number of instructions in the program by forcing more variable and temporary values to live in memory, and they require extra load and store instructions. Consequently, this design approach increases both the number of cycles for program execution and the power dissipated.

As the number of the registers increases, the marginal benefits of a two-operand instruction format decline. In particular, at least 16 general registers are required for good RISC performance. Three 4-bit register fields require at least 12 bits to encode. Bits for opcode and constant fields are also required. So 16-bit encoding, as used by some processors, is not sufficient for good performance.

---

## 14.6    INSTRUCTION WIDTH

Prior RISC architectures failed to achieve an appropriate balance between code size and performance because RISC ISA designers felt constrained to certain instruction sizes such as 16 and 32 bits. There are indeed advantages to using instruction sizes that are simple ratios to the data word width of the processor. However, relaxing the restriction somewhat has significant advantages.

Xtensa processors use a 24-bit fixed-length encoding (Figure 14.22) as a starting point; 24 bits are sufficient for achieving high performance while providing extensibility and room for powerful instructions that will decrease the number of instructions required to execute a program. The Xtensa ISA's 24-bit encoding represents a 25% reduction in instruction size relative to the more common RISC 32-bit instruction word, which reduces code size requirements relative to most 32-bit RISC instruction sets. Most important, 24 bits is simple to accommodate in a processor with 32-bit data path widths.

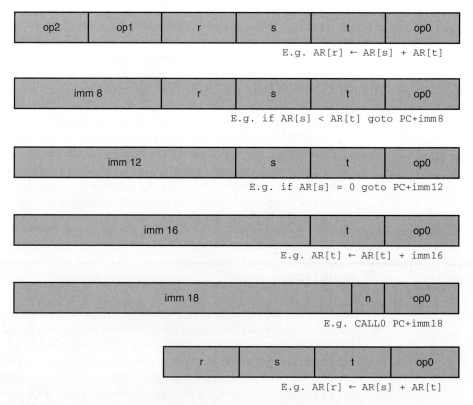

**FIGURE 14.22**    Xtensa Typical 16-/24-bit Instruction Types (Reprinted by permission of Tensilica Incorporated.)

The Xtensa architecture uses 4-bit register fields as shown Figure 14.23, the minimum required for acceptable performance and the maximum that fits well within a 24-bit instruction word. Many RISC instruction sets use 32 registers (5-bit register fields).

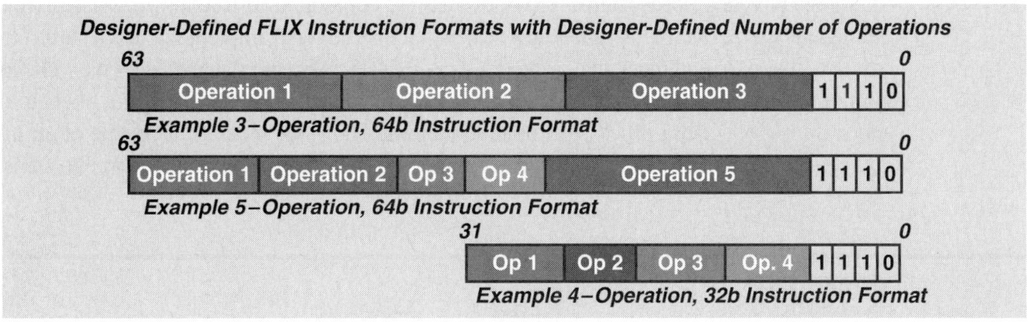

**FIGURE 14.23** The 32-/64-bit extended instructions (Reprinted by permission of Tensilica Incorporated.)

The difference in performance between 16 and 32 general registers (about 5%) is not as large as the difference between 8 and 16 general registers, and it is small enough that other features can be introduced to make up the lost performance (e.g., compound instructions and register windows). The resulting increase in the number of instructions needed to encode a program (also about 5%) is more than offset by the difference between 24-bit and 32-bit encoding (a reduction of 25%).

Note that many instruction sets with 5-bit register fields do not provide 32 general registers for compilation. Most dedicate a register to hold zero, even though the addition of a few extra instruction op codes can easily eliminate the need for a zero register (e.g., the Xtensa NEG instruction). Also, other registers are often given specific uses that can be avoided by including other features in the instruction set. For example, some RISC architectures dedicate 2 of its 31 general registers for exception handling and one more register for a global area pointer. So, in effect, the architecture provides the program with only 28 general registers for variables and temporary storage. That is only 12 more registers than an instruction set that uses 4-bit register fields.

The division of general registers into caller and callee saved registers by software convention is common and further restricts the utility of larger register files. The Xtensa ISA includes features that avoid this, which brings the effectiveness of the 16 registers almost to the level of 32 registers. The Xtensa ISA shows that a 24-bit encoding of a full-featured RISC instruction set is possible. In addition, the Xtensa ISA supports parallel operation by extension of the instruction width, as shown in Figure 14.23. The Xtensa ISA is a significant step forward for processor design.

## 14.7   COMPOUND INSTRUCTIONS

To improve performance and code size, the Xtensa ISA also provides instructions that combine the functions of multiple instructions typically found in RISC and other processor instruction sets into a single instruction. The first example of a compound instruction is a simple "left shift and add/subtract." High-end architectures like the SUN Sparc are examples of instruction sets that provide these operations.

Address arithmetic and multiplication by small constants often use these combinations, and providing these operations reduces the instruction count but potentially increases the processor clock period because of the additional series logic added to the computation pipeline stage. However, various implementations have shown that when the shift range is limited to 0 to 3, the extra logic is not the most critical constraint on the clock frequency. An instruction set that provides arbitrary shift and add, consequently, with a corresponding ISA will have degraded maximum clock frequencies.

Right shifts are often used to extract a field from a larger word. For an unsigned field extract, two instructions (either left shift followed by right shift, or right shift followed by an AND with a constant) are typically used. Xtensa provides a single compound instruction, EXTUI (extract unsigned immediate), to perform this function. The EXTUI instruction is implemented as a shift followed by an AND with a specified mask that is encoded in the instruction word using just 4 bits. The logical AND portion of the EXTUI instruction is so trivial that its inclusion in the ISA is not likely to increase the clock period of Xtensa processor implementations. The same would not be true of an instruction to extract signed fields, so there's no corresponding EXTSI instruction included in the Xtensa ISA.

## 14.8    BRANCHES

Most processor instruction sets, both RISC and otherwise (e.g., ARM, Intel Viiv, Freescale PowerPC, and Sun SPARC), use a compare instruction that sets condition code(s), followed by a conditional branch instruction that tests the condition code(s) for program flow control. Conditional branches constitute 10% to 20% of the instructions in most RISC instruction sets, and each is usually paired with a compare instruction. This style of instruction set is wasteful. Some instruction sets (e.g., ARM, MIPS, HP PA-RISC, Sun SPARC V9) provided a compound compare-and-branch facility of varying flexibility. The Xtensa ISA provides the most useful compound compare-and-branch instructions.

Choosing the exact set requires balancing the utility of each compare and branch with the op code space that it consumes, especially when 24-bit (as opposed to 32-bit) instruction encoding is the target. Other instruction sets fail this test. Compound compare-and-branch instructions reduce instruction count, when compared with instruction sets that have separate compare-and-branch instructions, and even when compared with the partial compare-and-branch instructions. Some Xtensa processor implementations may require an increase in clocks per instruction to implement some compound compare-and-branch instructions, but the overall performance effect of these compound instructions is still positive.

The Xtensa ISA's compare-and-branch instructions also support comparisons to immediate values and use clever encoding of constants to increase their utilization. The BEQI, BNEI, BLTI, and BGEI instructions also use a 4-bit field that encodes various common constants. The BLTUI and BGEUI instructions use a different encoding, as unsigned comparisons have a different set of useful values.

The Xtensa processor's compound compare-and-branch instruction set packs all these immediate values into a single instruction word, resulting in smaller fields. These instructions combine the comparison opcode, two source-register fields, and an 8-bit PC-relative offset target specifier into a 24-bit instruction word. The 8-bit relative target specifier will be too small in some infrequent cases, so the compiler or assembler compensates by using a conditional branch of the opposite nature around an unconditional branch with a longer range. The Xtensa ISA also provides a series of compound compare-and-branch instructions that test against zero, the most common case. These compound compare-and-branch instructions have a 12-bit PC-relative offset, which provides much greater range.

The Xtensa architecture adds another important and unique goal to instruction set design: complete support for extensibility that allows for the addition of new data types, implemented in new instructions and closely coupled coprocessors (Figure 14.24). The Xtensa ISA uses an additional method for allowing coprocessor conditional branches. The Xtensa ISA offers an option that adds 16 1-bit Boolean registers. The Xtensa ISA's BF (branch if false) and BT (branch if true) instructions test these Boolean registers and branch accordingly.

**FIGURE 14.24**  Extended
Instruction Set

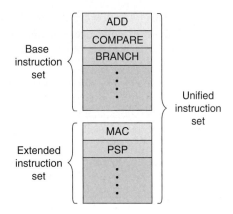

Designer-defined coprocessors can implement instructions that set the Boolean registers based on comparisons of their supported data types. All Xtensa coprocessors share the baseline ISA's Boolean register set and the BF and BT instructions. This approach makes efficient use of the Xtensa ISA's short, 24-bit instruction word. This scheme is a new variant of compare-and-branch condition codes found in many earlier processor ISAs. The use of single-bit (Xtensa, MIPS) instead of multibit comparison-result registers (most other ISAs) increases the number of comparison op codes required but decreases the number of branch op codes required. This ISA design approach also makes the introduction of a broad range of application-specific branches and conditional operations simple and efficient for users to implement, a very important feature for an ISA designed expressly for extensibility.

The Xtensa ISA also provides a general-purpose, zero-overhead loop feature similar to that found in some DSPs (digital signal processors). Most RISC processors use their existing conditional branch instructions to implement software loops. However, this op code economy increases program cycle count and consequently reduces execution speed. For many RISC ISAs, loop overhead consists of three instructions: add, compare, and conditional branch. The performance impact of the loop overhead is higher when the loop body is small. For small software loops, many compilers use an optimization called loop-unrolling (Figure 14.25) to spread the loop overhead over two or more loop iterations, but this approach duplicates the loop body and can significantly increase code size.

**FIGURE 14.25**  Unrolled Loop

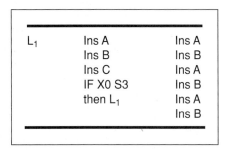

By contrast, many DSPs and some general-purpose processors provide other ways to perform certain kinds of loops. The first method is to provide an instruction that repeats the succeeding instruction a fixed number of times (e.g., TI TMS320C2x, Intel x86). For one-instruction loops, a repeat prefix instruction eliminates loop overhead and saves power by eliminating the need to repeatedly fetch the same instruction within the loop. Some ISAs with repeat instructions require that the processor not take an interrupt during the loop.

This limitation can impose unacceptable interrupt latency because loop execution may require many machine cycles to complete. An improvement on simple repeat prefix instructions is the ability to iterate a block of instructions multiple times with reduced or zero loop overhead (e.g., TI TMS320C5x). The Xtensa ISA provides this zero-overhead loop capability via its LOOP, LOOPGTZ, and LOOPNEZ instructions. The Xtensa ISA's LOOP instructions eliminate instruction execution cycles required for incrementing the loop index and for comparison and branch operations, and it avoids the taken-branch penalty that is typically associated with a compilation of loops based on conditional-branch instructions.

The Xtensa ISA demonstrates how a reduced-overhead looping capability can be integrated into a general-purpose processor ISA (as opposed to a DSP) to improve both execution performance and code size. Overall, the Xtensa architecture makes six important contributions to general branch instructions:

A choice of compare-and-branch instructions in a RISC ISA with the most useful comparisons

Compare-and-branch with encoded immediate values, including branch-on-bit instructions

Instruction formats with longer target specifiers for common cases (test against zero)

The encoding of all branch instructions in a 24-bit instruction word

Extensible support for branches on coprocessor Boolean registers (condition codes) with logical operations on Booleans

Zero-overhead loops that eliminate branch execution delay and reduce code size.

## 14.9    INSTRUCTION PIPELINE

The Xtensa LX incorporates a standard RISC five-stage pipeline as shown in Figure 14.26. It is composed of instruction fetch, register access, execute, data-memory access, and register writeback. Also, for more design flexibility, the Xtensa LX can support a seven-stage pipeline. This is useful, for example, when using slower memories to save power and silicon area. Higher-performance designs might have memories too fast for the five-stage pipeline.

**FIGURE 14.26**  Tensilica Xtensa LX2 Five-Stage Pipeline (Reprinted by permission of Tensilica Incorporated.)

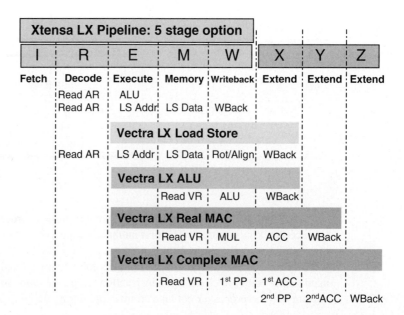

## 14.10    LIMITED INSTRUCTION CONSTANT WIDTH

No Xtensa baseline instruction is longer than 24 bits, so constant fields in the instruction word are constrained. The Xtensa architecture addresses this issue in several ways. The Xtensa ISA provides small constant fields to capture the most common constants. Xtensa instructions encode the constant value rather than specifying it directly. The encoded values are chosen from a wide array of program statistics as the N (e.g., 16) most frequent constants for each instruction type.

The Xtensa architecture uses this technique in the ADDI4 instruction, where the 16 values are chosen to be $-1$ and 1 to 15, rather than 0 to 15. Adding 0 is of no utility (there is a separate MOVE instruction), and adding $-1$ is common. The constants used in bitwise-logical operations (e.g., AND, OR, XOR) represent bit masks of various sorts and often do not fit in small constant fields. Bit patterns consisting of a sequence of 0s followed by a sequence of 1s, and a sequence of 1s followed by a sequence of 0s are quite common. For this reason, the Xtensa architecture has instructions that avoid the need for putting a mask directly into the instruction word. The EXTUI instruction (described earlier) performs a shift followed by a mask consisting of a series of 0s followed by a series of 1s, where the number of 1s is a constant field in the instruction.

Xtensa load-and-store instructions use an instruction format with an 8-bit constant offset that is added to a base address from a register. The Xtensa ISA both makes the most of these 8 bits and provides a simple extension method when 8 bits is insufficient. Xtensa load/store offsets are zero-extended rather than sign-extended because the values 128 to 255 are more commonly used by load and store instructions than the values $-128$ to $-1$. Also, the offset is shifted left appropriately for the reference size because most references are to aligned addresses from an aligned base register.

The offset for 32-bit loads and stores is shifted by 2 bits, the offset for 16-bit loads and stores is shifted by 1 bit, and the offset for 8-bit loads and stores is not shifted. Most loads and stores are 32 bit, and so this technique provides 2 additional bits of range. When the 8-bit constant offset specified in a load/store instruction (or an ADDI instruction) is insufficient, the Xtensa ISA provides the ADDMI instruction, which adds its 8-bit constant shifted left by 8 bits. Thus a two-instruction sequence has 16 bits of range, 8 bits from the ADDMI and 8 bits from the load/store or ADDI instruction.

## 14.11    SHORT INSTRUCTION FORMAT

The Xtensa ISA consists of a core set of instructions that must be present in all implementations of the instruction set and a set of optional instruction packages that may or may not be present in a given implementation. One of the most popular packages is the short instruction format package. It provides even further code size reductions by reducing the average number of bits per instruction (refer to Figure 14.27). When these short-format instructions are present, the Xtensa

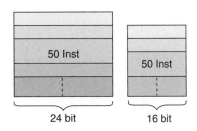

**FIGURE 14.27**    Instruction Width

ISA changes from a fixed-length (24-bit) instruction set, to one with two instruction sizes (24-bit and 16-bit). Note that the Xtensa architecture does not employ modes to add the 16-bit instructions to the ISA the way some other RISC processors do.

The Xtensa ISA's 24- and 16-bit instruction formats are operative simultaneously so there is zero overhead incurred in switching from one instruction format to another. Because the Xtensa short instruction forms are optional, these forms are used solely for improving code size; no new capabilities are added by the Xtensa ISA's 16-bit instructions. The set of instructions that can be encoded in 16 bits consists of the most statically frequent instructions that will fit. The most frequently used instructions in most instruction sets are loads, stores, branches, adds, and moves; these are exactly the instructions present in the Xtensa ISA's 16-bit instruction set.

Only the most frequent instructions need short encodings, so three register fields are still available (because the op code field is small), and narrow, encoded constant fields can capture a significant fraction of the uses. Approximately half of the Xtensa instructions needed to represent an application can be encoded in just 6 of the 16 op codes available in a 16-bit instruction encoding after three 4-bit fields are reserved for register-specifiers or constants.

## 14.12    REGISTER WINDOWS

Another important Xtensa architecture feature is the windowed register file, as shown in Figure 14.28. Register windows reduce code size and improve performance. Register windows are found on a few other processors, such as Sun's SPARC ISA. The name "register window" describes the typical implementation where the register field in the instruction specifies a register in the current window into a larger register file. Register windows avoid the need to save and restore registers at procedure entry and exit. Instead of saving and restoring registers on a stack, a processor with register windows merely changes a register-offset pointer, which hides some registers from view and exposes new ones.

**FIGURE 14.28** Register Windows (Reprinted by permission of Tensilica Incorporated.)

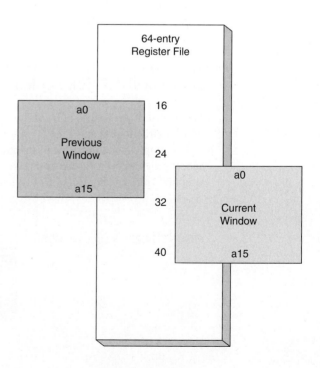

The exposed registers usually do not contain valid data and can be used directly. Register windows that overlap in their views of the physical register file between the caller and callee also avoid argument shuffling that can occur when arguments to procedures are passed in registers. Finally, register windows alter the breakeven point for allocating a variable or temporary to a register and thus encourage register use, which is faster and smaller than using a memory location.

Unlike SPARC's fixed-window overlap increment, the Xtensa ISA employs a variable increment for register windowing. This feature keeps implementation cost low by allowing a much smaller physical register file to be used. For example, many Sun SPARC ISA implementations use a physical register file of 136 entries, whereas Xtensa ISA implementations require a register file of only 64 entries to achieve similar performance. The Xtensa ISA specifies new methods to detect window overflow and underflow and to organize the stack frame.

## 14.13    XTENSA LX2 SUMMARY

The Xtensa architecture makes a number of fundamental contributions to embedded processor architecture, including

> A full 16-visible windowed register file, three-operand programming model in less than 32-bit instruction encoding for performance, generality, and code size.
>
> Rich selection of commonly occurring instruction combinations as compound instructions.
>
> Encoding of common immediate values for performance and code size.
>
> An unusually rich and powerful branch architecture, including compare and branch, bit-test branches, coprocessor condition codes and branches, and zero-overhead loops, for performance and code size.
>
> An available 16-bit instruction subset that can be freely intermixed with 24-bit base instructions for further code density improvement.
>
> Register windows to reduce load/store traffic and further improve performance and code density.

## QUESTIONS AND PROBLEMS

1. What is meant by concurrency in SoC design?
2. What is happening to design complexity versus design productivity?
3. List two key concerns for chip designers.
4. List three benefits of high silicon integration.
5. What are two key challenges system developers are facing?
6. What is the "programability overhead"?
7. What is the key goal of SoC development?
8. How are configurable, extensible processor cores generated?
9. List the key elements of the SoC design flow.
10. List four implications of automatic instruction set generation.
11. List the four key goals for the Xtensa architecture.
12. List the eight competing goals to balance in an ISA.
13. List six key features of RISC.
14. Why does instruction width matter?

15. What is a compound instruction?
16. Describe "register windowing."

## SOURCES

*Catching UP with Moore's law: How to fully exploit the benefits of nanometer silicon.* White Paper. Tensilica, Inc.

Dixit, Ashish. 2004. *Introducing the Xtensa LX Processor Generator.* Tensilica Inc.

Leibson, Steve. 2003. *Designing with configurable processors instead of RTL.* Tensilica Inc.

Rowen, Chris. 2002. *Reducing SoC simulation and development time.* Computer, December 2002, reprint.

Rowen, Chris. 2006. *The reinvention of the microprocessor.* Tensilica Inc.

*Xtensa architecture and performance.* White Paper. October 2005. Tensilica, Inc.

*Xtensa LX microprocessor overview handbook.* Data Book. Tensilica Inc.

# CHAPTER 15

# Digital Signal Processors

## OBJECTIVE: TO UNDERSTAND DSP CONTROLLER ARCHITECTURE

The reader will learn the basic architectures and features of the

1. Texas Instruments TMS320C55x.
2. Analog Devices ADSP-BF533 (Blackfin).

## 15.0 DSP OVERVIEW

Specialized digital signal processing devices increase the performance of digital signal processor (DSP) algorithms dramatically over general-purpose devices. As we have seen with the COTS and IP Core devices, they may have DSP extensions but still are fundamentally general purpose in nature. For applications where the primary function is DSP, devices are designed to meet this requirement.

## 15.1 TMS320C55x

The TMS320C55x (C55x) offers a cost-effective solution to the challenges in personal and portable processing applications as well as digital communications infrastructure with restrictive power budgets. Compared to a 120-MHz C54x, a 300-MHz C55x will deliver approximately 5X higher performance and dissipate one-sixth the core power dissipation of the C54x.

The C55x core's ultra-low power dissipation is achieved through attention to low-power design and advanced power management techniques. The C55x implements a high level of power-down configurability and granularity coupled with power management that occurs automatically and is transparent to the user.

The C55x core delivers twice the cycle efficiency of the C54x through a dual-MAC (multiply-accumulate) architecture with parallel instructions, additional accumulators, ALUs, and data registers. An advanced instruction set, a superset to that of the C54x, combined with expanded busing structure, supports the additional hardware execution units.

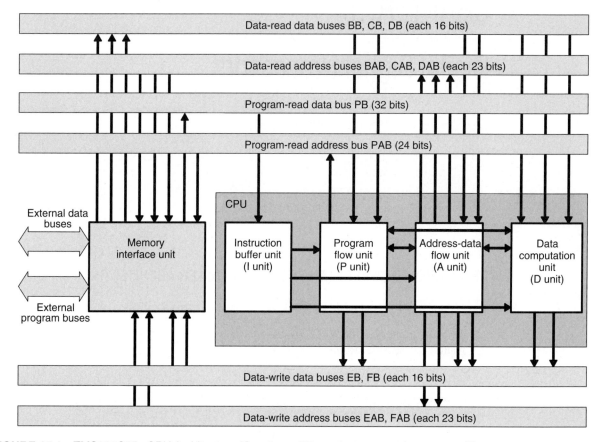

**FIGURE 15.1** TMS320C55x CPU Architecture (Courtesy of Texas Instruments Incorporated.)

The C55x instruction set is designed for increased code density to support lower system cost. The C55x instructions are variable byte lengths ranging in size from 8 bits to 48 bits. With this scalable instruction word length, the C55x reduces control code size per function for increased memory density. Reduced control code size means reduced memory requirements and lower system cost.

### 15.1.1 Characteristics of the TMS320C55x

The TMS320C55x delivers a combination of ultra-low power performance and low system cost to fuel the continued digitization and miniaturization of personal and portable applications. The C55x architecture and hardware design was developed with three interrelated objectives:

1. Ultra-low power
   Allows increased battery life for portable applications or greater channel density for power-efficient infrastructure systems.
2. Efficient DSP performance
   Allows more functionality to be added to systems or faster processing time for existing algorithms.
3. High in code density
   Requires less memory to perform given functions, which translates to lower system cost and/or smaller system size. All this as a result of more tightly packed code.

***15.1.1.1 Market Segments*** The C55x supports four basic market segments that require low power, low system cost, and high performance with different levels of importance. Those segments include applications such as

1. Applications that require much longer battery life while maintaining or slightly increasing performance. Examples include extending the battery life of today's digital cellular handsets, portable audio players, or digital still cameras from hours to days, or days to weeks, while maintaining the same level of functionality.
2. Applications that require much higher performance while maintaining or slightly increasing battery life. Examples include tomorrow's 3G wireless handsets or Internet appliances, which may converge audio, video, voice, and data into a single multifunction mobile product. Consumers have come to expect a certain level of battery life in standby and active modes and will not be willing to sacrifice this for more functionality.
3. Applications that require very small size, ultra-low power consumption, and low-to-medium levels of DSP performance. Examples include the personal medical market, whereby new advances in hearing aids and medical diagnostics require DSP capability but with battery life measured in weeks or months.
4. Power-efficient infrastructure applications (RAS, VOP, multiservice gateways, etc.) that need increased channel density while meeting stringent board-level power and space budgets.

***15.1.1.2 DSP Applications*** Generally speaking, the C55x is broadly targeted at the consumer and communication markets, which utilize DSP algorithms such as

> Speech coding and decoding
> Line or acoustic echo cancellation; noise cancellation
> Modulation and demodulation
> Image and audio compression and decompression
> Speech encryption, decryption
> Speech recognition, speech synthesis

## 15.1.2 Key Features of the C55x

The C55x incorporates a rich set of features that provide processing efficiency, low-power dissipation, and ease of use. These important features include

> A 32x 16-bit instruction buffer queue
> Two 17-bit x17-bit MAC units
> One 40-bit ALU
> One 40-bit barrel shifter
> One 16-bit ALU
> Four 40-bit accumulators
> Twelve independent buses:
>> Three data read buses
>> Two data write buses
>> Five data address buses
>> One program read bus
>> One program address bus
> User-configurable IDLE domains
> Buffers variable length instructions and implements efficient block repeat operations
> Execute dual MAC operations in a single cycle
> Performs high-precision arithmetic and logical operations

Can shift a 40-bit result up to 31 bits to the left, or 32 bits to the right

Performs simpler arithmetic in parallel to main ALU

Hold results of computations and reduce the required memory traffic

Provide the instructions to be processed as well as the operands for the various computational units in parallel—to take advantage of the C55x parallelism.

Improve flexibility of low-activity power management

### 15.1.3 Instruction Set Architecture

The C55x architecture achieves power-efficient performance through increased parallelism and complete focus on reduction in power dissipation.

The CPU supports an internal bus structure (Figure 15.2) composed of

One program bus

Three data read buses

Two data write buses

Additional buses dedicated to peripheral and DMA activity

These buses provide the ability to perform up to three data reads and two data writes in a single cycle. In parallel, the DMA controller can perform up to two data transfers per cycle independent of CPU activity.

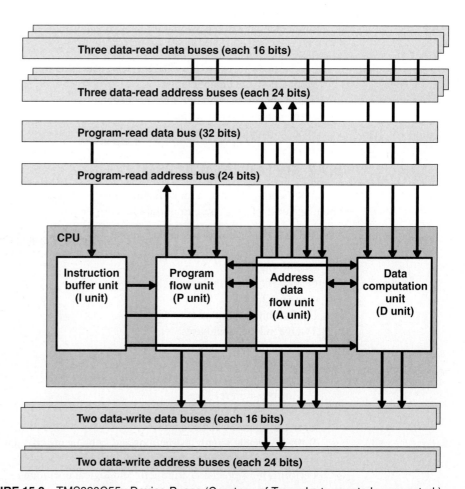

**FIGURE 15.2**　TMS320C55x Device Buses (Courtesy of Texas Instruments Incorporated.)

**15.1.3.1 Instruction Pipelining** The C55x DSPs perform instruction fetching, decoding, and execution in seven pipelined stages as shown in Figures 15.3(a) and 15.3(b). They are as follows:

1. Fetch stage reads program data from memory into the instruction buffer queue.
2. Decode stage decodes instructions and dispatches tasks to the other primary functional units.
3. Address stage computes addresses for data accesses and branch addresses for program discontinuities.
4, 5. Access1/Access 2 stages send data read addresses to memory.
6. Read stage transfers operand data on B bus, C bus, and D bus.
7. Execute stage executes operation in the A unit and D unit and performs writes on the E bus and F bus.

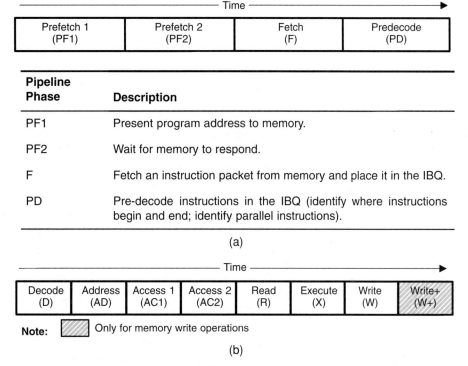

| Prefetch 1 (PF1) | Prefetch 2 (PF2) | Fetch (F) | Predecode (PD) |

| Pipeline Phase | Description |
| --- | --- |
| PF1 | Present program address to memory. |
| PF2 | Wait for memory to respond. |
| F | Fetch an instruction packet from memory and place it in the IBQ. |
| PD | Pre-decode instructions in the IBQ (identify where instructions begin and end; identify parallel instructions). |

(a)

| Decode (D) | Address (AD) | Access 1 (AC1) | Access 2 (AC2) | Read (R) | Execute (X) | Write (W) | Write+ (W+) |

**Note:** ▨ Only for memory write operations

(b)

**FIGURE 15.3** (a) TMS320C55x First Stages of the Pipeline; (b) TMS320C55x Second Stages of the Pipeline (Courtesy of Texas Instruments Incorporated.)

The C55x pipeline is protected, meaning it will automatically insert cycles as necessary to prevent pipeline conflicts. Pipeline protection cycles are inserted when:

An instruction is supposed to write to a location, but a previous instruction has not yet read that location (extra cycles are inserted so the read occurs first).

An instruction is supposed to read from a location, but a previous instruction has not yet written to that location (extra cycles are inserted so the read occurs first).

**15.1.3.2 CPU Features** The C55x CPU provides two multiply-accumulate (MAC) units, each capable of 17-bit x 17-bit multiplication in a single cycle. A central 40-bit arithmetic/logic unit (ALU) is supported by an additional 16-bit ALU. Use of ALUs is subject to instruction set control. This programmability provides the capacity to optimize parallel activity and power consumption. These resources are managed in the address data flow unit (AU) and data computation unit (DU) of the C55x CPU.

***15.1.3.3  Instruction Set***    The TMS320C55x is a low-power, general-purpose signal processing architecture with an instruction set optimized for efficiency, ease of use, and compactness. The C55x architecture supports a variable byte width instruction set for improved code density. The instruction buffer unit (IU) performs 32-bit program fetches from internal or external memory and queues instructions for the program unit (PU). The program unit decodes the instructions, directs tasks to AU and DU resources, and manages the fully protected pipeline. Configurable instruction cache is also available to minimize external memory accesses, thus improving data throughput and conserving system power.

A highly parallel architecture complements the C55x instruction set and enables increased code density while reducing the number of cycles required per operation. The union of an efficient, compact instruction set with a highly parallel architecture provides a high-performance signal processing engine while minimizing code size and power consumption.

The C55x instruction set includes a flexible set of orthogonal features to enhance ease of use and program efficiency. Powerful addressing modes, which include the absolute addressing mode, the register-indirect addressing mode, and the direct addressing mode (also known as displacement), greatly reduce the instruction count required for signal processing algorithms. A three-operand instruction format, with support for both memory and register references, also provides excellent instruction density.

All C55x instructions that move data support any of the major addressing modes and operand formats. This regularity is conducive to efficient high-level language compiler use and simplifies programming in assembly. The instruction set also includes syntax that allows the programmer or compiler to schedule multiple instructions for parallel execution. These instruction set features simplify the task of the programmer and optimize the efficiency of C55x code, resulting in shorter product development time.

A key to the processing power and superior code density of the C55x is its efficient implementation. This implementation uses variable length instruction encoding to achieve optimal code density and efficient bus usage. Multiple computational units are included to carry out computations in parallel, thereby reducing the number of cycles required per operation. The dual multiply-and-accumulate (MAC) units can perform two 17-bit x 17-bit MAC operations in a single cycle, while the 40-bit ALU can be used to operate on 32-bit data or can be split to perform dual 16-bit operations.

A second 16-bit ALU for general-purpose arithmetic further increases parallelism and adds flexibility. Based on a modified Harvard architecture, the C55x includes one program bus and three independent read data buses that can simultaneously bring data operands to the various computational units. The high degree of parallelism and efficient instruction encoding maximize the overall processor efficiency without sacrificing performance.

## 15.1.4  Primary Functional Units

The C55x architecture is built around four primary blocks: the instruction buffer unit (IU), the program flow unit (PU), the address data flow unit (AU), and the data computation unit (DU). These functional units exchange program and data information with each other and with memory through multiple dedicated internal buses.

Figures 15.4(a) and 15.4(b) show the principal blocks and bus structure in the C55x devices. Program fetches are performed using the 24-bit program address bus (PAB) and the 32-bit program read bus (PB).

The functional units read data from memory via three 16-bit data read buses named B-bus (BB), C-bus (CB), and D-bus (DB). Each data read bus also has an associated 24-bit data read address bus (BAB, CAB, and DAB). Single operand reads are performed on the D-bus. Dual-operand reads use C-bus and D-bus. B-bus provides a third read path and can be used to provide coefficients for dual-multiply operations.

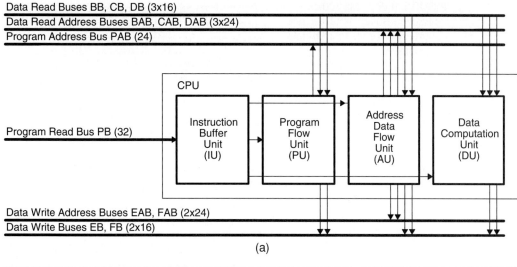

Data Read Buses BB, CB, DB (3x16)

Data Read Address Buses BAB, CAB, DAB (3x24)

Program Address Bus PAB (24)

Program Read Bus PB (32)

Data Write Address Buses EAB, FAB (2x24)

Data Write Buses EB, FB (2x16)

(a)

| Action Performed | Bus Used |
|---|---|
| Program Fetches | 24-bit Program Address Bus (PAB)<br>32-bit Program Read Bus (PB) |
| IU, AU, DU, and PU read from data memory | Three 16-bit Data Read Buses: BB, CB, and DB |
| Single-operand reads | D bus |
| Dual-operand reads | C bus and D bus |
| Coefficient reads for dual-multiply operations | B bus |
| Program and data writes | Two 16-bit Data Write Buses: EB and FB<br>Associated 24-bit Data Write Address Buses: EAB and FAB |

(b)

**FIGURE 15.4**    (a) TMS320C55x Functional Block Diagram; (b) TMS320C55x Bus Usage (Courtesy of Texas Instruments Incorporated.)

Program and data writes are performed on two 16-bit data write buses called E-bus (EB) and F-bus (FB). The write buses also have associated 24-bit data write address buses (EAB and FAB). Additional buses are present on the C55x devices to provide dedicated service to the DMA controller and the peripheral controller.

***15.1.4.1   Instruction Buffer Unit***   The instruction buffer unit receives program code into its instruction buffer queue and decodes instructions. The I unit then passes the appropriate information to the program flow unit, address data flow unit, and data computation units for execution. The CPU fetches 32-bit packets from memory into the instruction buffer queue (IBQ). The IBQ holds up to 64 bytes of instructions in queue to be decoded. The IBQ provides 6 bytes at a time to the instruction decoder, which then dispatches actions to the other primary functional units in the CPU.

The instruction buffer unit of the TMS320C55x handles the task of bringing the instruction stream from memory into the CPU. During each CPU cycle, the I unit receives 4 bytes of program code from the 32-bit program bus and decodes 1 to 6 bytes of code that were previously received in the queue.

The I unit then passes the decoded information to the P unit, the A unit, and the D unit for execution of the instructions. Figure 15.5 shows a block diagram of the I unit.

**FIGURE 15.5**  TMS320C55x
Instruction Buffer Unit Diagram
(Courtesy of Texas Instruments
Incorporated.)

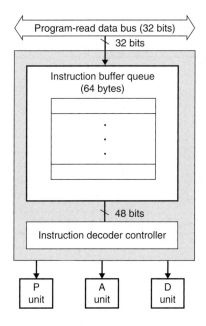

During the prefetch phase of the pipeline, the CPU fetches 32 bits of code from program memory and places it in the instruction buffer queue. When the CPU is ready to decode instructions, up to 6 bytes are transferred from the queue to the instruction decoder. The instruction buffer queue can hold up to 64 bytes of code at a time, optimizing the performance of the CPU by maintaining a continuous program flow.

The instruction buffer queue is also used in conjunction with the local repeat instruction to repeat or loop a block of code stored in the queue. This method of looping is extremely efficient in both performance and power dissipation because once the code is loaded into the queue, no additional memory fetches are required to execute the loop.

Another benefit of the instruction buffer queue is that it can perform speculative fetching of instructions while a condition is being tested for conditional program flow control instructions (conditional call, conditional return, or conditional goto). This capability minimizes overhead due to program flow discontinuities by preventing the need to flush the pipeline. Cycles that otherwise would have been lost to a pipeline flush are converted to useful processing cycles.

In the decode phase of the pipeline, the instruction decoder accepts up to 6 bytes of program code from the instruction buffer queue and decodes those bytes. Instructions are decoded in the order that they are received in the instruction buffer queue—the I unit does not perform dynamic scheduling. This results in predictable execution time, which is essential for designing real-time embedded systems.

The C55x instruction set has a variable length encoding, with instruction lengths varying from 1 to 6 bytes. Instead of encoding all instructions with the same number of bits, simple instructions are encoded with fewer bits than complex instructions. The instruction decoder identifies the boundaries of instructions so that it can decode 8-, 16-, 24-, 32-, 40-, and 48-bit instructions. This encoding method results in very high density program code and optimal use of program memory.

**15.1.4.1.1  Instruction Cache**  Figure 15.6 shows how the I-cache fits into the DSP system. CPU status register ST3_55 contains three cache-control bits for enabling, freezing, and flushing the I-Cache. To configure the I-cache and check its status, the CPU accesses a set of registers in the I-cache. For storing instructions, the I-cache contains one two-way cache. The two-way cache uses two-way set associative mapping and holds up to 16 K bytes: 512 sets, two lines per set, four 32-bit words per line. In the two-way cache, each line is identified by a unique tag.

**FIGURE 15.6** TMS320C55x
Instruction Cache (Courtesy of
Texas Instruments Incorporated.)

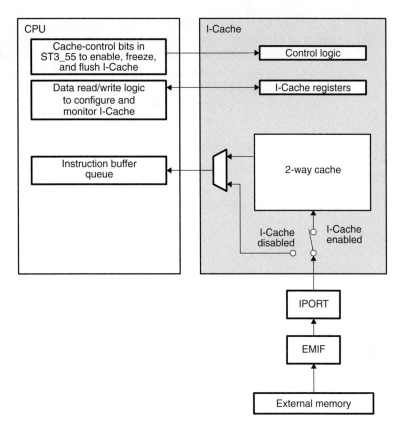

When fetching instruction code from external memory, the CPU sends a 32-bit access request to the instruction cache (I-cache), using its program read data bus (P bus). If the instruction cache is disabled, the request goes directly to the IPORT and then to the external memory interface (EMIF). The EMIF must read 32 bits from the external memory and then pass all 32 bits to the IPORT, which in turn sends the data to the CPU.

Two things could happen if the instruction cache is enabled. In the case of a cache hit, the CPU request will be immediately serviced by the instruction cache, and no data will be read from external memory. In the case of a cache miss, the instruction cache will request four 32-bit words from the EMIF through the IPORT. The EMIF will read four 32-bit words from external memory and then pass the data to the instruction cache through the IPORT. The instruction cache will then send the requested data to the CPU and update its memory contents.

The flexible C55x instruction cache also provides a configurable cache capability that can be used to optimize the cache operation for different types of code. Improving the cache hit ratio means fewer external accesses and less system power consumed. The burst-fill capability of the instruction cache can minimize external memory accesses and their associated loss of performance and power efficiency.

***15.1.4.2 Program Flow Unit*** The program flow unit receives instructions from the I unit and coordinates program flow actions, including the following:

Interpreting conditions for conditional instructions
Determining branch (goto) addresses
Initiating interrupt servicing when an interrupt is requested
Managing single- and block-repeat operations
Managing execution of parallel instructions

The C55x program flow unit, or P unit, controls the sequence of instructions executed in a program. It generates the addresses for instruction fetches from program memory and directs operations such as hardware loops, branches, and conditional execution. This unit also includes the logic for managing the instruction pipeline and the four status registers used to control and monitor various features of the CPU. The components of the P unit enable the superior cycle efficiency of the C55x. Figure 15.7 shows a block diagram of the P unit.

**FIGURE 15.7**  TMS320C55x Program Flow Unit Diagram (Courtesy of Texas Instruments Incorporated.)

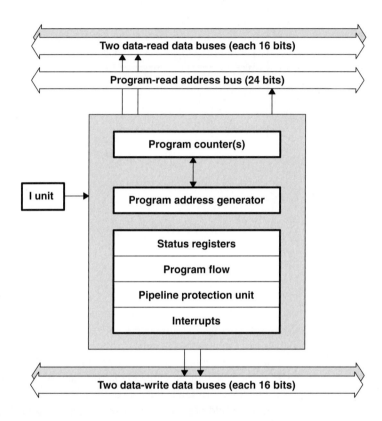

Within the P unit, the program address generation logic generates 24-bit addresses for instruction fetches from program memory. There are no alignment restrictions on code placement within memory, because the P unit supports byte addressing. The 24-bit address gives the C55x a program range of 16 M bytes to accommodate large programs.

The P unit normally generates sequential addresses using the program counter to keep track of the execution point within a program. However, this logic also generates nonsequential addresses for program control operations, such as the following:

Branches
Calls
Returns
Hardware looping (repeats)
Conditional execution
Interrupt servicing

The P unit is highly optimized for efficient execution of program flow operations with minimal impact on pipeline performance. The address generation logic of the P unit is completely independent of the other units within the CPU. Because of this, the target address of a branch can be calculated, and the condition for a conditional branch can be tested early in the pipeline to minimize branch latency.

This parallelism also enables the execution of program-control instructions in the same execute phase of the pipeline as data processing instructions. This greatly enhances C55x performance over previous architectures that only allow delay slots as a means for improving branch performance. Other features of the P unit that enhance program control performance include speculative branching logic and a separate program counter dedicated for fast returns from subroutines or interrupt service routines.

The looping capabilities provided by the P unit include repetition of a single instruction or a block of instructions. Three levels of hardware loops are possible on the C55x by nesting a block repeat operation within another block repeat operation and including a single repeat in either or both of the repeated blocks. The P unit also includes hardware to support conditional repeats.

A major benefit that the P unit provides is dedicated logic for pipeline protection. In addition to handling control hazards, the P unit provides full protection against write-after-read (WAR) and read-after-write (RAW) data hazards. When such data hazards occur in a C55x instruction stream, the pipeline protection logic inserts cycles to maintain the intended order of operations and correct execution of the program.

**15.1.4.3  *Address Data Flow Unit***    The address data flow unit generates the addresses for read and write accesses to data space. This unit contains all the logic and registers necessary to generate the addresses for the three data-read address buses and the two data-write address buses. It also contains a general-purpose 16-bit arithmetic logic unit (ALU) with shifting capability. Figure 15.8 shows a block diagram of the A unit.

**FIGURE 15.8**  TMS320C55x Address Flow Unit Diagram (Courtesy of Texas Instruments Incorporated.)

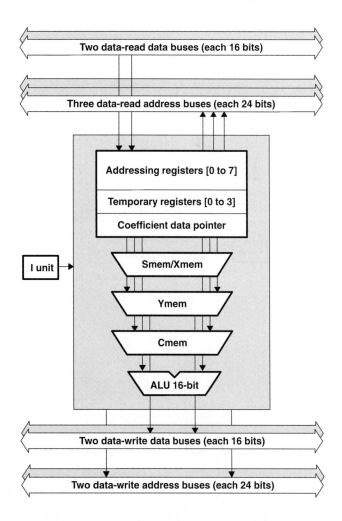

There are eight auxiliary registers for use as address pointers and coefficient data pointer registers to provide a dedicated pointer to a coefficient table. The registers for control of circular addressing are also managed by the A unit. The A unit also contains a 16-bit ALU capable of performing arithmetical, logical, shift, and saturation operations.

The 16-bit ALU allows simpler arithmetic operations to be performed in parallel with more complex operations performed in the D unit. It accepts immediate values from the I unit and communicates bidirectionally with memory, the A-unit registers, the D-unit registers, and the P-unit registers. Within the A unit, the ALU can manipulate four general-purpose 16-bit registers or any of the address-generation registers. The four general-purpose registers enable improved compiler efficiency and minimize the need for memory accesses.

Either the general-purpose ALU, or one of the three addressing register ALUs (ARAUs) can modify the nine addressing registers used for indirect addressing. The three ARAUs provide independent address generators for each of the three data-read buses of the C55x. This parallelism allows two 16-bit operands and a 16-bit coefficient to be read into the D unit during each CPU cycle. The A unit also includes dedicated registers to support circular addressing for instructions that use indirect addressing. Up to five independent circular buffer locations can be used simultaneously with up to three independent buffer lengths. There are no address alignment constraints for these circular buffers.

**15.1.4.4  *Data Computation Unit***    The data computation unit is the primary part of the CPU where data is processed. Three data-read buses feed the two multiply-and-accumulate (MAC) units and the 40-bit ALU, and intermediate results can be stored in one of four 40-bit accumulator registers. The parallelism of this unit minimizes the cycle count required per task to provide efficient execution of signal processing algorithms. Figure 15.9 shows a block diagram of the D unit.

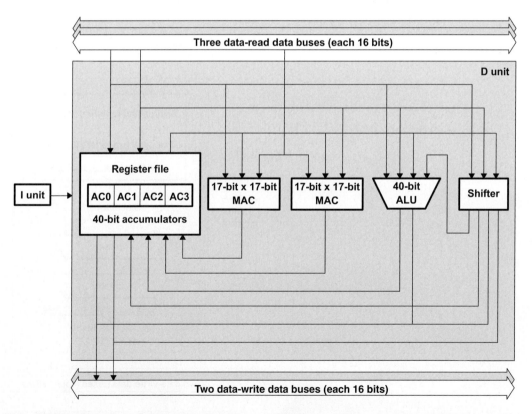

**FIGURE 15.9**    TMS320C55x Data Computation Unit Diagram (Courtesy of Texas Instruments Incorporated.)

The key to the computational power of the C55x is the dual MAC architecture. A MAC unit consists of a multiplier and a dedicated adder with saturation logic. In a single cycle, each MAC unit can perform a 17-bit by 17-bit multiplication and a 40-bit addition or subtraction with optional 32-/40-bit saturation. The three data-read buses can be used to carry two data streams and a common coefficient stream to the two MAC units. The results from the MAC units can be placed in any of four 40-bit accumulators within the D unit. This dual mac data computation unit (D unit) capability greatly increases the C55x performance for executing block filtering algorithms as well as for other signal processing applications.

The D unit also includes a 40-bit arithmetic logic unit (ALU) that is completely separate from the MAC units. The D unit ALU can perform arithmetical or logical operations on 40-bit values from the accumulators, or it can be used to perform dual 16-bit arithmetic operations simultaneously.

In addition to accepting inputs from the 40-bit accumulator registers of the D unit, the ALU can accept immediate values from the I unit and communicates bidirectionally with memory, the A-unit registers, or the P-unit registers.

A powerful barrel shifter complements the MACs and ALU of the D unit. This shifter can shift 40-bit accumulator values up to 31 bits to the left or up to 32 bits to the right. It accepts immediate values from the I unit and communicates bidirectionally with memory, the A-unit registers, and the P-unit registers. In addition, it can supply a shifted value to the D-unit ALU as an input for further calculation.

The results of computational operations in the D unit are written to memory by the two 16-bit data-write buses. These buses, together with the A-unit address generation logic, can perform two 16-bit writes or a single 32-bit write to memory in one CPU cycle. This data throughput is essential for supporting the real-time processing speed provided by the D unit.

#### 15.1.4.4.1 Bit Shifter
The D-unit shifter performs the following actions:

Shifts 40-bit accumulator values up to 31 bits left or up to 32 bits right. The shift count can be read from a temporary register or can be supplied as a constant in the instruction.
Shifts 16-bit register, memory, or I/O space values up to 31 bits left or up to 32 bits right.
Shifts 16-bit immediate values up to 15 bits left. The shift count is contained in the instruction as a constant.
Normalizes accumulator values
Extracts and expands bit fields and performs bit counting
Rotates register values
Rounds and/or saturates accumulator values before they are stored to data memory

#### 15.1.4.4.2 Arithmetic Logic Unit
The D unit contains a 40-bit ALU that accepts values from the I unit and communicates calculation results to all the other primary functional units. The functions of the D-unit ALU include the following:

Performs additions, subtractions, comparisons, rounding, saturation, Boolean logic operations, and absolute value calculations
Performs two arithmetical operations simultaneously when a dual 16-bit arithmetic instruction is executed
Tests, sets, clears, and complements D-unit register bits
Moves register values data computation unit (DU or D unit)

#### 15.1.4.4.3 Multiply-Accumulate Units
Two MAC units support multiplication and addition/subtraction. In a single cycle, each MAC can perform a 17-bit x 17-bit multiplication (fractional or integer) and a 40-bit addition or subtraction with optional 32-/40-bit saturation.

The accumulators receive all the results of MAC operations. The presence of three dedicated read buses and two dedicated write buses provides the capability for sustained, single-cycle dual-MAC operations.

**15.1.4.4.4  Registers**  The D unit contains four 40-bit accumulators (AC0–AC3) that are used as source/destination for calculations performed by the MAC units and the ALU.

The D units also contains two 16-bit transition registers (TRN0, TRN1) that hold the transition decision path to new metrics to perform the Viterbi algorithm.

## 15.1.5 Device Special Features

The C55x incorporates many features to enhance functionality in DSP embedded system design. These features enable the system designer to optimize parameters for maximum performance at minimal power dissipation levels.

**15.1.5.1  Low-Power Dissipation**  A series of process, design, and architectural enhancements collectively enable high levels of power reduction. These design enhancements to the C55x not only achieve ultra-low power but also greatly increase performance. This allows for significantly greater functionality in mobile devices with DSP requirements. Maximum operating frequency can maintained with as low as 1.1 volts for significantly extended battery life in applications such as medical devices.

## 15.1.6 Low-Power Design

The 40-bit ALU is consistent with the architecture of the C54x generation, and is used for primary computational tasks. The C55x architecture implements an additional 16-bit ALU, which can be used for smaller arithmetic and logic tasks. The flexible instruction set provides the capability to direct simpler computational or logical/bit-manipulation tasks to the 16-bit ALU, which consumes less power. This redirection of resources also saves power by reducing cycles per task because both ALUs can operate in parallel.

**15.1.6.1  Memory Accesses**  Memory accesses, both internal and external, can be a major contributor to power dissipation. Minimizing the number of memory accesses necessary to complete a given task furthers the goal of minimizing power dissipation per task. The C55x reduces the number of fetches necessary to provide instructions to the CPU. On the C55x, program fetches are performed as 32-bit accesses.

In addition, the variable-byte-length instruction set means that each 32-bit instruction fetch can retrieve more than one instruction. Variable length instructions improve code density and conserve power by scaling the instruction size to the amount of information needed. This alliance of instruction set design and architecture minimizes the power necessary to keep the application running at top performance.

**15.1.6.2  Automatic Power Mechanisms**  The C55x core processor actively manages power consumption of on-chip peripherals and memory arrays. This resource power management is fully automatic and transparent to the user. It is performed without any impact on the software or the computational performance of the application. It is another contributor to power reduction without impact on performance.

When individual on-chip memory arrays are not being accessed, they are automatically switched into a low-power mode. When an access request arrives, the array returns to normal operation, without latency in the application, and completes the memory access. If no further accesses to that array are requested, the array returns to a low-power state until it is needed again.

### 15.1.6.3  Low-Power Enhancements

The processor provides a similar control to on-chip peripherals. Peripherals can enter low-power states when they are not active and the CPU does not require their attention. The peripherals also respond to processor requests and exit their low-power states without latency. This power management occurs in addition to the software controllable low-power states provided by the IDLE domain control of the peripherals.

A critical component of power conservation is minimizing the power used when an application is in an idle or low-activity state. The C55x generation improves the flexibility of low-activity power management through the implementation of user-controllable IDLE domains. These domains are sections of the device, which can be selectively enabled or disabled under software control. When disabled, a domain enters a very low-power IDLE state in which register or memory contents are still maintained. When the domain is enabled, it returns to normal operation. Each of the domains can be separately enabled or disabled providing the application the capability to manage low-activity power situations as efficiently as possible.

On initial C55x devices, the sections of the device configured as separate IDLE domains are the CPU, the DMA, the peripherals, the external memory interface (EMIF), the instruction cache, and the clock generation circuitry.

### 15.1.6.4  Power Conservation

Three features are included to minimize power usage:

Software-programmable idle domains that provide configurable low-power modes
Automatic power management
Advanced low-power CMOS process

### 15.1.6.5  Idle Domains

The flexible architecture of C55x devices provides a means to dynamically conserve power through software-programmable idle domains. Blocks of circuitry on the device are organized into idle domains. Each domain can operate normally or can be placed in a low-power idle state. The idle control register (ICR) determines which domains will be placed in the idle state when the execution of the next IDLE instruction occurs. The six domains are the following:

CPU domain
DMA domain
Peripherals domain
Clock generator domain
Instruction cache domain
EMIF domain

Each domain can be placed in a low-power state when its capabilities are not required. This control provides the user the capability to dynamically modify the power consumption of the device based on activity. Note that when each domain is in the idle state, the functions of that particular domain are not available. An exception to this exists in the peripheral domain. In the peripheral domain, each peripheral has an idle enable bit, which controls whether or not the peripheral will respond to the changes in the idle state.

Thus, peripherals can be individually configured to idle or remain active when the peripheral domain is idled. The idle state can be exited by modifying the ICR (if the CPU and clock generation domains were not idled) or by an external interrupt.

### 15.1.6.6  Advanced Technology

In addition to the power dissipation reductions achieved by the architectural and instruction set enhancements, the C55x generation of processors further challenges the barriers to power reduction through advanced low-voltage CMOS technologies.

C55x devices are based on a power-efficient CMOS technology that supports devices running at 1.5 V and 0.9 V. These low-voltage processors still maintain the capability to interface directly to other standard 3.3 V CMOS components.

## 15.1.7 Processor On-Chip Peripherals

Figure 15.10 lists on-chip processor peripherals available for members of the TMS320C55x generation of fixed-point digital signal processors (DSPs). Different combinations of peripherals can be selected based on the target application. On a given device, some peripherals may share pins, making the peripherals' use mutually exclusive.

**FIGURE 15.10** TMS320C55x Available Peripherals (Courtesy of Texas Instruments Incorporated.)

| Peripheral |
| --- |
| Analog-to-digital converter (ADC) |
| Clock generator with PLL |
| Direct memory access (DMA) controller |
| External memory interface (EMIF) |
| Host port interface (HPI) |
| Instruction cache |
| Inter-integrated circuit (I2C) module |
| Multichannel buffered serial port (McBSP) |
| MultiMediaCard/ SD card controller |
| Power management/ Idle configurations |
| Real-time clock (RTC) |
| Timer, general-purpose |
| Timer, watchdog |
| Universal Asynchronous Receiver/Transmitter (UART) |
| Universal Serial Bus (USB) module |

***15.1.7.1 On-Chip Memory*** The on-chip memory is composed of three subsections. Dual-access RAM (DARAM) supports two memory accesses per cycle. Up to 128k words (256k bytes) of DARAM is available. Single-access RAM (SARAM) supports one memory access per cycle for a total of 32k words (64k bytes). Finally a ROM capability is included that provides nonvolatile storage for program instructions or data values.

***15.1.7.2 Analog-to-Digital Converter*** The ADC (see Figure 15.11) converts an analog input signal to a digital value for use by the DSP. The ADC can sample one of up to four inputs (AIN0–AIN3) at a time and generates a 10-bit digital representation (ADCDATA) of the samples. The maximum sampling rate of the ADC is 21.5 kHz. This performance makes the ADC suitable for sampling analog signals that change at a slow rate.

The ADC could be used to sample the voltage across a potentiometer on a user interface panel or to monitor the supply voltage drop on a battery-operated circuit. It is not intended for sampling data for signal processing. Note that the ADC is only available on the TMS320VC5509A.

**FIGURE 15.11** TMS320C55x
Analog-to-Digital Converter
(Courtesy of Texas Instruments
Incorporated.)

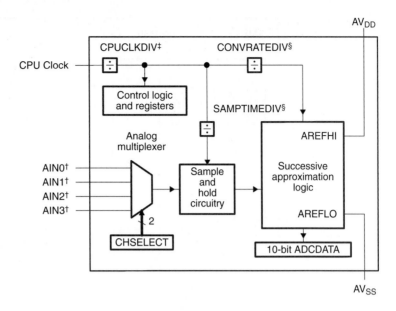

### 15.1.7.3 DSP Clock Generator
The DSP clock generator supplies the DSP with a clock signal that is based on an input clock signal connected at the CLKIN pin. Included in the clock generator is a digital phase-lock loop (PLL), which can be enabled or disabled. You can configure the clock generator to create a CPU clock signal that has the desired frequency (Figure 15.12).

**FIGURE 15.12** TMS320C55x
Clock Generator (Courtesy of
Texas Instruments Incorporated.)

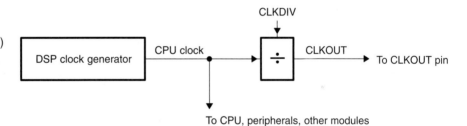

The clock generator has a clock mode register, CLKMD, for controlling and monitoring the activity of the clock generator. For example, you can write to the PLL ENABLE bit in CLKMD to toggle between the two main modes of operation:

In the bypass mode, the PLL is bypassed, and the frequency of the output clock signal is equal to the frequency of the input clock signal divided by 1, 2, or 4. Because the PLL is disabled, this mode can be used to save power.

In the lock mode, the input frequency can be both multiplied and divided to produce the desired output frequency, and the output clock signal is phase-locked to the input clock signal. The clock mode is entered if the PLL ENABLE bit of the clock mode register is set and the phase-locking sequence is complete. (During the phase-locking sequence, the clock generator is kept in the bypass mode.)

### 15.1.7.4 DMA Controller
Acting in the background of CPU operation, the DMA controller can transfer 32-bit data values among internal memory, external memory, and on-chip peripherals in the same clock cycle. It is essentially acting in a co-processor mode to the CPU. Figure 15.13 shows the basic block diagram.

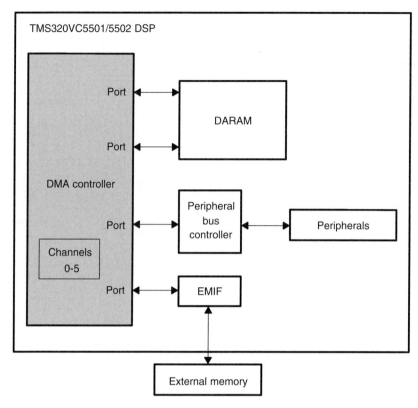

**FIGURE 15.13**    TMS320C55x DMA Controller (Courtesy of Texas Instruments Incorporated.)

The DMA controller has the following important features:

Operation that is independent of the CPU.

Four standard ports: Two for internal dual-access RAM (DARAM), one for external memory, and one for peripherals.

Six channels, which allow the DMA controller to keep track of the context of six independent block transfers among the standard ports.

Bits for assigning each channel a low priority or a high priority.

Event synchronization. DMA transfers in each channel can be made dependent on the occurrence of selected events.

An interrupt for each channel. Each channel can send an interrupt to the CPU on completion of certain operational events.

Software-selectable options for updating addresses for the sources and destinations of data transfers.

A dedicated idle domain. You can put the DMA controller into a low-power state by turning off this domain. Each multichannel buffered serial port (McBSP) on the C55x DSP has the ability to temporarily take the DMA domain out of this idle state when the McBSP needs the DMA controller.

If the CPU and the DMA controller simultaneously request access to the same DARAM block in internal memory, CPU requests always have priority over DMA requests. The DMA requests to a DARAM block will be serviced when there are no more CPU requests. Refer to the device-specific data manual for specific information on the start and end addresses for each DARAM block.

The DMA controller has six paths, called channels, to transfer data among the four ports (two for DARAM, one for external memory, and one for peripherals). Each channel reads data from one port (from the source) and writes data to that same port or another port (to the destination). Each channel has a first in, first out (FIFO) buffer that allows the data transfer to occur in two stages (see Figure 15.14):

*Port read access.* Transfer of data from the source port to the channel FIFO buffer.
*Port write access.* Transfer of data from the channel FIFO buffer to the destination port.

The FIFO buffer in each channel is eight 32-bit words deep.

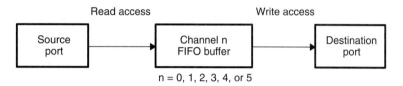

**FIGURE 15.14**   The Two Parts of a DMA Transfer (Courtesy of Texas Instruments Incorporated.)

***15.1.7.5   External Memory Interface***   Figure 15.15 illustrates how the EMIF is interconnected with other parts of the DSP and with external memory devices. The connection to the peripheral bus controller allows the CPU to access the EMIF registers.

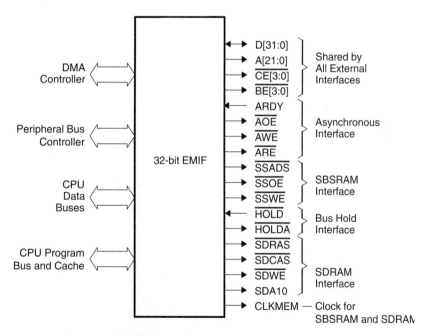

**FIGURE 15.15**   TMS320C55x External Memory Interface (Courtesy of Texas Instruments Incorporated.)

***15.1.7.6   I²C Module***   The I²C module supports any slave or master I²C-compatible device. Figure 15.16 shows an example of multiple I²C modules connected for a two-way transfer from one device to other devices.

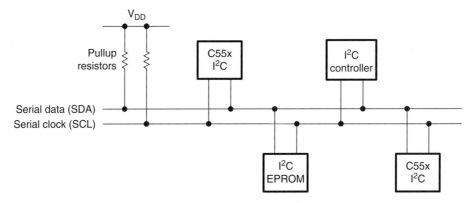

**FIGURE 15.16** I²C Device Interconnect (Courtesy of Texas Instruments Incorporated.)

Each device connected to an I²C-bus, including any C55x DSP connected to the bus with an I²C module, is recognized by a unique address. Each device can operate as either a transmitter or a receiver, depending on the function of the device. A device connected to the I²C-bus can also be considered as the master or the slave when performing data transfers.

A master device is the device that initiates a data transfer on the bus and generates the clock signals to permit that transfer. During this transfer, any device addressed by this master is considered a slave. The I²C module supports the multimaster mode, in which one or more devices capable of controlling an I²C-bus can be connected to the same I²C-bus.

For data communication, the I²C module has a serial data pin (SDA) and a serial clock pin (SCL), as shown in Figure 15.17. These two pins carry information between the C55x device and other devices connected to the I²C-bus. The SDA and SCL pins both are bidirectional. They each must be connected to a positive supply voltage using a pull-up resistor. When the bus is free, both pins are high. The driver of these two pins has an open-drain configuration to perform the required wired-AND function.

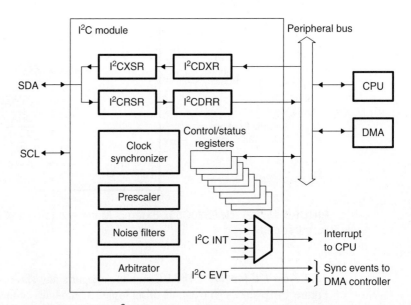

**FIGURE 15.17** TMS320C55x I²C Block Diagram (Courtesy of Texas Instruments Incorporated.)

### 15.1.7.7 Multimedia/SD Card Controller

As shown in Figure 17.18, the MMC controller passes data between the CPU or the DMA controller on one side and one or more a memory cards on the other side. The CPU or the DMA controller can read from or write to the control and status registers in the MMC controller. As necessary, the CPU and/or the DMA controller can store or retrieve data in the DSP memory or in the registers of other peripherals. The CPU can monitor data activity by reading the status registers and responding to interrupt requests.

The DMA controller can be notified of data reception/transmission status with the two DMA events. Data transfers between the MMC controller and a memory card can use one bidirectional data line (for the MMC protocol) or four parallel data lines (for the SD protocol). If multiple cards are connected, the MMC controller uses commands of the MMC/SD protocol to select and communicate with one card at a time. Note that the MMC/SD controller is only available on the TMS320VC5509A.

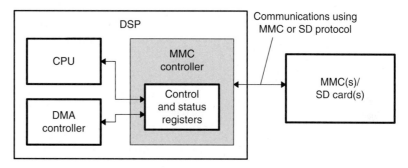

**FIGURE 15.18**   TMS320C55x MMC Controller (Courtesy of Texas Instruments Incorporated.)

### 15.1.7.8 Programmable Timers

The DSP contains three programmable timers, as shown in Figure 15.19. Two are general-purpose (GP) timers, Timer0 (TIM0) and Timer1 (TIM1). Each of the GP timers can be configured in one of three modes using the timer mode bits in global timer control register 1 (GCTL1): 64-bit timer, dual 32-bit timers chained, or dual 32-bit timers unchained.

The two GP timers do not support the watchdog timer mode. At reset, the two GP timers are configured as 64-bit timers. The third timer can be configured as either a GP timer or a watchdog timer.

At reset, the third timer is configured as a 64-bit GP timer. Each timer can be driven by an external clock at the timer pin or by an internal clock. When the internal clock is selected, the timer clock is generated as shown in Figure 15.18. The DSP clock generator receives a signal from a clock source as described in the device-specific data manual. One of the clocks produced by the DSP clock generator is the fast peripherals clock (SYSCLK1). SYSCLK1 is a divided-down version of the CPU clock, as described in the device-specific data manual. A clock divider inside the timer divides SYSCLK1 by a preset divisor to produce the timer clock.

### 15.1.7.9 UART

The UART peripheral (refer to Figure 15.20) is based on the industry-standard TL16C55x0 asynchronous communications element, which in turn is a functional upgrade of the TL16C450. Functionally similar to the TL16C450 on power up (single-character or TL16C450 mode), the UART can be placed in an alternate FIFO (TL16C55x0) mode. This relieves the CPU of excessive software overhead by buffering received and transmitted characters. The receiver and transmitter FIFOs store up to 16 bytes, including three additional bits of error status per byte for the receiver FIFO.

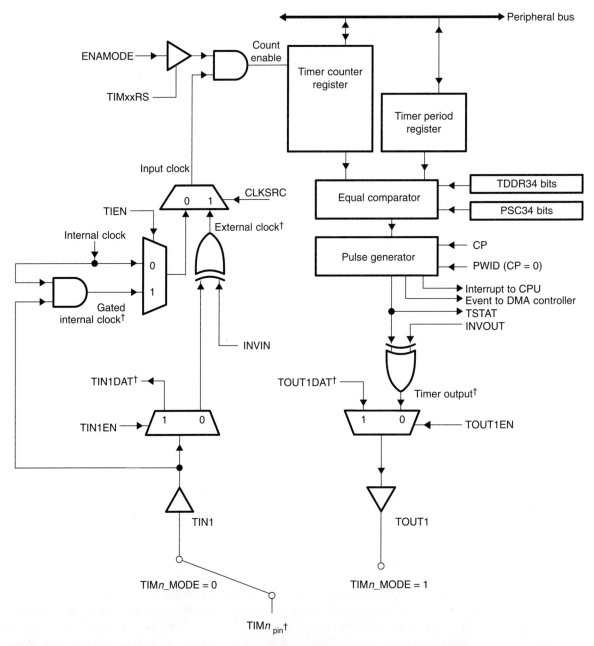

**FIGURE 15.19** TMS320C55x Generation of Internal Timer Clock (Courtesy of Texas Instruments Incorporated.)

The UART performs serial-to-parallel conversions on data received from a peripheral device and parallel-to-serial conversion on data received from the CPU. The CPU can read the UART status at any time. The UART includes control capability and a processor interrupt system that can be tailored to minimize software management of the communications link. The UART includes a programmable baud generator capable of dividing the UART input clock by divisors from 1 to 65535 and producing a 16. reference clock for the internal transmitter and receiver logic.

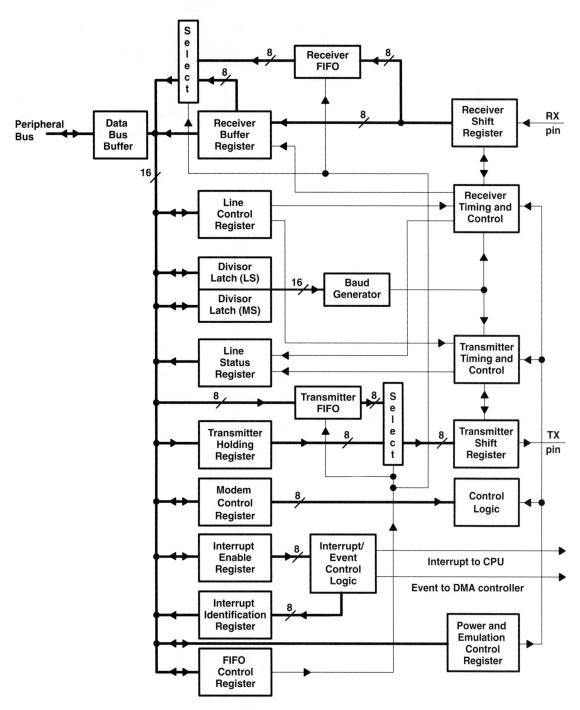

**FIGURE 15.20**  TMS320C55x UART Functional Diagram (Courtesy of Texas Instruments Incorporated.)

***15.1.7.10  USB Module***  The USB module is a USB 2.0-compliant, full-speed (12 Mbps) slave module (Figure 15.21) available on the TMS320VC5509A. The USB module has 16 endpoints. Two control endpoints (for control transfers only) are OUT0 and IN0. Fourteen general-purpose endpoints (for other types of transfers) include OUT1–OUT7 and IN1–IN7. Each of these endpoints can support bulk, interrupt, and isochronous transfers.

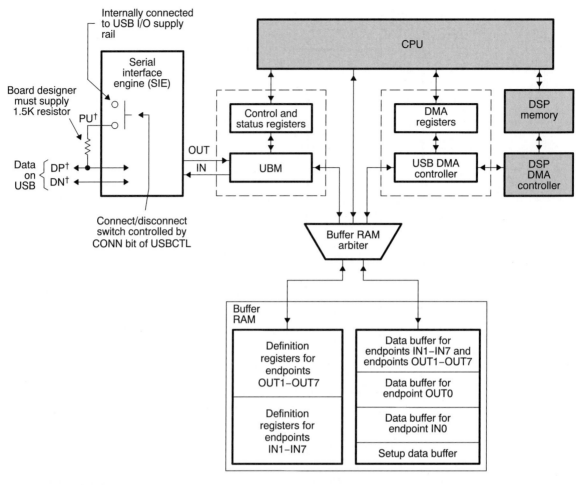

**FIGURE 15.21**    TMS320C55x USB Block Diagram (Courtesy of Texas Instruments Incorporated.)

An optional double-buffer scheme for fast data throughput is also supported, including a dedicated DMA channel. The DMA controller inside the USB module can pass data between the general-purpose endpoints and the DSP memory while the CPU performs other tasks. (The USB DMA controller cannot access the control endpoints.)

### 15.1.8  Emulation and Test

On-chip scan-based emulation capability with program history, tracking of recent program counter values, and discontinuities (Trace FIFO) are supported. This enables a comprehensive real-time debug capability via an integrated development environment. Also incorporated is a noninvasive IEEE 1149. One (JTAG) boundary scan test capability.

## 15.2    ANALOG DEVICES ADSP-BF535 BLACKFIN PROCESSOR

The ADSP-BF535 processor is a member of the Blackfin processor family of products, incorporating the Micro Signal Architecture (MSA), jointly developed by Analog Devices, Inc. and Intel Corporation. The basic functional block diagram is shown in Figure 15.22.

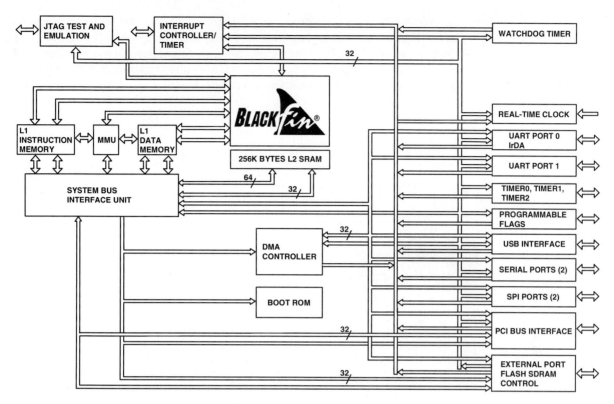

**FIGURE 15.22** Blackfin BF-535 Functional Block Diagram (Reprinted by permission of Analog Devices Incorporated.)

The architecture combines a dual MAC state-of-the-art signal processing engine, the advantages of a clean, orthogonal RISC-like microprocessor instruction set, and single-instruction, multiple data (SIMD) multimedia capabilities into a single instruction set architecture. By integrating a rich set of industry leading system peripherals and memory, Blackfin processors are the platform of choice for next-generation applications that require RISC-like programmability, multimedia support, and leading-edge signal processing in one integrated package.

### 15.2.1 Portable Low-Power Architecture

Blackfin processors provide advanced power management and performance. Blackfin processors are designed in a low-power and low-voltage design methodology and feature dynamic power management, the ability to independently vary both the voltage and frequency of operation to significantly lower overall power consumption. Varying the voltage and frequency can result in a substantial reduction in power consumption, by comparison to just varying the frequency of operation. This translates into longer battery life for portable appliances.

### 15.2.2 System Integration

The ADSP-BF535 Blackfin processor is a highly integrated system-on-a-chip solution for the next generation of digital communication and portable Internet appliances. By combining industry-standard interfaces with a high-performance signal processing core, users can develop cost-effective solutions quickly without the need for costly external components. The ADSP-BF535 Blackfin processor system peripherals include UARTs, SPIs, SPORTs, general-purpose timers, a real-time clock, programmable flags, watchdog timer, and USB and PCI buses for glueless peripheral expansion.

The ADSP-BF535 Blackfin processor peripherals are connected to the core high band-width buses (Figure 15.23), providing flexibility in system configuration as well as excellent overall system performance. The base peripherals include general-purpose functions such as UARTs, timers with PWM (pulse-width modulation) and pulse measurement capability, general-purpose flag I/O pins, a real-time clock, and a watchdog timer.

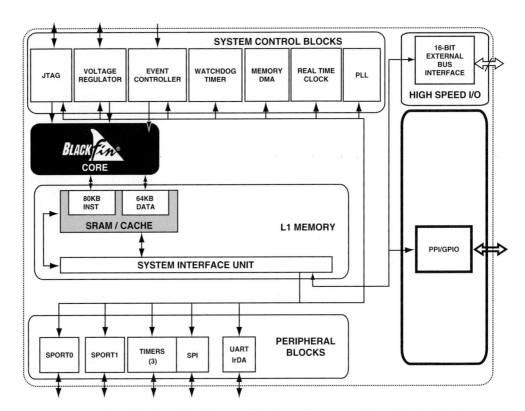

**FIGURE 15.23**    Blackfin Base Peripherals (Reprinted by permission of Analog Devices Incorporated.)

This set of functions satisfies a wide variety of typical system support needs and is aug-mented by the system expansion capabilities of the part. In addition to these general-purpose pe-ripherals, the ADSP-BF535 Blackfin processor contains high-speed serial ports for interfaces to a variety of audio and modem CODEC functions. It also contains an event handler for flexible management of interrupts from the on-chip peripherals and external sources. And it contains power management control functions to tailor the performance and power characteristics of the processor and system to many application scenarios.

The on-chip peripherals can be easily augmented in many system designs with little or no glue logic due to the inclusion of several interfaces providing expansion on industry-standard buses. These include a 32-bit, 33 MHz, V2.2 compliant PCI bus, SPI serial expansion ports, and a device type USB port. These enable the connection of a large variety of peripheral devices to tailor the system design to specific applications with a minimum of design complexity.

All the peripherals, except for programmable flags, real-time clock, and timers, are sup-ported by a flexible DMA structure with individual DMA channels integrated into the peripher-als. There is also a separate memory DMA channel dedicated to data transfers between the various memory spaces, including external SDRAM and asynchronous memory, internal Level 1 and Level 2 SRAM, and PCI memory spaces.

Multiple on-chip 32-bit buses, running at up to 133 MHz, provide adequate bandwidth to keep the processor core running along with activity on all the on-chip and external peripherals.

### 15.2.3 Processor Core

As shown in Figure 15.24, the Blackfin processor core contains two multiplier/accumulators (MACs), two 40-bit ALUs, four video ALUs, and a single shifter. The computational units process 8-bit, 16-bit, or 32-bit data from the register file. Each MAC performs a 16-bit by 16-bit multiply in every cycle, with an accumulation to a 40-bit result, providing 8 bits of extended precision.

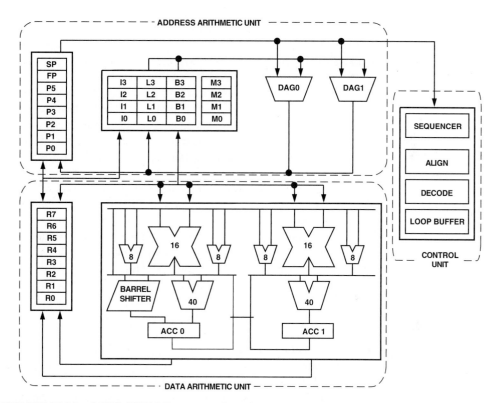

**FIGURE 15.24**  ADSP-BF533 Processor Core Architecture (Reprinted by permission of Analog Devices Incorporated.)

The ALUs perform a standard set of arithmetic and logical operations. With two ALUs capable of operating on 16- or 32-bit data, the flexibility of the computation units covers the signal processing requirements of a varied set of application needs. Each of the two 32-bit input registers can be regarded as two 16-bit halves, so each ALU can accomplish very flexible single 16-bit arithmetic operations. By viewing the registers as pairs of 16-bit operands, dual 16-bit or single 32-bit operations can be accomplished in a single cycle. Quad 16-bit operations can be accomplished simply, by taking advantage of the second ALU. This accelerates the per cycle throughput.

The powerful 40-bit shifter has extensive capabilities for performing shifting, rotating, normalization, extraction, and for depositing data. The data for the computational units is found in a multiported register file of sixteen 16-bit entries or eight 32-bit entries. A powerful program sequencer controls the flow of instruction execution, including instruction alignment and decoding. The sequencer supports conditional jumps and subroutine calls, as well as zero-overhead looping. A loop buffer stores instructions locally, eliminating instruction memory accesses for tightly looped code.

Two data address generators (DAGs) provide addresses for simultaneous dual operand fetches from memory. The DAGs share a register file containing four sets of 32-bit index, modify, length, and base registers. Eight additional 32-bit registers provide pointers for general indexing of variables and stack locations.

Blackfin processors support a modified Harvard architecture in combination with a hierarchical memory structure. Level 1 (L1) memories are those that typically operate at the full processor speed with little or no latency. Level 2 (L2) memories are other memories, on-chip or off-chip, that may take multiple processor cycles to access. At the L1 level, the instruction memory holds instructions only.

The two data memories hold data, and a dedicated scratch pad data memory stores stack and local variable information. At the L2 level, there is a single unified memory space, holding both instructions and data. In addition, the L1 instruction memory and L1 data memories may be configured as either static RAMs (SRAMs) or caches.

The memory management unit (MMU) provides memory protection for individual tasks that may be operating on the core and may protect system registers from unintended access. The architecture provides three modes of operation: user mode, supervisor mode, and emulation mode.

User mode has restricted access to certain system resources, thus providing a protected software environment, whereas supervisor mode has unrestricted access to the system and core resources.

The Blackfin processor instruction set has been optimized so that 16-bit op codes represent the most frequently used instructions, resulting in excellent compiled code density. Complex DSP instructions are encoded into 32-bit op codes, representing fully featured multifunction instructions. Blackfin processors support a limited multiple issue capability, where a 32-bit instruction can be issued in parallel with two 16-bit instructions, allowing the programmer to use many of the core resources in a single instruction cycle.

The Blackfin processor assembly language uses an algebraic syntax for ease of coding and readability. The architecture has been optimized for use in conjunction with the C/C++ compiler, resulting in fast and efficient software implementations.

### 15.2.3.1  *Instruction Pipeline*

The program sequencer determines the next instruction address by examining both the current instruction being executed and the current state of the processor. If no conditions require otherwise, the processor executes instructions from memory in sequential order by incrementing the lookahead address. The processor has a ten-stage instruction pipeline with the stages listed in Figure 15.25.

### 15.2.3.2  *Instruction Pipeline Flow*

Figure 15.26 shows a diagram of the processor pipeline. The instruction fetch and branch logic generates 32-bit fetch addresses for the instruction memory unit. The instruction alignment unit returns instructions and their width information at the end of the IF3 stage. For each instruction type (16, 32, or 64 bits), the instruction alignment unit ensures that the alignment buffers have enough valid instructions to be able to provide an instruction every cycle. Because the instructions can be 16, 32, or 64 bits wide, the instruction alignment unit may not need to fetch an instruction from the cache every cycle.

For a series of 16-bit instructions, the instruction alignment unit gets an instruction from the instruction memory unit once in four cycles. The alignment logic requests the next instruction address based on the status of the alignment buffers. The sequencer responds by generating the next fetch address in the next cycle, provided there is no change of flow. The sequencer holds the fetch address until it receives a request from the alignment logic or until a change of

| Pipeline Stage | Description |
|---|---|
| Instruction Fetch 1 (IF1) | Issue instruction address to IAB bus, start compare tag of instruction cache |
| Instruction Fetch 2 (IF2) | Wait for instruction data |
| Instruction Fetch 3 (IF3) | Read from IDB bus and align instruction |
| Instruction Decode (DEC) | Decode instructions |
| Address Calculation (AC) | Calculation of data addresses and branch target address |
| Data Fetch 1 (DF1) | Issue data address to DA0 and DA1 bus, start compare tag of data cache |
| Data Fetch 2 (DF2) | Read register files |
| Execute 1 (EX1) | Read data from LD0 and LD1 bus, start multiply and video instructions |
| Execute 2 (EX2) | Execute/Complete instructions (shift, add, logic, etc.) |
| Write Back (WB) | Writes back to register files, SD bus, and pointer updates (also referred to as the "commit" stage) |

**FIGURE 15.25** ADSP-BF533 Instruction Pipeline Stages (Reprinted by permission of Analog Devices Incorporated.)

**FIGURE 15.26** ADSP-BF533 Processor Pipeline (Reprinted by permission of Analog Devices Incorporated.)

| | Instr Fetch 1 | Instr Fetch 2 | Instr Fetch 3 | Instr Decode | Addr Calc | Data Fetch 1 | Data Fetch 2 | Ex1 | Ex2 | WB |
|---|---|---|---|---|---|---|---|---|---|---|
| | Instr Fetch 1 | Instr Fetch 2 | Instr Fetch 3 | Instr Decode | Addr Calc | Data Fetch 1 | Data Fetch 2 | Ex1 | Ex2 | WB |
| Instr Fetch 1 | Instr Fetch 2 | Instr Fetch 3 | Instr Decode | Addr Calc | Data Fetch 1 | Data Fetch 2 | Ex1 | Ex2 | WB | |

flow occurs. The sequencer always increments the previous fetch address by 8 (the next 8 bytes). If a change of flow occurs, such as a branch or an interrupt, data in the instruction alignment unit is invalidated. The sequencer decodes and distributes instruction data to the appropriate locations such as the register file and data memory.

### 15.2.4 Memory Architecture

The ADSP-BF535 Blackfin processor views memory (Figure 15.27) as a single unified 4 Gbyte address space, using 32-bit addresses. All resources, including internal memory, external memory, PCI address spaces, and I/O control registers, occupy separate sections of this common address space. The memory portions of this address space are arranged in a hierarchical structure to provide a good cost/performance balance with very fast, low latency memory as cache or SRAM very close to the processor and larger, lower cost, and lower performance memory systems farther away from the processor.

**FIGURE 15.27** Blackfin Memory Architecture (Reprinted by permission of Analog Devices Incorporated.)

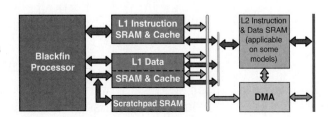

The L1 memory system is the primary highest performance memory available to the Blackfin processor core. The off-chip memory system, accessed through the external bus interface unit (EBIU), provides expansion with SDRAM, flash memory, and SRAM, optionally accessing more than 768 Mbytes of external physical memory.

The memory DMA controller provides high bandwidth data movement capability. It can perform block transfers of code or data between the internal L1/L2 memories and the external memory spaces (including PCI memory space).

### *15.2.4.1 Internal (On-Chip) Memory*

The ADSP-BF535 Blackfin processor has four blocks of on-chip memory providing high bandwidth access to the core. The first is the L1 instruction memory consisting of 16 Kbytes of four-way set-associative cache memory, as shown in Figure 15.28. In addition, the memory may be configured as an SRAM. This memory is accessed at full processor speed.

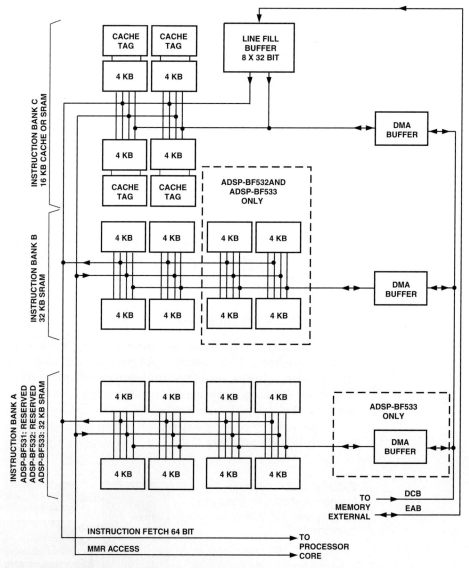

**FIGURE 15.28**    ADSP-BF533 L1 Instruction Bank Memory Architecture (Reprinted by permission of Analog Devices Incorporated.)

The second on-chip memory block is the L1 data memory (Figures 15.27), consisting of two banks of 16 Kbytes each. Each L1 data memory bank can be configured as one way of a two-way set-associative cache or as an SRAM and is accessed at full speed by the core.

The third memory block is a 4 Kbyte scratch pad RAM, which runs at the same speed as the L1 memories, but is only accessible as data SRAM (it cannot be configured as cache memory and is not accessible via DMA).

The fourth on-chip memory system is the L2 SRAM memory array, which provides 256 Kbytes of high-speed SRAM at the full bandwidth of the core and slightly longer latency than the L1 memory banks. The L2 memory is a unified instruction and data memory and can hold any mixture of code and data required by the system design. The Blackfin processor core has a dedicated low-latency 64-bit wide data path port into the L2 SRAM memory.

***15.2.4.2 PCI*** The PCI bus defines three separate address spaces, which are accessed through windows in the ADSP-BF535 Blackfin processor memory space. These spaces are PCI memory, PCI I/O, and PCI configuration. In addition, the PCI interface can either be used as a bridge from the processor core as the controlling CPU in the system or as a host port where another CPU in the system is the host and the ADSP-BF535 is functioning as an intelligent I/O device on the PCI bus.

When the ADSP-BF535 Blackfin processor acts as the system controller, it views the PCI address spaces through its mapped windows and can initialize all devices in the system and maintain a map of the topology of the environment. The PCI memory region is a 4 Gbyte space that appears on the PCI bus and can be used to map memory I/O devices on the bus.

The ADSP-BF535 Blackfin processor uses a 128 Mbyte window in memory space to see a portion of the PCI memory space. A base address register is provided to position this window anywhere in the 4 Gbyte PCI memory space while its position with respect to the processor addresses remains fixed.

The PCI I/O region is also a 4 Gbyte space. However, most systems and I/O devices only use a 64 Kbyte subset of this space for I/O mapped addresses. The ADSP-BF535 Blackfin processor implements a 64 Kbyte window into this space along with a base address register, which can be used to position it anywhere in the PCI I/O address space, while the window remains at the same address in the processor's address space.

PCI configuration space is a limited address space, which is used for system enumeration and initialization. This address space is a very low performance communication mode between the processor and PCI devices. The ADSP-BF535 Blackfin processor provides a one-value window to access a single data value at any address in PCI configuration space. This window is fixed and receives the address of the value, and the value if the operation is a write. Otherwise, the device returns the value into the same address on a read operation.

***15.2.4.3 I/O Memory Space*** Blackfin processors do not define a separate I/O space. All resources are mapped through the flat 32-bit address space. On-chip I/O devices have their control registers mapped into memory-mapped registers (MMRs) at addresses near the top of the 4 Gbyte address space. These are separated into two smaller blocks, one that contains the control MMRs for all core functions, and the other that contains the registers needed for setup and control of the on-chip peripherals outside the core.

The core MMRs are accessible only by the core and only in supervisor mode and appear as reserved space by on-chip peripherals, as well as external devices accessing resources through the PCI bus. The system MMRs are accessible by the core in supervisor mode and can be mapped as either visible or reserved to other devices, depending on the system protection model desired.

## 15.2.5 Event Handling

The event controller on the ADSP-BF535 Blackfin processor handles all asynchronous and synchronous events to the processor. The ADSP-BF535 Blackfin processor provides event handling that supports both nesting and prioritization. Nesting allows multiple event service routines to be active simultaneously.

Prioritization ensures that servicing of a higher-priority event takes precedence over servicing of a lower-priority event. The controller provides support for five different types of events:

*Emulation*. An emulation event causes the processor to enter emulation mode, allowing command and control of the processor via the JTAG interface.

*Reset*. This event resets the processor.

*Nonmaskable interrupt (NMI)*. The NMI event can be generated by the software watchdog timer or by the NMI input signal to the processor. The NMI event is frequently used as a power-down indicator to initiate an orderly shutdown of the system.

*Exceptions*. Events that occur synchronously to program flow, for example, the exception will be taken before the instruction is allowed to complete. Conditions such as data alignment violations, undefined instructions, and so on, cause exceptions.

*Interrupts*. Events that occur asynchronously to program flow. They are caused by timers, peripherals, input pins, explicit software instructions, and so on.

Each event has an associated register to hold the return address and an associated return-from-event instruction. The state of the processor is saved on the supervisor stack, when an event is triggered.

The ADSP-BF535 Blackfin processor event controller consists of two stages, the core event controller (CEC) and the system interrupt controller (SIC). The core event controller works with the system interrupt controller to prioritize and control all system events. Conceptually, interrupts from the peripherals enter into the SIC and are then routed directly into the general-purpose interrupts of the CEC.

### 15.2.5.1  *Core Event Controller (CEC)*

The CEC supports nine general-purpose interrupts (IVG15–7), in addition to the dedicated interrupt and exception events. Of these general-purpose interrupts, the two lowest priority interrupts (IVG15–14) are recommended to be reserved for software interrupt handlers, leaving seven prioritized interrupt inputs to support the peripherals of the ADSP-BF535 Blackfin processor.

Figure 15.29 describes the inputs to the CEC, identifies their names in the Event Vector Table (EVT), and lists their priorities.

**FIGURE 15.29**  ADSP-BF533 Core Event Controller (Reprinted by permission of Analog Devices Incorporated.)

| Priority (0 is Highest) | Event Class | EVT Entry |
|---|---|---|
| 0 | Emulation Test | EMU |
| 1 | Reset | RST |
| 2 | Non-Maskable | NMI |
| 3 | Exceptions | EVX |
| 4 | Global Enable | |
| 5 | Hardware Error | IVHW |
| 6 | Core Timer | IVTMR |
| 7 | General Interrupt 7 | IVG7 |
| 8 | General Interrupt 8 | IVG8 |
| 9 | General Interrupt 9 | IVG9 |
| 10 | General Interrupt 10 | IVG10 |
| 11 | General Interrupt 11 | IVG11 |
| 12 | General Interrupt 12 | IVG12 |
| 13 | General Interrupt 13 | IVG13 |
| 14 | General Interrupt 14 | IVG14 |
| 15 | General Interrupt 15 | IVG15 |

**15.2.5.2   *System Interrupt Controller (SIC)*** The system interrupt controller provides the mapping and routing of events from the many peripheral interrupt sources to the prioritized general-purpose interrupt inputs of the CEC. Although the ADSP-BF535 Blackfin processor provides a default mapping, the user can alter the mappings and priorities of interrupt events by writing the appropriate values into the interrupt assignment registers (IAR). Figure 15.30 describes the inputs into the SIC and the default mappings into the CEC.

**FIGURE 15.30** ADSP-BF533 System Interrupt Controller (Reprinted by permission of Analog Devices Incorporated.)

| Peripheral Interrupt Event | Peripheral Interrupt ID | Default Mapping |
|---|---|---|
| Real-Time Clock | 0 | IVG7 |
| Reserved | 1 | |
| USB | 2 | IVG7 |
| PCI Interrupt | 3 | IVG7 |
| SPORT 0 Rx DMA | 4 | IVG8 |
| SPORT 0 Tx DMA | 5 | IVG8 |
| SPORT 1 Rx DMA | 6 | IVG8 |
| SPORT 1 Tx DMA | 7 | IVG8 |
| SPI O DMA | 8 | IVG9 |
| SPI 1 DMA | 9 | IVG9 |
| UART 0 Rx | 10 | IVG10 |
| UART 0 Tx | 11 | IVG10 |
| UART 1 Rx | 12 | IVG10 |
| UART 1 Tx | 13 | IVG10 |
| Timer 0 | 14 | IVG11 |
| Timer 1 | 15 | IVG11 |
| Timer 2 | 16 | IVG11 |
| GPIO Interrupt A | 17 | IVG12 |
| GPIO Interrupt A | 18 | IVG12 |
| Memory DMA | 19 | IVG13 |
| Software Watchdog Timer | 20 | IVG13 |
| Reserved | 26–21 | |
| Software Interrupt 1 | 27 | IVG14 |
| Software Interrupt 2 | 28 | IVG15 |

**15.2.5.3   *Interrupt Event Control*** An interrupt is an event that changes normal processor instruction flow and is asynchronous to program flow. In contrast, an exception is a software-initiated event whose effects are synchronous to program flow. The event system is nested and prioritized. Consequently, several service routines may be active at any time, and a low-priority event may be preempted by one of higher priority.

The processor employs a two-level event control mechanism. The processor system interrupt controller (SIC) works with the core event controller (CEC) to prioritize and control all system interrupts. The SIC provides mapping between the many peripheral interrupt sources and the prioritized general-purpose interrupt inputs of the core. This mapping is programmable, and individual interrupt sources can be masked in the SIC.

The ADSP-BF535 Blackfin processor provides the user with a very flexible mechanism to control the processing of events. The basic interrupt processing flow diagram is show in Figure 15.31.

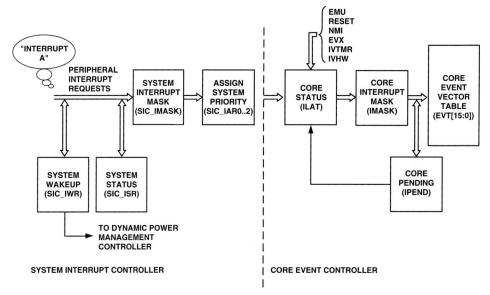

FIGURE 15.31   ADSP-BF533 Interrupt Processing Block Diagram (Reprinted by permission of Analog Devices Incorporated.)

In the CEC, three registers are used to coordinate and control events. Each of the registers is 16 bits wide, and each bit represents a particular event class. The SIC allows further control of event processing by providing three 32-bit interrupt control and status registers. Each register contains a bit corresponding to each of the peripheral interrupt events.

Because multiple interrupt sources can map to a single general-purpose interrupt, multiple pulse assertions can occur simultaneously, before or during interrupt processing for an interrupt event already detected on this interrupt input. The IPEND register contents are monitored by the SIC as the interrupt acknowledgment.

The appropriate ILAT register bit is set when an interrupt rising edge is detected (detection requires two core clock cycles). The bit is cleared when the respective IPEND register bit is set. The IPEND bit indicates that the event has entered into the processor pipeline. At this point, the CEC will recognize and queue the next rising edge event on the corresponding event input. The minimum latency from the rising edge transition of the general-purpose interrupt to the IPEND output asserted is three core clock cycles; however, the latency can be much higher, depending on the activity within and the mode of the processor.

### 15.2.6  DMA Controller

DMA transfers can occur between the ADSP-BF535 Blackfin processor's internal memories and any of its DMA-capable peripherals. Additionally, DMA transfers can be accomplished between any of the DMA-capable peripherals and external devices connected to the external memory interfaces, including the SDRAM controller, the asynchronous memory controller and the PCI bus interface, as shown in Figure 15.32.

DMA-capable peripherals include the SPORTs, SPI ports, UARTs, and USB port. Each individual DMA-capable peripheral has at least one dedicated DMA channel. DMA to and from PCI is accomplished by the memory DMA channel.

To describe each DMA sequence, the DMA controller uses a set of parameters called a descriptor block. When successive DMA sequences are needed, these descriptor blocks can be linked or chained together, so the completion of one DMA sequence auto initiates and starts the

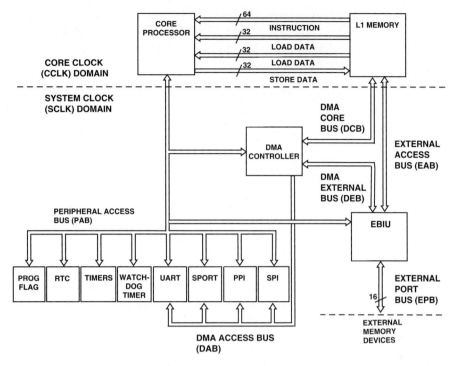

**FIGURE 15.32** ADSP-BF533 DMA Controller Interfaces (Reprinted by permission of Analog Devices Incorporated.)

next sequence. The descriptor blocks include full 32-bit addresses for the base pointers for source and destination, enabling access to the entire ADSP-BF535 Blackfin processor address space.

In addition to the dedicated peripheral DMA channels, there is a separate memory DMA channel provided for transfers between the various memories of the ADSP-BF535 Blackfin processor system. This enables transfers of blocks of data between any of the memories, including on-chip Level 2 memory; external SDRAM, ROM, SRAM, and flash memory; and PCI address spaces with little processor intervention.

### 15.2.7 External Memory Control

Figure 15.33 shows the external bus interface unit (EBIU) on the ADSP-BF535 Blackfin processor. It provides a high-performance, glueless interface to a wide variety of industry-standard

**FIGURE 15.33** ADSP-BF533 External Bus Interface Unit (Reprinted by permission of Analog Devices Incorporated.)

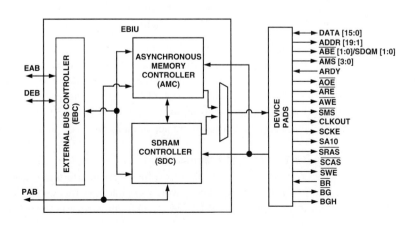

memory devices. The controller is made up of two sections: the first is an SDRAM controller for connection of industry-standard synchronous DRAM devices and DIMMs (dual inline memory module), whereas the second is an asynchronous memory controller intended to interface with a variety of memory devices.

***15.2.7.1    SDRAM Controller***    The SDRAM controller provides an interface for up to four separate banks of industry-standard SDRAM devices or DIMMs, at speeds up to fSCLK. Fully compliant with the PC133 SDRAM standard, each bank can be configured to contain between 16 Mbytes and 128 Mbytes of memory. The controller maintains all the banks as a contiguous address space so that the processor sees this as a single address space, even if different size devices are used in the different banks. This enables a system design where the configuration can be upgraded after delivery with either similar or different memories.

A set of programmable timing parameters is available to configure the SDRAM banks to support slower memory devices. The memory banks can be configured as either 32 bits wide for maximum performance and bandwidth or 16 bits wide for minimum device count and lower system cost.

All four banks share common SDRAM control signals and have their own bank select lines providing a completely glueless interface for most system configurations. The SDRAM controller address, data, clock, and command pins can drive loads up to 50 pF. For larger memory systems, the SDRAM controller external buffer timing should be selected, and external buffering should be provided so that the load on the SDRAM controller pins does not exceed 50 pF.

## 15.2.8  Asynchronous Controller

The asynchronous memory controller provides a configurable interface for up to four separate banks of memory or I/O devices. Each bank can be independently programmed with different timing parameters, enabling connections to a wide variety of memory devices including SRAM, ROM, and flash EPROM, as well as I/O devices that interface with standard memory control lines.

Each bank occupies a 64 Mbyte window in the processor's address space but, if not fully populated, these windows are not made contiguous by the memory controller logic. The banks can also be configured as 16-bit wide or 32-bit wide buses for ease of interfacing to a range of memories and I/O devices tailored either to high performance or to low cost and power.

## 15.2.9  PCI Interface

The ADSP-BF535 Blackfin processor provides a glueless logical and electrical, 33 MHz, 3.3 V, 32-bit PCI (peripheral component interconnect), Revision 2.2 compliant interface. The PCI interface is designed for a 3 V signaling environment. The PCI interface provides a bus bridge function between the processor core and on-chip peripherals and an external PCI bus. The PCI interface of the ADSP-BF535 Blackfin processor supports two PCI functions:

> A host to PCI bridge function, in which the ADSP-BF535 Blackfin processor resources (the processor core, internal and external memory, and the memory DMA controller) provide the necessary hardware components to emulate a host computer PCI interface, from the perspective of a PCI target device.
> A PCI target function, in which an ADSP-BF535 Blackfin processor-based intelligent peripheral can easily be designed to interface with a Revision 2.2 compliant PCI bus.

***15.2.9.1    PCI Host Function***    As the PCI host, the ADSP-BF535 Blackfin processor provides the necessary PCI host (platform) functions required to support and control a variety of off-the-shelf PCI I/O devices. These would include Ethernet controllers, bus bridges, and so on in a system in which the ADSP-BF535 Blackfin processor is the host.

Note that the Blackfin processor architecture defines only memory space (no I/O or configuration address spaces). The three address spaces of PCI space (memory, I/O, and configuration space) are mapped into the flat 32-bit memory space of the ADSP-BF535 Blackfin processor.

Because the PCI memory space is as large as the ADSP-BF535 Blackfin processor memory address space, a windowed approach is employed, with separate windows in the ADSP-BF535 Blackfin processor address space used for accessing the three PCI address spaces. Base address registers are provided so that these windows can be positioned to view any range in the PCI address spaces while the windows remain fixed in position in the ADSP-BF535 Blackfin processor's address range.

For devices on the PCI bus viewing the ADSP-BF535 Blackfin processor's resources, several mapping registers are provided to enable resources to be viewed in the PCI address space. The ADSP-BF535 Blackfin processor's external memory space, internal L2, and some I/O MMRs can be selectively enabled as memory spaces that devices on the PCI bus can use as targets for PCI memory transactions.

***15.2.9.2  PCI Target Function***   As a PCI target device, the PCI host processor can configure the ADSP-BF535 Blackfin processor subsystem during enumeration of the PCI bus system. Once configured, the ADSP-BF535 Blackfin processor subsystem acts as an intelligent I/O device. When configured as a target device, the PCI controller uses the memory DMA controller to perform DMA transfers as required by the PCI host.

## 15.2.10  USB Device

The ADSP-BF535 Blackfin processor provides a USB-compliant device type interface to support direct connection to a host system. The USB core interface provides a flexible programmable environment with up to eight endpoints. Each endpoint can support all the USB data types, including control, bulk, interrupt, and isochronous. Each endpoint provides a memory mapped buffer for transferring data to the application.

The ADSP-BF535 Blackfin processor USB port has a dedicated DMA controller and interrupt input to minimize processor polling overhead and to enable asynchronous requests for CPU attention only when transfer management is required. The USB device requires an external 48 MHz oscillator. The value of SCLK must always exceed 48 MHz for proper USB operation.

## 15.2.11  Real-Time Clock

The ADSP-BF535 Blackfin processor real-time clock (RTC) provides a robust set of digital watch features, including current time, stopwatch, and alarm, as shown in Figure 15.34. The RTC is clocked by a 32.768 kHz crystal external to the ADSP-BF535 Blackfin processor. The RTC peripheral has dedicated power supply pins, so that it can remain powered up and clocked, even when the rest of the processor is in a low power state. The RTC provides several programmable interrupt options, including interrupt per second, minute, or day clock ticks, interrupt on programmable stopwatch countdown, or interrupt at a programmed alarm time.

The 32.768 kHz input clock frequency is divided down to a 1 Hz signal by a prescaler. The counter function of the timer consists of four counters: a 6-bit second counter, a 6-bit minute counter, a 5-bit hour counter, and an 8-bit day counter.

When enabled, the alarm function generates an interrupt when the output of the timer matches the programmed value in the alarm control register. There are two alarms: one is for a time of day, the second is for a day and time of that day.

The stopwatch function counts down from a programmed value, with 1-minute resolution. When the stopwatch is enabled and the counter underflows, an interrupt is generated. Like the other peripherals, the RTC can wake up the ADSP-BF535 Blackfin processor from a low power state on generation of any interrupt.

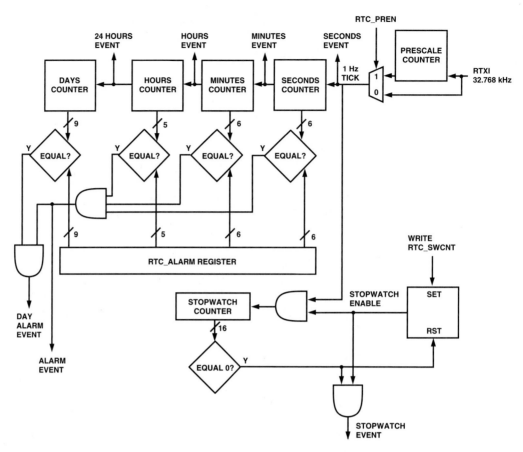

**FIGURE 15.34**    ADSP-BF533 Real-Time Clock (Reprinted by permission of Analog Devices Incorporated.)

## 15.2.12  Watchdog Timer

The ADSP-BF535 Blackfin processor includes a 32-bit timer, which can be used to implement a software watchdog function. A software watchdog can improve system availability by forcing the processor to a known state, via generation of a hardware reset, nonmaskable interrupt (NMI), or general-purpose interrupt, if the timer expires before being reset by software. The programmer initializes the count value of the timer, enables the appropriate interrupt, then enables the timer. Thereafter, the software must reload the counter before it counts to zero from the programmed value. This protects the system from remaining in an unknown state where software, which would normally reset the timer, has stopped running because of external noise conditions or a software error.

After a reset, software can determine if the watchdog was the source of the hardware reset by interrogating a status bit in the timer control register, which is set only on a watchdog generated reset. The timer is clocked by the system clock (SCLK) at a maximum frequency of fSCLK.

## 15.2.13  Timers

There are four programmable timer units in the ADSP-BF535 Blackfin processor. Figure 15.35 shows the basic block diagram of the timer function. Three of the general-purpose timers have an external pin that can be configured either as a pulse-width modulator (PWM) or timer output, as an input to clock the timer, or for measuring pulse widths of external events.

**FIGURE 15.35** ADSP-BF533 General-Purpose Timer (Reprinted by permission of Analog Devices Incorporated.)

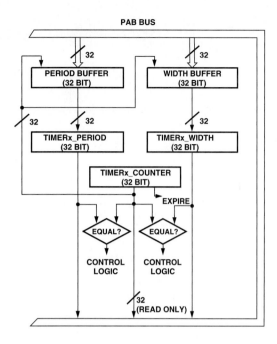

Each of the three general-purpose timer units can be independently programmed as a PWM, internally or externally clocked timer, or pulse-width counter. The general-purpose timer units can be used in conjunction with the UARTs to measure the width of the pulses in the data stream to provide an auto baud detect function for a serial channel.

The general-purpose timers can generate interrupts to the processor core providing periodic events for synchronization, either to the processor clock or to a count of external signals. In addition to the three general-purpose programmable timers, a fourth timer is also provided. This extra timer is clocked by the internal processor clock (CCLK) and is typically used as a system tick clock for the generation of operating system periodic interrupts.

## 15.2.14 Serial Ports

The ADSP-BF535 Blackfin processor incorporates two complete synchronous serial ports (SPORT0 and SPORT1) for serial and multiprocessor communications as part of the base peripheral functions.

A SPORT as shown in Figure 15.36 support these features:

*Bidirectional operation.* Each SPORT has independent transmit and receive pins.

*Buffered (8-deep) transmit and receive ports.* Each port has a data register for transferring data-words to and from other processor components and shift registers for shifting data in and out of the data registers.

*Clocking.* Each transmit and receive port can either use an external serial clock or generate its own, in frequencies ranging from (fSCLK/131070) Hz to (fSCLK/2) Hz.

*Word length.* Each SPORT supports serial data-words from 3 to 16 bits in length transferred in a format of most significant bit first or least significant bit first.

*Framing.* Each transmit and receive port can run with or without frame sync signals for each data-word. Frame sync signals can be generated internally or externally, active high or low, with either of two pulse widths and early or late frame sync.

*Companding in hardware.* Each SPORT can perform A-law or μ-law companding according to ITU recommendation G.711. Companding can be selected on the transmit and/or receive channel of the SPORT without additional latencies.

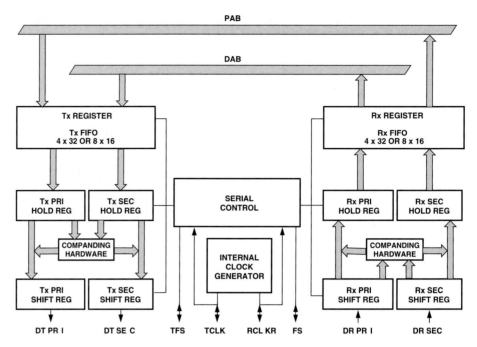

**FIGURE 15.36**    ADSP-BF533 SPORT Block Diagram (Reprinted by permission of Analog Devices Incorporated.)

*DMA operations with single-cycle overhead.* Each SPORT can automatically receive and transmit multiple buffers of memory data. The Blackfin processor can link or chain sequences of DMA transfers between a SPORT and memory. The chained DMA can be dynamically allocated and updated through the descriptor blocks that set up the chain.

*Interrupts.* Each transmit and receive port generates an interrupt on completing the transfer of a data-word or after transferring an entire data buffer or buffers through the DMA.

*Multichannel capability.* Each SPORT supports 128 channels and is compatible with the H.100, H.110, MVIP-90, and HMVIP standards.

## 15.2.15 Serial Peripheral Interface (SPI) Ports

The ADSP-BF535 Blackfin processor has two SPI compatible ports (Figure 15.37) that enable the processor to communicate with multiple SPI-compatible devices. The SPI interface uses three pins for transferring data: two data pins (master output-slave input, MOSIx, and master input-slave output, MISOx) and a clock pin (serial clock, SCKx). Two SPI chip select input pins (SPISSx) let other SPI devices select the processor, and fourteen SPI chip select output pins (SPIxSEL7–1) let the processor select other SPI devices.

The SPI select pins are reconfigured programmable flag pins. Using these pins, the SPI ports provide a full duplex, synchronous serial interface, which supports both master and slave modes and multimaster environments. Each SPI port's baud rate and clock phase/polarities are programmable, and each has an integrated DMA controller, configurable to support transmit or receive data streams. The SPI's DMA controller can only service unidirectional accesses at any given time.

During transfers, the SPI ports simultaneously transmit and receive by serially shifting data in and out on two serial data lines. The serial clock line synchronizes the shifting and

**FIGURE 15.37** ADSP-BF533 SPI Block Diagram (Reprinted by permission of Analog Devices Incorporated.)

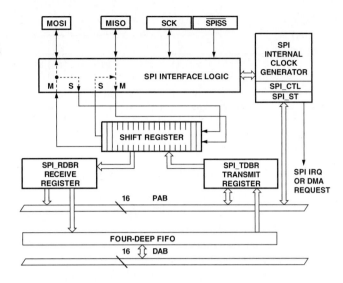

sampling of data on the two serial data lines. In master mode, the processor performs the following sequence to set up and initiate SPI transfers:

1. Enables and configures the SPI port's operation (data size and transfer format)
2. Selects the target SPI slave with an SPIxSELy output pin (reconfigured programmable flag pin)
3. Defines one or more TCBs in the processor's memory space (optional in DMA mode only)
4. Enables the SPI DMA engine and specifies transfer direction (optional in DMA mode only)
5. Reads or writes the SPI port receive or transmit data buffer (in non-DMA mode only)

The SCKx line generates the programmed clock pulses for simultaneously shifting data out on MOSIx and shifting data in on MISOx. In the DMA mode only, transfers continue until the SPI DMA word count transitions from 1 to 0.

In slave mode, the processor performs the following sequence to set up the SPI port to receive data from a master transmitter:

Enables and configures the SPI slave port to match the operation parameters set up on the master (data size and transfer format) SPI transmitter.

Defines and generates a receive TCB in the processor's memory space to interrupt at the end of the data transfer (optional in DMA mode only).

Enables the SPI DMA engine for a receive access (optional in DMA mode only).

Starts receiving data on the appropriate SPI SCKx edges after receiving an SPI chip select on an SPISSx input pin (reconfigured programmable flag pin) from a master.

In DMA mode only, reception continues until the SPI DMA word count transitions from 1 to 0. The processor can continue, by queuing up the next command TCB. A slave mode transmit operation is similar, except the processor specifies the data buffer in memory from which to transmit data, generates and relinquishes control of the transmit TCB, and begins filling the SPI port's data buffer. If the SPI controller is not ready to transmit, it can transmit a "zero" word.

## 15.2.16 UART Ports

The ADSP-BF535 Blackfin processor provides two full-duplex universal asynchronous receiver/transmitter (UART) ports (UART0 and UART1) fully compatible with PC-standard UARTs. The UART ports provide a simplified UART interface to other peripherals or hosts, supporting full-duplex, DMA-supported, asynchronous transfers of serial data. Each UART port includes support for 5 to 8 data bits; 1 or 2 stop bits; and none, even, or odd parity (Figure 15.38).

**FIGURE 15.38** ADSP-BF533 UART Word Format (Reprinted by permission of Analog Devices Incorporated.)

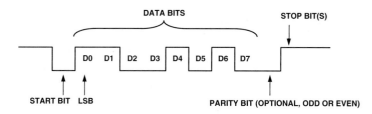

The UART ports support two modes of operation.

PIO (programmed I/O)—The processor sends or receives data by writing or reading I/O-mapped UATX or UARX registers, respectively. The data is double buffered on both transmit and receive.

DMA (direct memory access)—The DMA controller transfers both transmit and receive data. This reduces the number and frequency of interrupts required to transfer data to and from memory. Each UART has two dedicated DMA channels, one for transmit and one for receive. The DMA channels have lower priority than most because of their relatively low service rates.

Each UART port's baud rate, serial data format, error code generation and status, and interrupts are programmable:

Bit rates ranging from (fSCLK/1048576) to (fSCLK/16) bits per second
Data formats from 7 to 12 bits per frame
Both transmit and receive operations can be configured to generate maskable interrupts to the processor.

Autobaud detection is supported, in conjunction with the general-purpose timer functions. The capabilities of UART0 are further extended with support for the infrared data association (IrDA) serial infrared physical layer link specification (SIR) protocol.

### 15.2.17 Dynamic Power Management

The ADSP-BF535 Blackfin processor provides four power operating modes, each with a different performance/power dissipation profile. In addition, dynamic power management provides the control functions, with the appropriate external power regulation capability to dynamically alter the processor core supply voltage, further reducing power dissipation. Control of clocking to each of the ADSP-BF535 Blackfin processor peripherals also reduces power dissipation.

***15.2.17.1 Full On Operating Mode***   In the full on mode, the PLL is enabled and is not bypassed, providing the maximum operational frequency. This is the normal execution state in which maximum performance can be achieved. The processor core and all enabled peripherals run at full speed.

***15.2.17.2 Active Operating Mode***   In the active mode, the PLL is enabled, but bypassed. The input clock (CLKIN) is used to generate the clocks for the processor core (CCLK) and peripherals (SCLK). When the PLL is bypassed, CCLK runs at one-half the CLKIN frequency. Significant power savings can be achieved with the processor running at one-half the CLKIN frequency. In this mode, the PLL multiplication ratio can be changed by setting the appropriate values in the SSEL fields of the PLL control register (PLL_CTL). When in the active mode, system DMA access to appropriately configured L1 memory is supported.

### 15.2.17.3 *Sleep Operating Mode*

The sleep mode reduces power dissipation by disabling the clock to the processor core (CCLK). The PLL and system clock (SCLK), however, continue to operate in this mode. Any interrupt, typically via some external event or RTC activity, will wake up the processor. When in sleep mode, assertion of any interrupt will cause the processor to sense the value of the bypass bit (BYPASS) in the PLL control register (PLL_CTL). If bypass is disabled, the processor transitions to the full on mode. If bypass is enabled, the processor transitions to the active mode. When in sleep mode, system DMA access to L1 memory is not supported.

### 15.2.17.4 *Deep Sleep Operating Mode*

The deep sleep mode maximizes power savings even further by disabling the clocks to the processor core (CCLK) and to all synchronous peripherals (SCLK). Asynchronous peripherals, such as the RTC, may still be running, but will not be able to access internal resources or external memory. This powered-down mode can only be exited by assertion of the reset interrupt (RESET) or by an asynchronous interrupt generated by the RTC.

When in deep sleep mode, assertion of RESET causes the processor to sense the value of the BYPASS pin. If bypass is disabled, the processor will transition to full on mode. If bypass is enabled, the processor will transition to active mode. When in deep sleep mode, assertion of the RTC asynchronous interrupt causes the processor to transition to the full on mode, regardless of the value of the BYPASS pin. The DEEPSLEEP output is asserted in this mode.

## 15.2.18 Operating Modes and States

The processor supports the following three processor modes:

- User mode
- Supervisor mode
- Emulation mode

Emulation and supervisor modes have unrestricted access to the core resources. User mode has restricted access to certain system resources, thus providing a protected software environment. User mode is considered the domain of application programs. Supervisor mode and emulation mode are usually reserved for the kernel code of an operating system.

The processor mode is determined by the event controller. When servicing an interrupt, a nonmaskable interrupt (NMI), or an exception, the processor is in supervisor mode. When servicing an emulation event, the processor is in emulation mode. When not servicing any events, the processor is in user mode.

## QUESTIONS AND PROBLEMS

1. List six types of DSP consumer and communication markets.
2. How many MAC units are incorporated in the C55x?
3. Describe the seven pipelined stages of the C55x.
4. Describe the four primary functional blocks of the C55x.
5. List the five key areas of architectural parallelism in the C55x DU.
6. List six advanced functions for low-power utilization in the C55x.
7. List the five functions of the C55x program flow unit.
8. What three key functions are incorporated into the Blackfin processor architecture?
9. List eight of the system peripherals for Blackfin.
10. Describe the capabilities of the dual MAC units for the Blackfin.

11. What is meant by the L1/L2 memory system of the Blackfin?
12. List the five types of events supported by the event handler of the Blackfin processor.
13. What is the advantage of using a DMA controller in the Blackfin?
14. Describe how the watchdog timer (WDT) of the Blackfin operates.
15. List the four power management modes of the Blackfin processor.

## SOURCES

*TMS320C55x™ DSP peripherals overview, reference guide.* Preliminary Draft. TI Literature Number: SPRU317G. February 2004.

*TMS320C55x™ DSP, programmer's guide, preliminary draft.* TI Literature Number: SPRU376A. August 2001.

*TMS320C55x™ DSP, functional overview.* TI Literature Number: SPRU312. June 2000.

*TMS320C55x™ DSP, CPU reference guide.* Literature Number: SPRU371F. February 2004.

*TMS320C55x™ technical overview.* TI Literature Number: SPRU393. February 2000.

*TMS320C55x™ DSP algebraic instruction set reference guide.* TI Literature Number: SPRU375G, October 2002.

*TMS320VC5507/5509 DSP analog-to-digital converter (ADC) reference guide.* TI Literature Number: SPRU586B. June 2004.

*TMS320VC55x01/5502 DSP direct memory access (DMA) controller reference guide.* TI Literature Number: SPRU613G. March 2005.

*TMS320VC55x10 DSP external memory interface (EMIF) reference guide.* TI Literature Number: SPRU590. August 2004.

*TMS320VC55x03/5507/5509 DSP external memory interface (EMIF) reference guide.* TI Literature Number: SPRU670A. June 2004.

*TMS320VC55x01/5502/5503/5507/5509 DSP inter-integrated circuit (I2C) module reference guide.* TI Literature Number: SPRU146D. October 2005.

*ADSP-BF535 Blackfin® embedded processor, data sheet.* REV. A. 2005, Analog Devices.

*ADSP-BF533 Blackfin® processor hardware reference.* Revision 3.2. July 2006, Part Number 82-002005-01, Analog Devices.

*Getting started with Blackfin® processors.* Revision 1.0. February 2005, Part Number 82-000850-01, Analog Devices.

*TMS320VC55x01/5502 DSP instruction cache reference guide.* TI Literature Number: SPRU630C. June 2004.

*TMS320VC55x01/5502 DSP timers reference guide.* TI Literature Number: SPRU618B. April 2004.

*TMS320VC55x01/5502 DSP universal asynchronous receiver/transmitter (UART) reference guide.* TI Literature Number: SPRU597B, November 2002. Revised March 2004.

*TMS320VC55x09 DSP multimediacard / SD card controller reference guide.* TI Literature Number: SPRU593. June 2003.

*TMS320VC55x07/5509 DSP universal serial bus (USB) module reference guide.* TI Literature Number: SPRU596A. June 2004.

# INDEX